Manmade Closed Ecological Systems

Earth Space Institute Book Series

The Earth Space Institute Book Series covers different aspects of the future of space. These books aim to show how space has become a new tool in further developments for the benefit of mankind. In this context, space can now be considered as part of the human experience. Each volume in the series will be devoted to selected topics in engineering, legal and financial issues, and space-based sciences (extraterrestrial, earth observation, life sciences, materials science) intended for students, researchers and space professionals. The series will also feature historical and general overviews of interest to the public.

Volume 1
Space Science in China
Edited by Wen-Rui Hu

Volume 2
Der Mensch im Kosmos
Edited by Peter R. Sahm and Gerhard Thiele

Volume 3
Multiple Gravity Assist Interplanetary Trajectories
By Alexei V. Lubunsky, Oleg V. Papkov and Kostantin G. Sukhanov

Volume 4
Large Space Structures Formed by Centrifugal Forces
By V.M. Melnikov and V. A. Koshelev

Volume 5
East Asian Archaeoastronomy: Historical Records of Astronomical Observations of China, Japan and Korea
By Zhentao Xu, Yaotiao Jiang and David W. Pankenier

Volume 6
Space Debris: Hazard Evaluation and Mitigation
Edited by Nickolay N. Smirnov

Volume 7
Physics of Fluids in Microgravity
Edited by Rodolfo Monti

Volume 8
Space Radio Science
By Oleg I. Yakovlev

Volume 9
Manmade Closed Ecological Systems
By I.I. Gitelson, G.M. Lisovsky and R.D. MacElroy

Manmade Closed Ecological Systems

I.I. Gitelson, G.M. Lisovsky
Institute of Biophysics of the Russian Academy of Science, Krasnoyarsk, Russia

and R.D. MacElroy
NASA Ames Research Center, USA

Taylor & Francis
Taylor & Francis Group
LONDON AND NEW YORK

In memory of:
Sergey Pavlovich Korolev, the father of Soviet cosmonautics, whose vigorous and timely support of the Bios project made its realization possible in Krasnoyarsk, and Leonid Vasilyevich Kirensky and Ivan Aleksandrovich Terskov, founders of the Institute of Physics and Institute of Biophysics in Krasnoyarsk.

First published 2003
by Taylor & Francis
11 New Fetter Lane, London EC4P 4EE

Simultaneously published in the USA and Canada
by Taylor & Francis Inc,
29 West 35th Street, New York, NY 10001

Taylor & Francis is an imprint of the Taylor & Francis Group

Publisher's note:
This book was prepared from camera-ready-copy supplied by the authors.

Printed and bound in Great Britain by
MPG Books, Bodmin, Cornwall

British Library Cataloguing in Publication Data
A catalogue record for this book is available from the British Library

Library of Congress Cataloging in Publication Data
A catalog record for this title has been requested

ISBN 0-415-29998-5

Contents

FOREWORD

During the early twentieth century, rocket science was being developed by the Russian Konstantin Eduardovich Tsiolkovsky, the American Robert Hutchings Goddard, and the German Hermann Oberth, leading to dreams of space travel by both scientists and the public at large. Advances after World War II made it possible for space travel to be achieved, and at the height of the cold war near the end of the 1950s, the Soviet Union and the United States were engaged in a race to put an artificial satellite into orbit around the Earth. The Soviets achieved this goal on October 4, 1957 when they launched the 83.6 kg satellite called Sputnik 1. This spectacular success was followed a month later by the launch of Sputnik 2, which carried the dog Laika, the first living creature to be shot into space and orbit the Earth. The success of the Sputniks galvanized the United States into a frenzy of effort to achieve equally spectacular goals. The National Aeronautics and Space Administration (NASA) was established in 1958, but the Soviets continued to lead as they were the first to launch a space craft, Luna 1, free of Earth's gravity, and the first to pioneer manned Earth-orbital flight when Yury Gagarin circled the Earth in April 1961. Nevertheless, on July 20, 1969, Apollo 11 astronauts Neil Armstrong and Edwin ("Buzz") Aldrin set foot on the Moon, strengthening the dreams of a human-occupied lunar colony, dreams that had so often been fortified by wonderful works of science fiction.

Such dreams—and the similar dreams of prolonged space flight to Mars or other distant places—always included the challenge of human survival away from the amenities of mother Earth. One obvious answer was to "take a bunch of sack lunches," and so far, this has been the approach to manned space travel: carry enough food, water, oxygen, and other needed supplies to last the duration of the planned trip. A second answer is to recycle these supplies to reduce the load that must be carried at launch. The possibility of recycling was apparent to the early rocket scientists, and it was also apparent that, in addition to physicochemical means of recycling, we might use algae or higher plants to remove atmospheric carbon dioxide and release oxygen in the process of photosynthesis, also producing food and, in the case of higher plants, transpiring water vapor that could be condensed and utilized.

The possibilities of developing bioregenerative life support systems were apparent to both the Soviets and the Americans. In the United States, some scientists in the U.S. Air Force, various Universities, and the Aerospace Industry carried out some preliminary studies on how algae or higher plants might be used in such systems. Beginning in 1960, I had one of a series of NASA-supported projects to study plant life under extreme environmental conditions and to consider the use of higher plants in bioregenerative life support systems. By the end of the 1960s, however, NASA had lost interest in life support projects although I continued to study active plant life under the snow and related matters. Interest in bioregenerative life support would not be revived by NASA until the late 1970s.

Meanwhile, beginning in the early 1960s and continuing to the present, Russian scientists were studying bioregenerative life support, principally in Moscow and in Krasnoyarsk, which is located in Siberia. Because of the cold war, we in the United States, Europe, and Japan were unaware of the work in the Soviet Union until the mid-1980s when we began to hear tantalizing rumors about the things that had been done in Krasnoyarsk. The prime movers were Leonid Vasilyevich Kirensky and Ivan Aleksandrovich Terskov, both now deceased, but Iosef I. Gitelson, this book's senior author, was a member of the group that initiated the studies. Genry M. Lisovsky joined the group early in its history and has worked closely with Professor Gitelson for many years. The third author, Robert D. MacElroy, who is at the NASA Ames Research Center in California, administered the funding for university and other research on bioregenerative life support beginning in the early 1980s. Dr. MacElroy made it possible for Prof. Gitelson to spend many months at NASA Ames, where this book was prepared.

Thus the authors are particularly well qualified to write this book. Dr. MacElroy acknowledges that his contribution was relatively minor so the book is largely a summary of the work done in the Soviet Union (and now Russia), especially during those cold-war years when we in the West were unaware of the Soviet efforts—and ourselves largely inactive in the field. Prof. Gitelson, with the help of Dr. Lisovsky and his other colleagues in Krasnoyarsk, summarizes in wonderful detail the work done there. Clearly, the book could hardly have been written by anyone else! For us in the West, it is extremely important that the Soviet work be made part of the record, part of the history of Soviet efforts during the cold war and part of the record of thought about closed ecological systems, especially those that might be used in space exploration. Prof. Gitelson also summarizes some of the work done in Moscow where results were kept largely secret during much of the cold war and have still not been made widely available. In working with a group at the Institute of Biomedical Problems in Moscow to grow wheat in the Russian Space Station Mir, I was often amazed as members of this group told me about what they had done with life support systems, also beginning in the early 1960s. Prof. Gitelson is probably more aware of this work than anyone outside of the Moscow group; hence, his summary in this book is highly valuable.

Although these Russians had a commanding lead in life support research, there are now various groups in Europe, Japan, and the United States who are fast catching up. Thus, this book should be of great value to them. In 1992, I visited for the third time the Bios-3 facility in Krasnoyarsk to prepare a report for the Johnson Space Center, where they were in the initial stages of planning a similar facility. This report was expanded into an article coauthored by Prof. Gitelson and Dr. Lisovsky, so I thought I knew nearly all there was to know about Bios-3. What a surprise it was to read this book and to learn that what I knew barely scratched the surface of the extensive and detailed work that had been done by this group. If the current teams are to have as much success as the Krasnoyarsk researchers had, they will have to consider the problems at the same level of detail and intensity. Reading this book before such an undertaking could save much of this effort, which would allow the researchers to apply their creative talents to application of technologies that have developed since Bios-3 was built in 1972.

Researchers in the field are not the only ones who will find valuable material in this book. There are many students who hope to enter the field of bioregenerative life support, and for them this book should be required reading. Furthermore, the Krasnoyarsk researchers discovered that their work had much broader application than life support for space travel. The Earth's biosphere is a system closed with respect to matter and open with respect to energy—just as is an ideal bioregenerative life support system for space travel. For that matter, small closed systems might be used in polar, desert, or other climates. In any case, wherever they are studied and used, they can help us to understand the much larger biosphere of the Earth. The authors of this book are well aware of these implications of their work, and they discuss them in some detail.

In reading this book, I was highly impressed with the quality of Soviet science during the cold war. This book provides many insights into the creative perception, attention to detail, and deep knowledge of biology, agronomy, chemistry, physics, and engineering that the Krasnoyarsk researchers were able to apply in their design and construction of closed life support systems. Although these scientists were essentially isolated from the rest of the world, they managed to remain on the cutting edge of many branches of both science and engineering. Indeed, we have yet to catch up with all of what they accomplished with Bios-3. It makes some of us feel like beginners in the field. There are many examples, but one that I especially noted (because I had worked in a related field) was their perception of the human biological clock; they were aware of its intricacies even while much of the current knowledge was being developed in the West.

I would like to comment on how the various chapters impressed me: I greatly enjoyed the Introduction and the first five chapters, which review some history and discuss some general problems of life support. Then I encountered Chapter 6 on Principles of Controlled Cultivation of Microalgae. Because of the mathematics, I suppose, I found this chapter to be very rough going. At the same time, it provides excellent insight into the depth of analysis achieved by the Soviet researchers. It is essential reading for anyone interested in use of algae in life support, but others may want to skim this chapter and move on. Chapter 8 describes the use of microalgae in the early Krasnoyarsk efforts to achieve bioregenerative life support. Having never been much interested in the use of algae, I expected to encounter equally difficult going through that section. I was most pleasantly surprised to find myself fascinated by the descriptions of how algae had been used in the early Bios facilities. I even became convinced that microalgal systems might still have some application in life support for space travel—and perhaps other uses. I expected to enjoy Chapter 7 on the use of higher plants, which was my own research field, and I was not disappointed. As with the algal studies, I was deeply impressed with the level of detail that the Krasnoyarsk scientists had reached as they prepared for and then used agricultural crops in their systems. (The use of chufa, not currently a crop anywhere in the world, is a unique and valuable contribution to life support studies.) Chapter 9 examines in detail the three trials with human occupants in Bios-3. This is where the depth of the Krasnoyarsk studies is most apparent. Chapter 10 discusses the role of microbial life in bioregenerative systems. As with algae, I had little interest in the topic—until I read the chapter! The discussion makes a strong case for careful study of the microbial life of any closed system. I was fascinated! Unfortunately, however, I found Chapter 11 to be as difficult of a read as Chapter 6—again probably because of the mathematics.

There will be readers who will be entranced by this chapter, but if you fail to be so entranced, please skip to the Conclusion. Prof. Gitelson's conclusions are worth waiting for! His discussion is a wonderful addition to our understanding of bioregenerative systems. There are insights that go well beyond the thinking of many researchers who are presently working in the field. As noted above, the authors have not been oblivious to the implications of their work for ecology in general as well as for life support systems, and that is made clear in the Conclusions.

To summarize: I found this book to be a marvelous addition to the literature of closed ecological systems in general and bioregenerative life support systems for space exploration in particular. Actually, such literature is limited and mostly in the form of relatively brief papers, often in obscure publications. Only rarely has anyone attempted to create a book-length examination of the problem. Indeed, there is no other book like this one nor will there ever be! The Krasnoyarsk work was totally unique and highly significant. We should be grateful to Professor Gitelson and his colleagues for preparing this spectacular summary of what they accomplished!

Frank B. Salisbury
Prof. Emeritus of Plant Physiology
Utah State University, Logan, Utah, U.S.A.

In memory of
Leonid Vasilyevich Kirensky and
Ivan Aleksandrovich Terskov,
founders of
Institute of Physics and
Institute of Biophysics
in Krasnoyarsk,
initiators of this work

ACKNOWLEDGEMENTS

Contributions to the creation, investigation, and testing of the biological life support system Bios were made at different times by more than one hundred people—specialists in electronics, biologists, physicians, physicists, engineers, mathematicians—all united by their profound interest in this area of research. Having no opportunity to name everyone, the authors extend their sincere thanks to all of them. The Bios would not have been created and this book could not have been written but for their expertise, skills, courage, and strong motivation.

Some of them are gone now: B.G. Kovrov, a unique researcher of versatile talents, the chief designer of the Bios; F.Ya. Sidko, who thoroughly investigated optics of microalgae and elaborated a system of plant light nutrition for the Bios; G. Tereshkova, who took part in the development and testing of early Bios versions. We honor their memory.

I. Gitelson also owes a debt to the American National Research council and NASA Ames Research Center, Moffett field. A large part of this work was supported by a grant from the former and performed thanks to the hospitality of the latter.

We appreciate the invaluable assistance of G.E. Kasaeva, who for many years had been involved in the Bios research and who made all the pictures for the book and took part in designing it. We must also thank Darlene Reddaway (Stanford University) and Elena Krasova (Institute of Biophysics) for translating the book. D. Reddaway translated Introduction and Chapters 4, 5, 6, 7 (partially), and 9 (partially). E. Krasova translated Chapters 1, 2, 3, 7 (partially), 8, 9 (partially), 10, 11, and Conclusion. She was also responsible for the technical preparation of the text.

"If humanity will only understand the enormity of the present moment and not use either its reason nor labor to bring about its own self-destruction, then we will find that a great future is opening up to us in the geologic history of the biosphere. The onus is on us to make a spontaneous process into a conscious one. We must convert our living area—the biosphere—into the kingdom of human reason—the noosphere."

V. I. Vernadsky

INTRODUCTION

Experimental Modeling of the Biosphere and Noosphere

Let's take a moment to ponder the paradox of life on Earth. The Earth's resources are limited. Although at any given moment there is only enough oxygen in the atmosphere to support the breathing of all living things on Earth for about 2000 years, life has existed on our planet significantly longer—at least 4 billion years, and no natural causes can be envisioned that would suggest its cessation in the future. The paradox of life lies exactly in this: its resources are finite, but it itself is endless. Such a contradictory state of affairs is feasible only because the resources accessible to life can be used over and over again. That is, the material cycle has resolved this contradiction and has come to unite all things living on the Earth during the course of evolution. It is precisely the material cycle that forms the modern conception of the "biosphere."

Humanity's intensely developing industrial activity within the biosphere imparts a heightened poignancy to the problem of preserving the closed material cycle. In an effort to achieve various utilitarian goals, industry more frequently enters into conflict with the course of those natural processes that enable the biosphere's self-maintaining, constant material cycle—outside of which the existence of humanity and all life on the planet are rendered impossible.

Conditions on Earth extant at the beginning of the 21st century have forced humanity to find a way to render aggressively growing technological civilization compatible with nature. Recently this task has been discussed in terms of a scenario for the sustainable development of humanity.

At the same time, it was during this technological surge that human activity in the solar system first became viable.

Space engineering has resolved the task of sending humanity into space. And this in turn has given rise to another problem: how to support long living periods for humans, extending for months or even years. Solving this problem will enable the continuation of the lives of those humans who may become torn away from their native Earth, the very Earth to which they and all living creatures are tied, as if by an umbilical cord, for their material exchange.

At first glance, the problem of sustainably developing civilization on Earth and the problem of supporting human life in space seem quite removed from one another. However, their scientific and methodological bases are the same. The key to the solution of both problems is the principle of a balanced material cycle. In the Earth's biosphere the task is to preserve the material cycle. In a space settlement the material cycle must be created, and although the material cycles designed for use in a space settlement may be rather far from reproducing an exact representation of the biosphere, they must be its functional equivalent.

In solving both of these problems then, it is essential to develop a thorough understanding of the one characteristic that makes the biosphere and its simulations unique—isolation. The isolation of the biosphere is such an essential and distinctive trait that it fully justifies the development of a special branch of ecology that would study the mechanisms that enable the self-maintenance and sustainability of the biosphere's closed material cycle, as well as the emergence, evolution, and function of these mechanisms.

In the second half of the 20th century two different lines of scientific development converged, those of ecology and space biology, and a new field of study appeared: the creation of closed manmade ecological systems. This new field has also been referred to as "biospherics," a term which has not yet found general acceptance. Biospherics is devoted to two things: 1) modeling closed ecosystems in order to discover the mechanism that enables sustainable existence of the Earth's closed ecosystem and 2) developing closed human life support systems capable of sustaining distant human space flights and which can also help to enhance the quality of life for those inhabiting unfavorable environmental conditions on Earth.

The idea that the biosphere's material cycle is closed traces its roots back to the 19th century and to the work of Liebig and Haeckel. Vernadsky elaborated this idea to a much fuller extent in the first half of the 20th century. Today the concept of the biosphere as an essentially closed material cycle on Earth has become relatively trivial and has been included in textbooks on ecology. This notwithstanding, modern ecologists, and especially those in the Western scientific community, are not sufficiently acquainted with the whole scope of Vernadsky's work on the biosphere, which has been deeply and thoroughly developed.

An aspect of Vernadsky's work that is especially relevant to our present study is even less well-known to the scientific community. This is the idea of the noosphere, that is, the governing of the biosphere by human reason. (In Greek, *noos* means reason.) In essence, it is precisely the idea of the noosphere that is so unwittingly and so heatedly discussed today in the widest scientific and social circles in its modern guise as the problem of "sustainable development" (Forrester, 1971; Koptyug, 1997).

According to Vernadsky and Le Roy (Le Roy, 1928), the noosphere was to become a new stage in the evolution in the Earth's biosphere. Only in the 20th century did humans become fully aware of the need to rationally govern the biosphere. Although humans have interacted with the biosphere from their first appearance on Earth, they have only recently made legitimate attempts to rationally control it, and have violated cyclicity of its natural processes instead. Today, humanity plays an increasingly greater role in the biosphere by virtue of its entire cognitive activity with respect to technology and science, and to a growing extent in the future humans will become responsible for the restoration and control of these processes. We humans must be willing to accept this responsibility if we are to defuse the conflict between our need to develop civilization and our need to preserve living nature, and we must discover how humanity and the biosphere can coadapt.

Vernadsky's contemporary Teilhard de Chardin (1943, 1987) formulated an essentially similar idea, in which humans assumed a governing role in the biosphere. Lovelock (1972, 1990, 1991) further expounded this concept in his representations of "Gaia." The idea subsequently acquired a mystical hue, not so much on the part of the original thinkers who authored it as much as with their followers. As the mystical interpretation came to overshadow the original idea, a cautious, skeptical approach to the noosphere was adopted by some careful-thinking scientists (Kolchinsky, 1990) even though the concept of the human factor in the biosphere as espoused by Vernadsky was elaborated from a point of positive knowledge and was completely lacking in mystical flourish. The subsequent development of society and its relationships with nature in the second half of the 20th century attest to the visionary depth and productivity of this idea, and they give sufficient cause for shedding any prejudice against the noosphere that may have appeared due to the influence of any spurious mystical offshoots.

In Vernadsky's definition, the noosphere is just as solid a scientific concept as is the biosphere. To an ever-growing extent during the noospheric stage in the evolution of the biosphere, humanity's conscious governing influence supplements the stochastically-acting natural regulation of the planet's biotic cycle. Human influence is manifested in its proclivity to regulate both energy streams and technological streams of matter while it attempts to use natural resources in a non-destructive way. Learning to wisely exert human influence may become the most important task for the 21st century.

What kind of knowledge must humanity attain in order to rationally govern the biosphere?

The basis for controlling the biosphere in a reasonable manner is a sufficient and exact knowledge of its laws and the means for describing all possible changes that can occur within it that would at the same time not violate its main quality: the material cycle. For it is precisely the material cycle that is so vital to the co-evolution of humanity and the biosphere, and it is the human capacity to control the material cycle that in the end will facilitate the transformation of the biosphere into the noosphere. The study of the biosphere as an independent scientific object demands the development and use of methods and research that are commensurate with 1) the enormity of its size and mass; 2) the multitude of its intricately interwoven physical, chemical, and biological processes; and 3) the formidable inertia of those processes that enable the material cycle on Earth.

One of these methods is the method of "empirical generalization." Within the framework of national and international scientific programs, "empirical generalization" as a method can be seen as developing over the last decades by virtue of the intense accumulation of factual material having to do with 1) the disposition of living matter within the bounds of the biosphere, 2) the productivity of various biomes, 3) the way pollution influences the environment, and so forth. For the first time remote sensing from space has allowed us to see the biosphere as a real and integral object and to chart the dynamics of biospheric processes. This factual material has formed the richest of databases, and it has enabled theoretical conceptualization and the construction of various hypotheses in the field of biospherics to occur.

However, observational methods alone are not sufficient to solve such crucially important problems of human existence as the biosphere's sustainability. These methods also cannot define the bounds of allowable interference with the biosphere in such a way as to predict their consequences. For the first time, humanity has set before itself a noospheric task: the elaboration of a strategy of sustainable development (the well-known series of publications as well as the Club of Rome, 1960–1970; Meadows *et al.*, 1974; Mesarovich and Pestel, 1974; Forrester, 1971; Koptyug, 1997). In order to solve this problem, biospheric science needs a powerful investigative instrument. The only applicable tool is experiment.

As computer science developed in the 1970s, a new method for studying the biosphere became available: the construction of large-scale mathematical models representing the biosphere and the anthropogenic influences on it based on information gathered by empirical methods (Forrester, 1971; Moiseev, 1979). No one doubted the wisdom of prohibiting direct experimentation on the biosphere: biospheric proportions are too gigantic, experiments are indescribably expensive, and more importantly, the consequences for humanity and the biosphere itself are potentially dangerous and irreversible.

Mathematical modeling was the irreplaceable alternative. According to mathematicians, such modeling represented the sole means available to predict possible changes in the biosphere's state resulting from humanity's large-scale influence on it.

American and Russian scientists convincingly demonstrated the prognostic force of mathematically modeling biospheric processes under significant anthropogenic influences when they researched the possible consequences of nuclear war (Moiseev, 1979; Turco *et al.*, 1983). The fact that their two separately developed models of nuclear winter are principally similar is objective testimony that "computer experiments" are highly effective when applied to biospheric studies. As long as the primary information invested in such an experiment is complete enough, an accurate prognosis can be made.

In time it became apparent that observations of Earth from space may quickly increase the available data on the biosphere and make for sound mathematical models, but they are limited by the Earth's present state of near equilibrium. Large-scale (catastrophic in their results) human influences on the biosphere that are amenable to mathematical modeling do not exhaust all real threats of damage to the biosphere's self-support mechanisms, and cannot even predict the potentially destructive forces that can dissolve these mechanisms. "The danger of some of the changes evoked by humanity is in their elusiveness. They are very easy to overlook, and the situation can get to the point where it is already too late to correct them." (Polunin and Greenwald, 1993).

It is characteristic of "living things" in the biosphere, that is individual specimens, species, and biocoenoses, to react to what would seem to be rather small changes in the environment (the appearance of a pollutant, the introduction of new species, etc.). Since we do not know many of the boundary conditions associated with many biospheric parameters, we cannot predict which small changes are critical. That is, we cannot predict which small changes are capable of escalating and disrupting the ecosystem's homeostasis to the point where a considerable restructuring of its organization can be expected in time, with the possibility that the ecosystem itself may undergo destruction.

These kinds of processes are not readily apparent to the researcher who is focusing on separate organisms, populations, or biocenoses. This is because open natural ecosystems are only a part of the biosphere's single closed material exchange, and at the same time they are dampened by this exchange. Since most traditional experiments in ecology are conducted in open natural ecosystems, we can draw the conclusion that very few experiments in that field have investigated the role that the biosphere plays in the biological exchange of matter and the anthropogenic influences upon it. It is necessary to conduct specific "biospheric" experiments under conditions that reproduce the distinguishing characteristic of the biosphere—the closedness of its material cycle.

In order to produce a salient prognosis then, it is necessary to study the behavior of the ecosystem when it is beyond the bounds of its norms, that is, when it experiences extreme states, all the while ensuring that the system reproduces a closed material cycle. Researchers must have access to models of closed ecosystems that they can push beyond the limits of stability in order to accelerate knowledge growth in this field. Such closed ecosystems would also form the basis for mathematical modeling of non-equilibrium and catastrophic biospheric states.

In nature, however, and at least on our planet, no ecological system other than the biosphere can be termed as truly closed.

A question arises: can humans artificially create a similar object? The solution to such a problem might become a test of the completeness of our ecological knowledge.

Attempts to create hermetically-sealed microbiospheres have been made since the 1920s. Researchers have had experience creating such "microbiospheres" based on the combination of unicellular algae and bacteria (Kovrov and Fishtein, 1980; Folsom and Hanson, 1986). For example it has been demonstrated that after keeping 90 such systemic variants in sealed test tubes in the presence of light, there were 6 cases in which the material cycle proved to be completely closed or so near to it that for the duration of the following 8 years there was no accumulation of deadlock products nor a change in the population counts that constituted the biocenosis of these microecosystems (Kovrov, 1992).

Constructing closed manmade ecosystems by synthetic means has not been limited to the microbic level. As early as the last decades of the 19th century, K.E. Tsiolkovsky was already conjecturing about the possibility of designing a biological life support system that would sustain human life in space, but a detailed scientific elaboration and experimental study began only on the threshold of humanity's imminent expansion into open space during the 1950s.

Experimental studies in the 1950s began to explore the possibility of creating closed ecosystems inhabited by humans. This trend developed with varying intensity in the United States, Japan, Germany, England, France, and the Soviet Union (continuing in Russia today). It led to the elaboration of the first human life support prototypes existent outside of the natural biosphere. In these prototypes, humans participated in the material cycle of a closed manmade ecosystem that they themselves controlled. As opposed to the biosphere that is not controlled by anyone, it turned out that a small closed ecosystem can be sustainable only under such control (Kirensky *et al.*, 1967; Gitelson *et al.*, 1975; Lisovsky, ed., 1979; Gitelson *et al.*, 1989).

As so often happens in science, the original and fundamental question of how to create a closed manmade ecosystem proved theoretical. Its corollary—how to create a biological life support system for prolonged human expeditions into space—served as the driving force for experimental investigation. As researchers grappled with the idea of space travel in the first stormy developmental years of cosmonautics, they found it necessary to construct a manmade biosphere. We must confess that even the authors of this book were initially drawn into the experimental solution of this problem, and we did not immediately grasp the full magnitude of this problem, confining our understanding of the problem to the design of life support systems.

For some reason practical cosmonautics has left the data yielded by these efforts untouched to this day. Now it is abundantly clear that these efforts will prove fundamentally significant in returning to the original question and solving global biospheric problems.

The creation of manmade, closed ecological systems allows us, in the same way as does a magnifying glass: 1) to approximate the remote consequences of weak changes in the environment, 2) to illuminate subtle mechanisms of homeostasis violation in the material cycle (or, conversely, of homeostasis support), and 3) to predict the possible character and evolution of such systems. These physical experiments are of crucial importance because they represent a new method for researching the biosphere.

The creation of closed systems differs from that of open systems by its fundamentally different approach. Closed systems are artificially synthesized from a small number of species (at least two). In ideal situations these species might be the metabolic antipodes of one another: "reducing agent–oxidizing agent," "photothron–heterothron," and so forth. The lower the organizational level of the species included in the system and the more primitive their demands on the surrounding environs, the greater the chances of more closely aligning their mass and energy exchange, and the nearer we are to a fully closed manmade ecosystem. Closed microecosystems can serve as convenient instruments in researching the ecology of microorganisms, in evaluating the influence that penetrating radiation and various physical fields have on the ecosystem, and in studying the evolution of both biocenoses and microecosystems themselves.

In this respect, a closed manmade system is more a model of the noosphere than the biosphere. We can begin to study and elaborate an algorithm for the controlling human function in the noosphere, the same function that humanity must assume in a coming stage of the biosphere's evolution. Human participation in controlling closed manmade ecosystems is a fitting school for their future governing of the Earth's biosphere.

As experiments have shown, a life support system based on a biological material exchange is fully realizable and contains possibilities for further improvement. This kind of closed human ecosystem can become a model not only of an earthly noosphere, but also of daughter noospheres that can be used to help humanity render space inhabitable without threatening other bodies of the solar system through the intrusion of earthly substances and processes. As such, it will allow humans to exist in space, while requiring only the input of energy and not allowing the release of metabolites into the surroundings.

We cannot envision a time when complete closure of exchange will become necessary for practical purposes under conditions anywhere on Earth. But partially closed technologies for regulating the atmosphere, water, and vegetable nutrients can radically enhance the quality of life in extreme regions—in the Arctic, Antarctica, in deserts, and in high mountain settlements. Another aspect of closed life support technologies is that they enable minimizing the environmental pollution that results from human waste products and those of their domestic animals. It is not at all unlikely that these technologies will find applications on Earth before human settlements appear on Mars or the Moon. The transition to essentially closed nonpolluting, non-deadlock life support technologies will be a noticeable step on the path toward the sustainable development of the noosphere.

Our book is devoted to the theory and practice of the creation and research of closed ecological systems. In this book, the authors have set themselves the task of summarizing the first developmental period in the study of closed manmade ecosystems. The scientific and practical significance of this period becomes all the more apparent as humanity paves a path into space and as our interaction with living nature on Earth is simultaneously strained. The authors of this book have participated in this work in Russia (Gitelson and Lisovsky) and in the United States (MacElroy) from its very beginnings.

History has somehow determined that work on closed ecosystems enjoys particularly good support in the Soviet Union. It was exactly here that the enormous scientific authority of predecessors in the field could be felt. It was here that Tsiolkovsky's best scientific dreams about space travel could be dreamt, and that Vernadsky's ideas on the noosphere could be developed. It was here that such farseeing pioneers of cosmonautics as Korolev garnered support. To no less a degree, these works were supported by the Academy of Sciences, which has a longstanding tradition of supporting fundamental research.

These works were never counted as secret in the Soviet Union, but the general isolation of this country from the international arena meant that the majority of these works (although by no means all of them) were published only in the Soviet Union in the 1960s and 1970s, and naturally, in Russian. This explains (although perhaps it does not excuse) why these works were not sufficiently well-known to the worldwide scientific community outside of Russia. Now we have the chance to rectify this situation.

We consider it high time to analyze prior achievements in order to lay a basis for future discovery. Materials describing experiments made on prolonged human habitation periods in closed biological systems which were conducted in the Soviet

Union are mainly described in three monographs that were published in Russian and for that reason have been practically inaccessible to the English-language reader. These works make up a significant portion of the present book. The articles summarized in this book are scattered in various editions, sometimes one-time editions of small circulation. These publications are primarily in Russian and not easily accessible, although they number in the hundreds.

These, in short, are the compelling reasons that convinced the authors and drove them to write this book with the hope that it will be useful. It is also the reason that the book, for the most part, is based on works that were done in the Soviet Union, and that even now are continuing to be done in Russia over the last decade.

The reference literature quoted in our book may not comply with customary expectations. The Western reader is used to seeing a list that normally includes works from the last five, or at the most, ten years. In our book, a great number of works that are 20 to 30 years old, or older, are quoted and discussed alongside those of recent years.

We have found this necessary for two reasons. First of all, work on manmade bioregenerative human life support systems reached its peak in the 1960s and 1970s when the projects at hand were flights to solar system planets and the establishment of bases on the Moon and on Mars. The research conducted then had immediate application even though the development of cosmonautics in both Russia (Soviet Union) and the USA proceeded more slowly than expected due to internal and external causes. Attention to and financial support of work on human life support systems for remote expeditions into space gradually slackened. Accordingly, the rate at which this work was executed also tapered off.

Even so, an enormous amount of work has been accomplished toward the fashioning of closed human life support systems, and this progress has been substantial. For example, in 1954 Myers conducted an experiment in which a single mouse was able to breathe thanks to air exchange with a *Chlorella* culture, but already in the 1960s and 1970s people were spending years inhabiting experimental bioregenerative life support systems. The first professional journal on this topic, *Life Support and Biosphere Science*, first came into print only five years ago and has not yet won a stable readership. We might mention that some good survey works have been compiled: Eckart's useful, but brief reference volume on life support systems in space (Eckart, 1996), and the earlier Soviet–American joint edition on space medicine and biology (1979 and 1993, second edition, Life Support and Habitability Manual, ed. by R. Binot, 1991; Fundamentals of Space Life Sciences, vol. 1 and 2, ed. by Susanne Churchill, 1997).

We must keep in mind that the human cognitive progress is unstoppable, and that the human drive to know the universe is indestructible. Undoubtedly, the scientific community will return to the results of this research, and we can surmise that humanity will once again return to the Moon, and that it will build bases on Mars. We can even predict that humanity will engage in constant activity in the farthest reaches of space.

There is reason to believe that in the first decades of the 21st century expeditions to Mars will have taken place and that there will be outposts established on the Moon. And what is even more important, the strategy for making short "hops" into near space will be replaced by a strategy directed at “free space travel” for distant and prolonged expeditions into the solar system.

And when that happens, the problem of creating closed ecosystems of prolonged duration will once again become vital.

In order to avoid wasting time, labor, and resources on a repetition of those things achieved by predecessors in the field (including their mistakes), we must know their works, and not forget them.

Those problems that can be solved using closed ecosystems or their elements are diverse, and the authors hope that their work will prove profitable to the ecologists and space biologists, and to the doctors and engineers who will be creating spaceships designed for distant travel. To no lesser extent, the authors hope that this book will benefit those who are working on means to better the conditions of human habitation in unfavorable ecological environments, and that it will engage all those who are interested in this scientific trend and the role they can play in enhancing human life on our planet and in its dissemination into space.

* * *

Before our readers proceed to the main part of the book, we would like to provide them with some recommendations, which are not to be necessarily followed but rather are useful directions that a pilot gives to the captain.

First of all, the readers should be aware of one specific difficulty in describing a closed ecosystem. Such an ecosystem is like a “black hole” in space—it is closed upon itself, and whatever part of it you describe first, the description cannot be complete without the knowledge of other parts. While writing the book, we invariably felt this difficulty. The authors’ position resembled that of a dog playing with a big round ball—it is too large to put it into the mouth as it is, but when torn into pieces, it is no longer round.

A characteristic example of this difficulty is the description of microflora in a closed ecosystem. Every component, including humans, contains the microflora as an important part, but, on the other hand, the microflora of a closed ecosystem is an integral whole. Two ways were possible: either to include the description of the microflora of each component in the relevant chapter or to devote a whole chapter to the microflora of the closed ecosystem. To emphasize the integrity of the microworld of the closed ecosystem, we chose the second way. Therefore, the microflora of the closed ecosystem is described in Chapter 10, which inevitably makes the description of other components in the previous chapters incomplete. The reader should remember that the missing details can be found later, in Chapter 10.

Another recommendation concerns Chapters 6 and 11. They contain much more mathematics than the rest of the book. In Chapter 6 mathematical formulae provide the foundation for the system of microalgal cultivation we have devised. Unless you intend to reproduce this system, you can skip the formulae—the text can be understood without them.

The same holds true for Chapter 11, unless the reader is a specialist in devising life support systems for space applications or a mathematician dealing with ecosystem modeling.

We wrote this book firmly believing that the society of the twenty-first century would be increasingly interested in the problems of the biosphere and closed ecosystems. Therefore, we tried to write it in such a way that even the slight knowledge of biology could be sufficient to understand the whole book.

It was a difficult problem to decide what units we should use to express power, light intensity, etc. Different authors used different units in different times. In our writing we adhere to SI, but retain the units used by other authors, when referring to their data.

CHAPTER 1

The Earth's Biosphere as a Closed Ecosystem

We can define the biosphere as a dynamic system with an essentially closed material cycle kept in motion by solar energy, in accordance with the program built into the totality of genomes of all species inhabiting the planet, and regulated by the totality of their enzymes.

The noticeable drawback of this definition is that it does not include explicitly abiogenic physical and chemical processes. They certainly occur, but they had been going on even before the advent of life. As for the appearance of the habitable close-to-surface part of the planet, it has been formed and is being transformed in the course of evolution, under the pressure of life. It is this case that we try to reflect in our definition.

There are two ways in which "biosphere" is defined. One of them is to define biosphere as a receptacle of life on the planet rather than the pattern of life. In this case the word "sphere" is a geometric term describing Earth's shape. The other tendency is to imply by "biosphere" life as a global phenomenon.

We adhere to the latter approach. To define the biosphere spatially, the term "limit" of life or of the biosphere would suffice.

The term "biosphere" was coined by E. Suess, an Austrian geologist, as early as the nineteenth century (1875). In scientific literature it was used in reference to the inhabited part of the planet, i.e. to describe the limits to which life extends, but not life itself.

Later, in Vernadsky's works (Vernadsky, 1927, 1937), the concept of biosphere acquired a much more profound meaning. He used it to refer to the totality of all living nature together with the accumulated excreta of all living organisms, called bio-inert bodies, contacting metamorphosed or inorganic rocks. Thus, the concept of biosphere has acquired a definite scientific meaning, the property of the object that can be subjected to analysis by rigorous methods.

In the hierarchy of ecological systems the biosphere is the most important and includes all other ecosystems as its component parts.

We try to summarize the main properties of the biosphere:

Integrity The biosphere possesses properties of a single system as the organisms included in it are all related to each other through external material exchange.

Closure Material exchange of the basic biogenic elements, carbon, in particular, is almost absolutely confined to the biosphere.

Cyclicity Material exchange in the biosphere goes on in essentially closed cycles. For different elements the degree to which the cycle is open can vary, but it does not exceed a fraction of a percent for a cycle.

Steady state Rates of cyclic processes are in mutual equilibrium, so that current values of live biomass and essential biogenic element concentrations of the environment vary about equilibrium constants over long geological periods.

Energy dependence The biosphere, the system with the properties described above, which is far from equilibrium in terms of thermodynamics, can exist only because solar radiation energy in the optical range flows through it continuously. This energy is captured in photosynthesis and then, after a number of transformations in living and bio-inert bodies, is dissipated into heat.

These functional properties of the biosphere ensure potentially everlasting life under the conditions of a limited quantity of substrate on the planet.

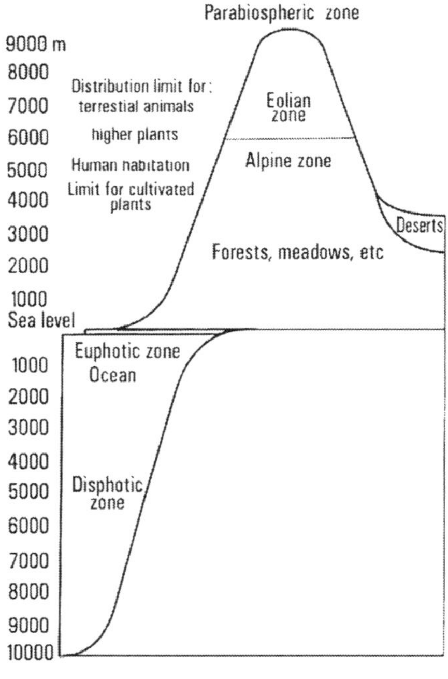

Figure 1.1 This scheme shows vertical extention of the biosphere. As the Earth's shell, the biosphere is of somewhat irregular shape, being surrounded by an indeterminate "parabiospheric" region in which life exists only in a dormant state as fungal spores and bacteria. The euphotic, or illuminated, zone of water bodies can span from several upper centimeters (in a very rapid river) to 100 m and more (in the most transparent ocean waters). (Quoted on J. Hatchinson "Biosphere," the Russian edition "Mir," Moscow, 1972 on The Biosphere Scientific American, Vol. 223, No 3, September 1970. With changes.)

Initially, steady state was not an inherent quality of the biosphere. A reducing atmosphere was characteristic of the early stages of evolution. The establishment of atmospheric O_2/CO_2 equilibrium first started only as a result of reciprocally balanced material exchange between autotrophic and heterotrophic life forms as late as in the Cambrian geological period of the Paleozoic, 600 million years ago. Apparently, cycling was not needed and did not occur in the period of the primary expansion of life, in the Archeozoic, billions of years ago. It developed as an adaptation to the limited resources of the environment and is, consequently, the

Table 1.1 Major quantitative parameters of present-day Earth's biosphere.

Parameter	Value
Biosphere area (km^2)	$509 \cdot 10^6$
including land	$148 \cdot 10^6$
including world ocean	$361 \cdot 10^6$
Mass of living material (t)	$1840 \cdot 10^9$
including plants (%)	99
including animals with microorganisms (%)	1
Ocean inhabitants	0.1
Land inhabitants	99.9
Elemental composition of biosphere (% by mass)	
Oxygen	75
Hydrogen	10
Carbon	3
Nitrogen	0.3
Mass of living material (%)	< 1
Mass of bio-inert material (%)	> 99
Average amount of solar energy incident upon Earth ($kcal \cdot cm^2 \cdot year^{-1}$)	167.00
Solar radiation use efficiency for biosphere (%)	0.13
Bio-inert mass in soil humus (t)	$2400 \cdot 10^{16}$
Net primary production (kcal/year)	$67 \cdot 10^{16}$
Land (t)	$40 \cdot 10^9$ C
Ocean (t)	$30 \cdot 10^9$ C

product of co-adaptation of the species comprising the biosphere and the evolution of the biosphere as a whole. Strictly speaking, it was only since that time that life became potentially everlasting. Apropos, the idea of a single living body covering the planet was also elaborated by J. Lovelock (1972) under the name "Gaia." Stripped of its mystical hue, Lovelock's idea is identical to Vernadsky's concept of biosphere, as the author himself mentioned in the latest editions (1991). It seems, though, that to use the word "biosphere" would be more proper not only because of its priority but also because it has a more accurate and specific meaning in natural science.

Consider the main quantitative parameters of the biosphere.

In space it extends down to 11 km, the maximal depths of the ocean, and to about the same distance up into the atmosphere (Figure 1.1). The Earth's surface is almost everywhere, except maybe for some central parts of the Antarctic Continent, covered in a relatively thin layer where life exists in the form of discrete organisms. The highest live biomass concentration and biological productivity can be found at phase interfaces: land–atmosphere and land–water. The major quantitative characteristics of the modern biosphere are listed in Table 1.1.

The current state of the biosphere is characterized by life under severe limitations that hold it in steady state. An example is provided by the situation in a tropical ocean. Moving along the equator in the Pacific Ocean, far away from the shores, over tens of thousand miles you can only see boundless deep blue, nearly violet water, a sure sign of the absence of life—phytoplankton. Measurements of productivity prove this. The production rate is hundreds of times lower than the average level for the ocean, like in a desert (Vinogradov and Shushkina, 1987). Why is this so under such optimal, "heavenly" conditions? Bright Sun pours energy onto the tropical ocean throughout the year, the water temperature is always at an optimal 25–28°C, and drought cannot occur. What can hinder the emergence of life? Analysis of water samples taken from the subsurface layer shows an almost total absence of biogenic elements—nitrogen, phosphorus, and iron in particular.

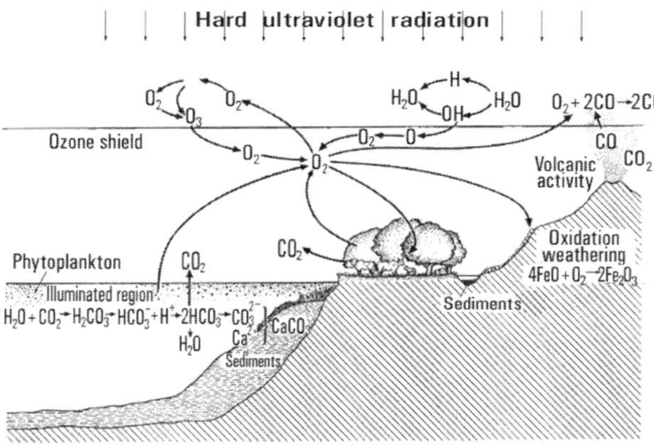

Figure 1.2 Oxygen cycling is rather intricate since it appears in many chemical forms besides the main form (O_2) and is a constituent of many compounds—water as well as other inorganic and organic substances. This simplified scheme shows some pathways of oxygen cycling on Earth. (Quoted on P. Cloud and A.Gibor "Biosphere," the Russian edition "Mir," Moscow, 1972 on The Biosphere Scientific American, Vol. 223, No 3, September 1970. With changes.)

The water samples collected from below a depth of 200–500 m contain sufficient amounts of biogenic elements to keep the life flourishing like in a tropical forest. But almost no solar light penetrates so deep into the water, and so life here is limited by energy deficiency. This is very typical of life—limiting factors can be different in different regions, but everywhere, under the pressure of life, they are used up. Only due to material cycling is a continuity of life maintained. An intricate food web uses the activity of its enzymes to keep biogenic elements cycling. Every element has its own pathways of cycling in the biosphere. Their schematic representation is given in Figures 1.2 and 1.3, showing patterns of oxygen and carbon cycling.

The rate of element cycling can be judged from the facts that, for instance, an individual CO_2 molecule stays about 10 years in the atmosphere, and an O_2 molecule more than a thousand years. It means that within the period of the biosphere's existence each atom involved in cycling has many times been a component of various bodies of the biosphere, animate and inanimate. A balance of cycling rates providing rather steady living conditions for all inhabitants of the planet is seen in the steadiness of the concentrations of the respiration gases O_2 and CO_2 in the atmosphere. For the Quaternary Period of the Cenozoic Era, that is for the last tens of million years, this steadiness has certainly been maintained.

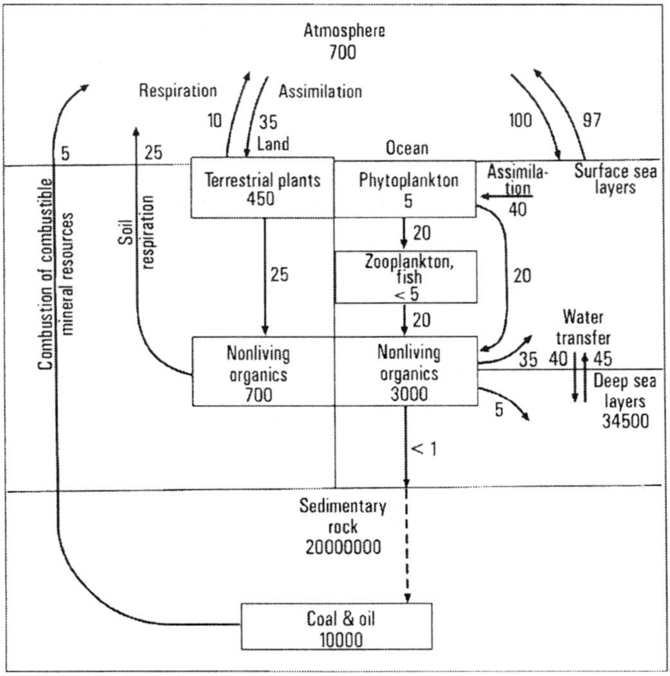

Figure 1.3 Carbon cycling in the biosphere consists of two different cycles, terrestrial and marine, with an interface between ocean and atmosphere. The ocean cycling is essentially autonomous. Carbon dioxide dissolved in sea-water is assimilated by phytoplankton, and oxygen enters the solution. Zooplankton and fish consume the carbon fixed by phytoplankton and use oxygen in respiration. As a result of the decomposition, organics of the carbon dioxide assimilated by phytoplankton returns to water. All values are given in billion tons. Annual combustion of about 5 billion tons of combustible mineral resources must result in a 0.7% increase in the amount of atmospheric CO_2, i.e. to 320 million $^{-1}$ (current content) almost 2 million $^{-1}$ must be added annually. Actually, an annual increase of mere 0.7 million $^{-1}$ is recorded. It means that two thirds of the carbon dioxide released in combustion quickly leaves the atmosphere, absorbed either by ocean water or by terrestrial flora. (Quoted on B. Bolyn "Biosphere," the Russian edition "Mir," Moscow, 1972 on The Biosphere Scientific American, Vol. 223, No 3, September 1970. With changes.)

It is this steadiness that guarantees sustainability and continuity of life on the planet and, consequently, the well being of humankind. Humanity's intensely developing industrial activity within the biosphere imparts a heightened poignancy to the problem of preserving the closed material cycle. In an effort to achieve various utilitarian goals, industry more and more often enters into conflict with the course of those natural processes that enable the biosphere's self-maintaining, constant material cycle—outside of which the existence of humanity and all life on the planet are rendered impossible. The scientific community has come to realize that it is vitally important for humanity to comprehend the mechanisms of biospheric cycling, keeping it closed and sustainable. In recent decades it has become clearly evident that technological activity is upsetting this equilibrium on a global scale. It is conspicuously manifested in the trend of increasing atmospheric CO_2 concentration. The alarmed world community has become more interested in the laws of existence of the biosphere and the limits of its sustainability. After millions of years of a relatively balanced existence, the biosphere is now getting out of this balance as a result of human activity. The current state of the biosphere and its dynamics cannot be described without accounting for the human impact factor, the serious interference in the biospheric cycle that is caused by technological processes devised by human intellect. Hence, even the current state of the biosphere can be regarded as a transition to Vernadsky's Noosphere. Table 1.2 presents some data confirming that this is not an unsubstantiated statement but rather a reflection of twenty-first-century reality. Based on the data of Table 1.2, we can compare the intensities of natural and technogenous flows of the principal respiration gases to have an idea of the magnitude of the latter.

Table 1.2 O_2 and CO_2 fluxes in the biosphere.

Process	Quantity
Atmospheric O_2 mass (t)	$1.0 \cdot 10^{16}$
Annual O_2 input to the atmosphere as a result of photosynthesis (t)	$5.0 \cdot 10^{11}$
Annual O_2 fixation in fuel combustion ("technological inspiration" of O_2) (t)	$9.0 \cdot 10^{12} - 40.0 \cdot 10^{9}$
Atmospheric CO_2 mass (t)	$2.2 \cdot 10^{12}$
CO_2 assimilation (t/year) by photosynthesis	$33.0 \cdot 10^{10}$
CO_2 emission by industry ("technological expiration") (t/year)	$16 \cdot 10^{9}$
Overall annual respiration (t)	$1.24 \cdot 10^{2}$
CO_2 release from Earth's tectonic sources (t)	$1 \cdot 10^{8}$

The CO_2 exchange is even more intense. Annual CO_2 assimilation in photosynthesis amounts to $3.3 \cdot 10^{11}$ t of $2.2 \cdot 10^{12}$ t contained in the atmosphere, i.e. 0.1

(otherwise reported as 0.01) of the atmospheric content. However this is counted, CO_2 completes its cycle in the atmosphere in less than a century, or over a million cycles during the Cenozoic Era, the balance being retained.

Human impact exercised through technological processes can be estimated by its contribution to CO_2/O_2 balance.

The annual inflow of oxygen into the atmosphere is one ten thousandth of the mass of atmospheric O_2, i.e. it takes only 10000 years for photosynthesis to regenerate an amount of oxygen equaling its atmospheric content. The CO_2 concentration in the atmosphere has been at approximately the same level for at least the whole of the Cenozoic Era, i.e. about 100 million years. These numbers imply that within this period oxygen has recycled in the atmosphere not less than 10000 times.

Technological production of CO_2 is only 20 times lower than consumption of it in photosynthesis, while O_2 technological consumption makes up a tenth of its evolution by photosynthesis. (A somewhat discrepant O_2/CO_2 ratio, i.e. the respiratory quotient of the biosphere, can be accounted for by lack of precision in global measurements.) Anyway, with the present-day level of industry, the contribution of human activity to O_2 and CO_2 exchange in the atmosphere amounts to 10–20%, which is obviously significant. As estimated by N. Moiseev (1979), even now combustion annually takes about 10 times more oxygen than all Earth plants taken together release into the atmosphere in photosynthesis. According to this calculation, in 300 years people will have breathing problems caused by a 3% drop in atmospheric oxygen partial pressure. Actually, asphyxia will not constitute a threat to humanity so soon. First, the ocean acts as a buffer: the oxygen dissolved in the ocean waters is in dynamic equilibrium with the atmosphere. Second, this estimate also needs revision. It does, however, suggest some imbalance in the biosphere caused by technological activity. It should be remembered, though, that concentration changes indicative of such imbalance of respiration gases in the atmosphere have occurred many times through the geological history of our planet. Yet the current changes occur at an unprecedented rate, leaving no time for evolutionary adaptation to them. It has already upset the balance of the biosphere, judging from the rate of the annual increment of atmospheric CO_2 concentration. The famous diagram of CO_2 concentration measurements made by the observatory located on Volcano Mauna Loa (Hawaii) demonstrates it clearly: the mean annual atmospheric CO_2 concentration increased by 2.6% during ten years (between 1959 and 1968). A longer series of measurements in Figure 1.4 shows that since the middle of the eighteenth century the CO_2 concentration has been steadily rising at an increasing rate. This value enables the permissible limits of the technosphere to be estimated.

Thus, the dynamics of the modern biosphere cannot be assessed ignoring the human impact factor.

Methods of remote sensing from space first made possible direct observation of dynamic processes in the biosphere, e.g. seasonal movement of the emergence of life across land and ocean, industrial impact on the ecosystem. The significance of this new vision, made available for humans, cannot be overestimated. However, even this does not enable us to determine the limits of biosphere sustainability.

This can be done only by studying its response to extreme conditions, challenging its very existence, that is, by experiment.

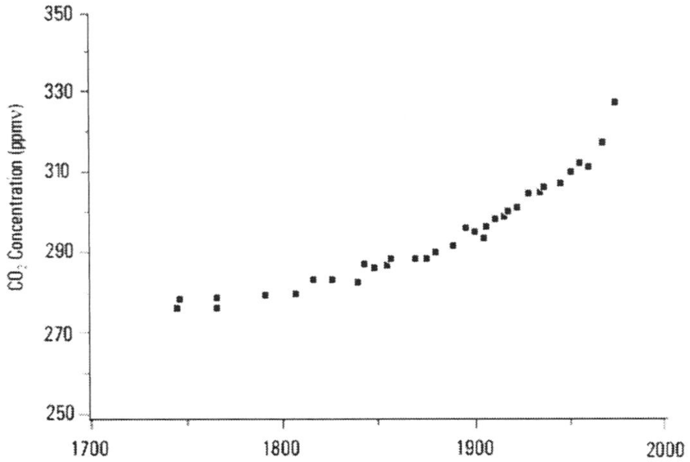

Figure 1.4 Atmospheric CO_2 concentrations since about 1750 derived from the Siple ice core in West Antarctica. (Quoted on B. Moore III and D.S. Barlet Earth and the Biosphere: Planetary Metabolism as a Paradigm for Global Biology in Space Biology and Medicine. *In* Space and its Exploration, Joint US/Russian Publication, 1993, p. 230.)

Let us turn to the tool for experimental investigation of the biosphere—modeling.

Models can be used if they adequately simulate closure—the basic specific quality of the biosphere.

There are two approaches to modeling—mathematical and physical. Let us examine both as applied to the biosphere.

A most powerful stimulus to the development of mathematical modeling was provided by the activities of the Club of Rome, which drew attention to the problem of the limits of technosphere expansion and the human population increase on the planet (Peccei, 1977). Works on system dynamics were developed to be applied to problems of the biosphere (Meadows *et al.*, 1974; Mesarovich and Pestel, 1974; Forrester, 1971).

Mathematical modeling of the biosphere as a single ecosystem of the highest hierarchic level has been most fully developed in the works of N.N. Moiseev and his school (Moiseev *et al.*, 1980).

It is not by chance that this line of inquiry has progressed in Russian science, Vernadsky's biospheric idea having deep roots in the philosophy of natural scientists there (see, e.g. Sukachev, 1944; Timofeev-Resovsky and Tyuryukanov, 1966; Yanshin, 1963; Yanshina, 1994; Shvarts, 1976; Korogodin, 1991; Kolchinsky, 1990; Gitelson *et al.*, 1995).

Figure 1.5 shows the block design of their model.

The model was used to run a number of computer simulations. Let us mention a few conclusions of interest. The scenario was typical of the currently separated world, in which one simulated region takes measures to halve pollution and to reduce the consumption of resources that cannot be regenerated, while the other does nothing of the kind. Computations showed that this scenario could defer but not prevent an outcome that would be catastrophic for the biosphere as a whole. The conclusion suggests the inevitability of world economics consolidation and absolute pre-eminence of global human interests above national ones. It is essential that this conclusion is not political but purely scientific.

Another scenario of the model, preservation of the existing trends of human impact, must cause, as the model predicts, serious cataclysms in climatic conditions and human population, as well as the limitation of the population by the quantity of food.

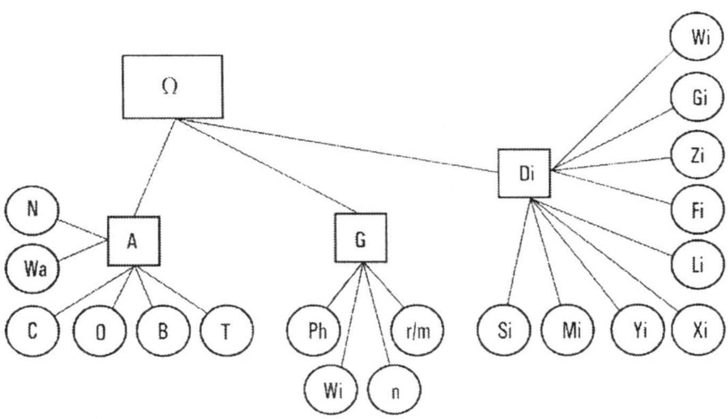

Figure 1.5 Biogeocenotic block diagram of the biosphere model.
Ω—biosphere; A—atmosphere: N—nitrogen (%), W_A—water vapor (mm), C—carbon dioxide (%), O—oxygen (%), B—turbidity (t/km^2), T—temperature (°C); G—ocean: Ph—phytoplankton (t/km^2), W_o—water (m), n—biogens (mg/m^3), r/m—necton (t/km^3); D_i—land region (i = 1): S_i—humus (t/km^2), M_i—mineral resources (t/km^2), X_i—agricultural vegetation (t/km^2), Li—forest (t/km^2), Yi—other vegetation (t/km^2), F_i—animal food (t/km^2), Z_i—pollutants (t/km^2), G_i—population (persons/km^2), W_i—water (m). (Quoted on N.N. Moiseev *et al.* "Human and Biosphere," MSU, No 4, 1980, p. 230.)

The third scenario is a global reduction in the intensity of technogenous impact on the natural environment. In this case, having lived through the period of two-century fluctuations, the biosphere reaches a state compatible with an indefinitely long existence of humans.

Though these prognoses are only relatively accurate, they vividly demonstrate the potentials of the mathematical modeling of biosphere dynamics.

However, mathematical models are barely if at all able to answer the question about the mechanisms responsible for the main quality of the biosphere—the closure of the material cycle, and to comprehend the laws of existence of this unique natural formation.

To resolve this problem, experiments should be performed on real, physically existing closed ecosystems. Since the biosphere is recorded as the only system of such a kind in nature, the models have to be created artificially.

The first attempts to create such systems were made almost a century and a half ago. In 1851 R. Warington described a balanced aquatic microecosystem. Later such systems received the name microcosms. Their chief advantage is that they can be used to conduct numerous versions of experiments with identical microbiospheres. Microcosms for various experiments were created by Odum and Hoskin (1957), Myers (1958), Beyers (1962), Golueke and Oswald (1963).

The use of microcosms yielded the most fruitful results in the investigation of ecosystem successions (Margalef, 1963) and fluctuations, as well as in the creation of the first models of life support systems for space applications. Myers was concerned with such systems several years before the first space flights (Myers, 1954, 1958).

In most of the above-mentioned works, however, microecosystems were not gas-closed, i.e. strictly speaking, they were not microcosms.

Investigations of entirely sealed microcosms were started and pursued simultaneously and independently by Clare Folsom, Professor of Microbiology at Hawaii University at Manoa (1986), and B. Kovrov, a Professor of the Institute of Physics (Biophysics), in Krasnoyarsk, Siberia (1992).

Some of the microcosms in sealed-off vessels, which they created in the 1970s, are still alive.

We can state that the laboratory tool, an object necessary for the investigation of the laws of biosphere existence and evolution, has been devised. However, a substantial part of the work remains to be done.

Laboratory microbiospheres have given rise to the line of creating experimental closed human ecosystems that can be named manmade noospheres. The chapters to follow are all devoted to the creation and experimental investigation of such systems.

CHAPTER 2

Manmade Closed Ecosystems and the Methodology of their Creation

Before considering the problem of creating artificial closed ecological systems, we have to define the notions and terms to be used since this field of knowledge still lacks a completely consistent nomenclature.

By an ecological system we mean some structural and functional whole including organisms and their habitats, in which processes of biological material and energy exchange go on, to support sustainable functioning of the self-contained system.

Besides a wide variety of natural ecosystems, humans can create different artificial systems. An artificial system designed to support the existence of a certain species (e.g. human) that dictates the qualitative specificity of all processes in the system can be termed this species' life support system.

An artificial ecological system can be conditionally subdivided into constituents—links. By a link we mean a species or a group of species occupying a structurally isolated part of the system and fulfilling the principal function of energy and material exchange in this part. The species whose life support is the purpose of the artificial ecological system can be termed the governing link, the other links all forming the environment for it.

No ecological system can function sustainably without energy exchange with the outside world, i.e. in thermodynamic isolation. Moreover, such a system cannot exist only due to an energy cycle, if only for the reason that in all transformation processes, primarily in biological links of the system, much of the energy is degraded to heat. Therefore, real ecosystems must always be open with regard to energy exchange. However, the specific energy exchange (i.e. power of the energy flowing through the system per, say, member of the governing link or unit mass of primary production of autotrophs) can be different for different systems, which is an essential parameter characterizing the efficiency of the system.

Ecosystems can be classified by the character of the material exchange with the outer environment. They can be divided into open ecosystems, existing due to outside mass exchange, and sealed ones, able to function for short to long intervals without an inflow of substances from the outside and without removal of substances from the system. An example of open natural ecosystems is provided by many biogeocenoses, and an example of open artificial ecosystems is provided by continuous flowing cultures of microorganisms.

Systems sealed with respect to material exchange can be subdivided into two extreme types: systems with stores of all the needed substances and systems in which the needed substances are regenerated in the cycles within the system. Systems of the first type can exist by utilizing the stored substances and building up the stores of the used substances, as was done in the first habitable spacecraft. The life of such closed systems is naturally limited, depending on the relationship between the quantity of the stored substances and the rate at which the substances are used.

Systems of the second type are closed ecological systems. Here, cycling of biogenic elements is organized within the system: some links use substances for biological exchange at a certain rate; other links regenerate them from the end products of their exchange to the initial state at the same average rate; then the regenerated substances are used in the same cycles of biological exchange. If the material cycle is perfectly coordinated, such systems can exist for infinitely long periods of time. The creation of such systems may be considered the ultimate goal of researchers. Since the middle of the 20th century, systems of this kind have been called sealed (Myers, 1954; Oswald, 1965; Shepelev, 1966; and others), closed (Nichiporovich, 1963; Gitelson *et al.*, 1964; Seryapin *et al.*, 1966; and others), self-contained (Henry, 1966), systems of material cycling (Yazdovsky, 1966), etc. The term "ecosystems with a closed material cycle" seems to be the most accurate. However, for brevity, we prefer a synonymous term, "closed ecosystems." In our opinion, such terms as "sealed" and "self-contained" cover a wider range of systems, including "closed" systems as a special and the most interesting case.

Unfortunately, in the literature written in English little or no distinction is made between the terms "sealed system" and "closed system," thus obscuring the fundamental property of closed systems: a material cycle within the system based on the perfect system that would exist to infinity, with limited amounts of substances involved in the exchange.

A closed material cycle can be attained by biological means, by physicochemical means, or by both. Since absolute closure of a material cycle is desirable but hardly attainable, it makes sense to apply the term "closed" systems not only to ideal systems, closed fully, but also to partially closed ones, in which the requirements of the links are satisfied both by the material cycle and by the stored substances.

Partial closure of the system can be estimated quantitatively, though somewhat conventionally. If the system is designed for human life support, it must satisfy human food, water, and respiration oxygen requirements. On a weight basis (kg), the daily requirements of one person are on average the following:

Oxygen	0.9	Food (dry weight)	0.6
Drinking and hygiene water	5.5	Total	7.0

For an artificially created, closed system including a human, it is customary to use the cumulative index of the system closure (R), characterizing the reduction in the portion of the expended mass of substances (or one separate substance) from the store, involved in the material cycle per unit time (Voronin and Polivoda, 1967; Gitelson *et al.*, 1975; Bartsev *et al.*, 1996). This index is defined as the ratio between the mass of the stores expended by the system and the mass of all substances (or one separate substance) consumed by a human in the system over the same time:

$$R = 1 - \frac{m}{M},$$

where m is the mass of the stored substances involved in the biological exchange which are expended daily by the system, and M is the mass of all substances used daily by a human. If the system does not need any stored substances ($m = 0$), then $R = 1$, which means complete closure of the system. R cannot be more than 1 but it can be less than 0, if the expenditure of stores in a life support system, e.g. for raising plants or animals, is larger than the mass of substances needed for the direct satisfaction of human requirements.

Sometimes it is not the total closure of the system, with respect to overall mass exchange, that is estimated, but a differential closure, with respect to a certain phase of matter (gas, liquid) or with respect to a separate substance. The systems that are fully closed by one, several, but not all differential indices are referred to as partially closed according to the general classification.

It should be noted that such estimation of system closure (let us term it "systemic" and designate this R_s) aims to reflect the achievements of researchers in reducing the weight of the pre-stored substances needed to provide human life support. This is rather a "technical economical" than an "ecological" characterization of the system, as we earlier defined it, without, as it appears, proper justification (Gitelson *et al.*, 1975, p.13–14).

Estimation of the system closure by the saving in the storage weight by no means reveals the nature of the principal ecological relations in the system—inter-link trophic chains realized in the cycling of biogenic elements; it does not determine the amounts of substances actually involved in the biological cycle; and in some cases such estimation even yields "negative closure"—an ecological nonsense.

It may be quite reasonable to estimate the system closure by the reduction in the needed mass of the stored substances versus the overall human requirement for them while developing and evaluating the efficiency of human life support systems for, say, space applications. However, such estimation is insufficient, to say the least, and in some instances unacceptable for comprehending processes in experimental closed systems in terms of ecology. For example, a human or humans should not be regarded as the obligatory (especially sole) heterotrophic link of experimental closed systems intended for simulating the biological material cycle in the biosphere or separate ecosystems. Of the most serious interest for researchers can and will be closed systems including various heterotrophs, without direct human participation in the biological cycle. Furthermore, in a system including a human it would be hardly reasonable to regard substances for satisfying some human requirements, such as water for washing, shower, doing the dishes, etc. as components of the biological material cycle. These substances amount to approximately half of all the stores used by a human, depending, however, on social rather than biological human living standards. Besides, oxygen and hydrogen (the biogenic elements composing water) are little involved in the processes of biological exchange, only in water photolysis in the course of photosynthesis and in water formation in the processes of heterotrophic metabolism.

For qualitative estimation of the degree of closure of the biological material cycle it seems to us reasonable to accept the approach proposed by Lisovsky and Tikhomirov.

In accordance with this approach, the closure of the biological cycle is estimated separately for each of the principal biogenic elements. In this way we can see the inner structure of the closure, see elements limiting the entire cycle of the system, and reveal the relations within the system in which this limitation is realized. The index of cycle closure for any biogenic element in the system (R_c—the cycle closure) is the ratio of the quantity of this element passing in unit time through the biological cycle forward and backward between heterotrophs and autotrophs to the total quantity of this element utilized in the system by both links over the same time:

$$R_c = \frac{2m_{min}}{M_a + M_h},$$

where m_{min} is the minimum of the two inflows of the element to the autotrophic or heterotrophic link, the inflow of one of the links determining the total carrying capacity of the element cycling through both of the links of the closed system; M_a and M_h are the requirements of autotrophs and heterotrophs for this element.

If several species of autotrophs and heterotrophs are closed in the system, exchange processes between them are considered in a generalized form, as relations between all of the autotrophs and all of the heterotrophs.

It is evident that to estimate the closure of the biological material cycle in the system, other input data than those used to estimate "systemic" closure are needed, characterizing the saving in weight of the stored substances expended by the system versus the cumulative weight of the store. The closure of the biological material cycle R_c is estimated based on the quantitative parameters of the flows of each of the principal biogenic elements involved in the two-way material exchange within the system between autotrophs and heterotrophs as well as on the total amount of these elements utilized by the system links in unit time. Using these values, we can easily calculate the overall estimate of the closure of the whole biological material cycle in the system, but such an estimate carries little information for comprehending the *structure* of the closure.

Sustainability, "vitality" of closed systems will be crucially dependent on the promptness of the control and accuracy in coordination of all processes maintaining the biological material cycle. In physicochemical closed systems it is inevitable that control should be chiefly exercised by specialized technical systems. In biological systems control can be based on the stochastic principle, as in natural ecosystems: biocenoses and the biosphere. It will be shown below, however, that the stochastic principle of control alone is insufficient for artificial closed biological systems. The necessity for a system of instrumental control as an integral part of an artificial closed biological system provides justification for many authors to use the term "biological technical systems" as reflecting not only ecological but also the organizational principles of the creation and operation of such systems.

Open life support systems, mostly based on the stores of all substances required by humans, have been the first systems to be used in space and still have wide application. The principle of creation, ways of technical realization, and results of the functioning of such systems in the conditions of real space flights are described in detail in the literature (Chizhov and Sinyak, 1973; Adamovich, 1975; Sauer and Shea, 1981; Popov and Bychkov, 1994).

An ecological system functioning on the basis of the stores of substances required by a human can exist, in principle, until the quality of the stored products degrades. However, there is another limiting factor for the ecological systems designed for space application: the weight characteristics of the system. Simple calculation shows that, for example, the crew of a spacecraft consisting of three members will need a store of the net weight 1890 kg for three months of operation of the life support system. Stores weighing that much present no difficulty for up-to-date space engineering. However, missions to other planets or stays of large crews in orbital stations will demand life support systems that can function over months and years and the weight of the storage will increase accordingly. If, for instance, life support is provided for a crew of five members during three years, only the net weight of the store will exceed 37 tons (including nearly 5 tons of oxygen). Rapid advances in space engineering, though, remove these weight limitations, at the same time posing the problem of creating an adequate environment in which humans could stay outside of the Earth for a long time. Researchers have to find the ways of creating self-contained homes, which could become the daughter formations of our planet and its biosphere, with a natural environment for humanity to spread into outer space.

That is why many researchers are searching for ways to create ecological systems with an organized biological material cycle. Should completely closed ecological systems be created, humanity could truly and biologically tear ourselves away from the Earth's biosphere, and infinitely long human life support outside of the Earth could become feasible.

Creation of an ecological system with humans as the governing link presupposes that other links of the system (physicochemical, biological) should reclaim oxygen, water and food from the consumed and processed end products of human metabolism: gaseous, liquid and solid wastes.

It was long ago proposed to study and evaluate physicochemical and chemical methods of producing oxygen by water electrolysis and CO_2 reduction in experimental closed life support systems (Voronin and Polivoda, 1967). The same is true of regenerating water required by humans from the sources within the system: condensate of moisture exuded by humans, urine, and hygiene sewage water. There are some physicochemical methods of treating contaminated water that yield satisfactory results (Sinyak and Chizhov, 1964; Seryapin *et al.*, 1966; Khomchenko *et al.*, 1971). A number of experimental systems with physicochemical methods of maintaining gas and water cycles have been designed and constructed. Tests of such systems including humans proved the feasibility of prolonged application (up to a year), maintaining a partially closed gas and water exchange (Olcott and Conner, 1968; Jackson *et al.*, 1968; Burnazyan *et al.*, 1969; Pecoraro and Morris, 1972).

The significant advances made in this field suggest that physicochemical processes should be widely used and investigated in artificially created closed systems. On the other hand, we should not underestimate the difficulties that will face researchers who take this approach. Regeneration of atmosphere and water by physicochemical methods brings into the system such undesirable factors as high temperature and pressure, explosive or toxic intermediate products, caustic compounds, etc. It is hardly a coincidence that until now both in the USA and in Russia spaceship crews have been provided with oxygen and water stores rather than with physicochemical systems to maintain the cycling of these substances (Bluth and Helppie, 1986; Humphries *et al.*, 1994).

The prospects for using chemical synthesis for food regeneration from human wastes are quite dim. Considerable work has been done along this line, much progress has been gained in the synthesis of carbohydrates and fats, and even specific schemes have been proposed for synthesizing separate substances for life support systems. On the whole, however, this problem is far from being solved. There is little hope that development of methods and technology of chemical synthesis of full-value food, acceptable for closed systems, can develop at the same pace as does space engineering stimulating experimental work on creating artificial ecological systems.

Processes of autotrophic biological synthesis possess a unique capacity for utilizing human gaseous and solid wastes and, at the same time, recycling them to generate molecular oxygen, water and organic nutrients, at normal temperature and pressure. Besides performing a number of functions in the closed system, living organisms can reproduce themselves, too, which is the prerequisite to their continuous and infinitely long functioning as environment-forming links. Physicochemical links of closed systems will hardly ever possess such properties. For this reason alone, in creating closed systems for continuous and prolonged operation, biological synthesis seems to be a more promising way than physicochemical systems. However, it is still early to give preference to biological or physicochemical systems for long-duration life support because both of them are in the initial stages of development. For the time being higher optimism about one or the other way of creating closed systems is most often based on approximate knowledge of its advantages and even worse knowledge of its potential disadvantages. It is only comprehensive prolonged theoretical and, what is more important, experimental investigations of both physicochemical and biological closed systems that with time will allow a well-grounded estimation of the true advantages and disadvantages of either variant, resulting most probably in creating combined biological physicochemical systems. Since the main goal of this book is to analyze the feasibility of creating a closed system with a biological material cycle, from here on we deal with this system only.

Different researchers have opposite opinions regarding the feasibility of the creation of closed biological systems. It is often argued that nature gives us an example of such a system, the Earth's biosphere. Even in 1958 A. Taylor asserted that it is undoubtedly feasible to develop a closed cycle of regeneration of the system elements, as such a cycle already exists in nature, and the problem is reduced to the development of technical aspects (Taylor, 1958). Discussing this problem,

Gyurdzhian (1961), Shepelev (1963) and some other authors assumed that it is actually a creation of the model of Earth's living nature. All human material and energy relations must fit into this model of the biosphere.

On the other hand, Violette and co-authors (1959) made an attempt to estimate theoretically the feasibility of creating a closed biological system as a model of the biosphere and discussed various schematic variants. Taking into account that real energy and mass exchange between humans and the biosphere is very complex and that the processes in the biosphere effecting this energy and mass exchange are complex, branched, and often extremely inertial, these authors arrived at a conclusion that a closed biological human life support system could not be designed in the near future and rated the projects associated with it as utopian.

S.A. Baranov (1967), basing his ideas on U. Liebig's theory of the "return" of mineral elements in agriculture, believes anthropogenic cenoses rather than the biosphere to be a prototype of closed ecological systems. V.P. Dadykin (1970a) is of the opinion that an artificial system can and must be calculated and organized by the laws governing processes in natural biocenoses. Anthropogenic and natural cenoses, however, are open systems; the material cycle within them is quantitatively imbalanced with respect to many elements and represents only partially the general material cycle of the closed system: the biosphere. A more correct opinion is expressed by O.G. Gazenko (1968): theoretical prerequisites based on the knowledge of open Earth biocenoses are too general and insufficient to solve such a specific and qualitatively peculiar problem as the creation of closed ecological life support systems.

It is the opinion of a number of authors (Lisovsky *et al.*, 1965, 1967; Shepelev, 1966) that it is not quite valid to approach the problem of creating closed biological systems as a problem of simulating the Earth's biosphere as a whole. Analysis of human–biosphere relations and biogeochemical processes in the biosphere proves that they do not conform to the requirements that a closed biological system must meet.

First, in the biosphere, part of the products of biogeochemical transformations drop out of biological cycles, forming excess buffer deposits and deadlock, excluded-from-the-cycle, products. Examples include oxygen in the atmosphere, carbon in coal and peat, carbon and calcium in limestone, etc. The amount of biogenic elements which drop out of the cycling is tens and hundreds of times larger than the amount of those involved in exchange processes. On the other hand, exchange processes in the biosphere continuously involve new, earlier inert masses of crustal material and substances released from inside the earth. Thus, at least part of the biogeochemical processes in the biosphere does not follow the pattern of an artificially closed cycle.

Second, in a real situation, physical and chemical environmental conditions, under which organisms and their communities live on Earth, vary with time. These variations are partly caused by cosmic and geological processes and forces independent of living organisms but mostly by the factors resulting from the vital activity of organisms (exhaustion of food reserves, biogenic processes of soil formation, influence on atmospheric composition, etc.). Therefore, the conditions under which mass exchange goes on are actually unstable both in the biosphere as a whole and in separate parts of it and so the mass exchange itself is unstable, too.

Third, the species of living organisms involved in biogeochemical processes also change on a historical scale. Among the factors causing evolutionary processes in the biosphere, one is very significant—the relationship between organisms and the outer environment, which is of positive feedback in character: processes of the vital activity of organisms themselves cause a number of changes in the environment, leading to evolutionary emergence of new varieties of living organisms. They again cause new, specific changes in the environment affecting the progress of evolution, etc. Thus, inherent in the very nature of biogeochemical processes occurring in the biosphere are conditions bringing about evolutionary changes in biological links and the biosphere as a whole, in this way determining changes in the paths of mass exchange and energy exchange processes.

The biosphere as a whole cannot be taken as the ideal biological system since not only biological but also abiotic physical and chemical processes play a significant part in its mass exchange. Neither is it the ideal completely closed system as some exchange processes in the biosphere continuously involve in the biological cycle various amounts of substances that have not participated in the cycle before. In terms of the above classification, the biosphere can be defined as a partially closed combined biological physicochemical system with stores. In designing an artificial system of the same type, the biosphere may serve as an original that can in principle be simulated by humans. To simulate the biosphere, it is not enough to reproduce qualitatively the same processes of mass and energy exchange in an artificial ecological system. An even more important condition of the vitality and sustainability of the system is the quantitative coordination of all the processes by regulating their rates. It should be remembered that artificial closed human life support systems are created to achieve specific aims, which was not the case in the formation and development of the biosphere. Besides, though feasible for people (and even for humanity as a whole), the creation of artificial closed systems is limited by their dimensions (weight, size, energy consumption), incommensurable with the dimensions of the biosphere. All these limitations are of principal importance. Due to them the biosphere as the original for simulation decreases in importance as the principal mechanisms responsible for its sustainability become inefficient in artificial closed systems.

In the biosphere mass and energy exchange rates are regulated only by the stochastic equalization of the results of numerous differently directed processes uncoordinated in time and space (biosynthesis and destruction of organic matter, birth and death of individual organisms, emergence and dying-out of species, etc.). Stochastic regulation can be illustrated by "waves of life" manifested in recurrent numerical fluctuations of the species trophically or otherwise related in biotic communities (Chetverikov, 1905; Volterra, 1931). Obviously, in artificial systems designed for human life support, survival and the number of humans cannot be dependent on merely stochastic means of coordinating mass and energy exchange rates in the system.

Relative stability of the biosphere despite temporal and spatial fluctuations in various exchange processes, i.e. efficiency of stochastic control, is determined by the buffer properties of the biosphere.

They are manifested, first, by the multitude of individuals of each species, i.e. by repeated duplication of each program of material and energy transformation, second, by the multitude of biological species able, to some extent, to replace each other in processes of material and energy exchange; third, by enormous amounts of reserve masses of substances involved in the biological cycle. These mechanisms are responsible for the high probability and reliability of all processes maintaining the material cycle in the biosphere.

The dimensions of an artificial closed system are limited, which significantly diminishes and sometimes excludes the possibility of ensuring relative stability of the system using buffer properties.

The stochastic principle of control will not do for artificial closed systems in addition because the biosphere has been continuously and progressively varying with time since the moment it originated, i.e. it has been evolving. The evolution proceeds ignoring vital requirements of any biological link similar to the governing link in artificial closed systems. On our planet it was not the biosphere that formed and developed adapting itself to certain species of living organisms, but species of organisms and biocenoses formed, owing to natural selection, in accordance with changing conditions of the part of the biosphere surrounding them. True enough, with the advent of human society people started controlling evolution of the biosphere, deliberately developing it along certain lines. Until now, though, humanity, with its great progress in science and technology, has been far from successful in making the evolution of the biosphere serve its interests. Neither has it achieved favorable developments of any significance in the biosphere on a planetary scale, though there are very many examples of unfavorable developments on a global scale.

In the artificial closed system designed for life support of a particular governing link spontaneous evolution of the other biological links, or any one of them can reduce the biological reliability of the system for such evolution can result in the disruption of coordinated interlink relations and the degeneration of the system as a whole.

Unlike the evolving biosphere, whose biological links are subject to transforming selection in various directions, the artificially created closed human life support system must be genetically stable. To do this requires continuous stabilizing selection in every biological link of the system. The only direction of natural species selection that is not prohibited in such an artificial system may be selection aimed at increasing exchange rates without changing the quality of exchange.

Examples of stabilizing selection can be found both in natural cenoses and in human activity aimed at preserving the desirable properties of the selected strains, cultivars, and breeds of organisms. In both cases, however, stabilizing selection is only a special case among other forms of selection. It reflects a temporary equilibrium state of the system, which generally changes and develops. In the artificial closed system stabilizing selection must become one of the obligatory constant conditions ensuring the vitality of the system.

In short, while the biosphere continuously and progressively changes thanks to evolution, the artificial closed system, by means of "evolution aimed at the absence of evolution" must be maintained stable for at least the lifetime of the governing link.

Summing up all arguments about the biosphere as the original for creating the artificial closed ecological system, we can conclude that at most it gives us an example of the elementary biological and physicochemical processes, which allows us to regard the fully closed life support system as principally feasible. However, for the selection of links realizing these processes and the arrangement and organization of these processes into an integral, stable artificial system that will, in the ideal case, function for an infinitely long period of time, the heuristic significance of the natural material cycle is not very high.

Imposing limits on the action of the stochastic principle of control, reduction of buffer capacity to the minimum, prohibition of transforming directions of evolution in artificial closed systems—all these requirements call for such principles of functioning and control that cannot be "borrowed" from the natural system, the biosphere.

Functional stability of energy and mass exchange in the artificial closed system, the needed accuracy of coordination, and promptness of regulation of processes in it are attainable only by employing a deterministic structural control system. Natural analogs in which this principle of control is effected are open systems—organisms.

The same principles serve as a basis for most control systems applied in engineering. In our opinion, by extending these principles to controlling artificial closed systems we can get a clue of how to solve the problem of their sustainability. Should this problem remain unsolved, creation of such systems would be possible only in a fantasy world.

To create artificial closed systems combining properties of two levels of biological organization—the pattern of mass exchange of the biosphere and the pattern of control of the organism—is a new scientific problem, unprecedented either in nature or in human activity.

The introduction of the principles of deterministic control into closed ecological systems gives us grounds to regard such systems not as experimental models of the biosphere but rather as experimental models of the future "noosphere," the sphere of life on Earth controlled by human intellect with the goal of keeping the biosphere in the conditions suitable for human life. Such models may allow substantiating experimentally the permissible limits of human controlling effect on processes of biological material cycles in the biosphere and finding boundary conditions of maintaining these cycles sustainable in complex interrelations of the biosphere and technosphere. Creation of models of the future noosphere including humans as the main link will be a general solution of the problem of human existence outside of the Earth, in daughter "mininoospheres," as long as necessary.

Consideration of the biosphere as a whole in the context of the problem we face shows that the natural system with the material cycle cannot be copied and reproduced as an artificial closed system. Naturally, since the 1950s many researchers have proposed different variants of theoretically conceivable partially and fully closed ecological systems (Brockmann *et al.*, 1958; Tisher, 1961; Nichiporovich, 1963; and others).

However, feasibility and viability of such systems cannot even be predicted theoretically because the character of metabolism and environment-forming functions of each organism in the closed system essentially depend on the selection of organisms it associates with. Our knowledge of the full range of metabolic products of different organisms and regularities of its fluctuations under the influence of possible changes in living conditions is extremely small. That is why theoretical models of closed systems usually take into account possibilities of interlink coordination only with respect to the most "bulky" metabolic products expressed as their biochemical "total."

These models cannot adequately take into consideration the mutual physiological and biochemical effects of the biological links of the system, produced by various metabolic products, for instance, micro-metabolites. First, the full range of micro-metabolites has not been sufficiently studied even in such a thoroughly investigated object as a human, even less profoundly in various microbes, plants, and animals. Second, because accumulation of knowledge in this field has just started, very little is known definitely of the biological role metabolites play in the interrelations of organisms. Third, even if we have adequate data on the peculiarities of the metabolism of individual organisms, the information obtained will not be sufficient to predict the mutual effects of these organisms should they exist in a single closed system. A.L. Kursanov (1966) pointed out that the physiology of a plant growing separately is different from that of a plant in a plant community since the very fact of co-existence of plants leaves an imprint on their growth and development. The same is absolutely true for animals and microorganisms.

Thus, even in open ecosystems, the mutual physiological biochemical effect of organisms is a real factor of their formation. In a closed system the contribution of such effects can be many times larger and influence most significantly the growth of co-existing organisms and the viability of the system as a whole. To estimate adequately the character, strength and remote consequences of the mutual effects of organisms in a closed system, it is not sufficient to study the links separately; it is essential to study the links actually functioning in the closed system.

We give this example to demonstrate that theoretically designed closed systems can only be used as a guide, though a very important one, to the approach to the solution of the problem. The state of the work in the field of biology allows a true solution of the problem in the context of science and, especially, for application, only in experiment.

Efforts to solve the problem experimentally can be based on different strategies.

On the one hand, it is possible to create a complex system in which the governing link will exist in the conditions maximally close to natural ones in terms of the number and nature of relations between this link and other links. An example of such an approach can be provided by the well-known experiments in Biosphere 2 (Allen, 1991; Nelson, 1993; Nelson *et al.*, 1993).

On the other hand, there is a possible way to break the historically developed natural relations among biological links and to create new bonds and relations, determined by necessity rather than tradition.

The latter approach allows closing of the system at the initial stages with the minimal number of links involved—probably, two: autotrophic and heterotrophic. It should be noted that each of the links will be represented not by separate species but by communities consisting of the "leading" species (human, *Chlorella*, wheat, etc.) and many species of associated microorganisms for it is beyond reason to hope now for bacterial sterility of humans and, consequently, of the system as a whole. It is this way, that of increasing complexity, that we prefer in our experimental work.

The creation of systems using the approach of increasing complexity presupposes, first of all, detailed knowledge of the outer exchange of each link included in the system. Humans had existed in the biosphere long before they found out, for instance, what chemical elements and molecules they need, but knowledge of this is an obligatory and preliminary condition of human existence in the artificial closed system. This is equally valid for the range of substances flowing into or out of the link, as in the closed system every link, the human link included, is a necessary condition for the existence of any other link.

Nature has a wide assortment of potential environment-forming links among which one can choose to create different variants of systems with various combinations of energy and mass exchange relations between links and to pursue a wide range of research. The choice of the governing link of the system is predetermined by the goal of creating it—to provide human life support. Hence it follows that human energy and mass exchange characteristics, human requirements on all parameters of the habitat in the artificial closed system, will primarily dictate the specificity of constructing the system and must be examined before examining the energy and mass exchange characteristics of other links of the human life support system.

CHAPTER 3

Creation of Manmade Closed Ecosystems for Human Life Support: History of Development

Space research has the virtue of being international. It is obviously important for humanity as a whole, and extremely costly to conduct. But from the outset, under the conditions of a cold war and very severe economic competition, there was cooperation and exchange of scientific observations and results. We can state with pride and satisfaction that in the field of space biology and medicine in particular, cooperation and data exchange between the USA and the USSR was regular and continuous from the early 1960s through the 1980s. This is specifically true for work on biological life support systems.

However, close cooperation in creating artificial closed ecological systems—clearly a radically new issue in the realm of the natural sciences—was not as intensive as it should have been. This is particularly sad when the importance of this line of investigation is considered, not only for prospective space applications but also for solving a fundamental global problem: preservation of our planet's biosphere.

This lack of close cooperation was probably due to a combination of reasons, including the political isolation of the countries from each other, real differences in research traditions, variations in approaches to the problems, and to the availability and costs of applicable technologies. However, the problem of creating closed ecosystems for use on Earth and for space applications really concerns humanity as a whole. For this reason we hope that in the 21st century, in the face of new ecological and geopolitical realities, the scientific communities dealing with this problem will be more far-sighted and integrated than in the past. At the moment, it still makes sense to describe the history of this field of knowledge individually for the USA, the USSR (now Russia), the European Community, and Japan.

3.1 IN THE USA

In the United States interest in using biological systems for life support began in the early 1950s. For several years (until 1958) the U. S. Air Force supported contracts with at least one private research company (Arthur D. Little, Co., Cambridge, Massachusetts) to grow green algae in closed systems. The intention of the research was apparently to explore the use of algae to produce oxygen for human airplane pilots. It was assumed that the technology would be eventually useful for extended flight airplanes that would fly above the atmosphere, and that the system could eventually be used in space.

Partially in response to the launch of Sputnik by the USSR in 1957, the National Aeronautics and Space Administration (NASA) was created by an act of Congress in 1958. Almost from its beginning NASA had an interest in various approaches to human life support. Because the Agency was formed from existing aeronautics laboratories (National Advisory Committee for Aeronautics (NACA)), and also had an extensive military tradition, physicochemical life support systems (such as those needed for airplanes and in submarines) were the predominant focus for life support research for space transportation. Studies of closed ecosystems were discussed at a NASA meeting at Ames Research Center in 1965, the proceedings of which were published in 1968.

Attempts to create experimental closed ecosystems that include microalgae have been made since the late 1950s. Experiments were done using small test animals, such as mice, rats or monkeys (Bowman, 1953; Gafford and Craft, 1959; Myers, 1958). However, balanced systems were not easily achieved, primarily because so little was known about the cultivation of microalgae, and no special apparatus had been developed for maintaining the system.

For example, Bowman and Thomas detected an increase in the carbon dioxide concentration in the atmosphere in their closed system, but with no decrease in oxygen concentration. Bates, who described the accumulation of carbon monoxide in the atmosphere of a closed system containing a monkey, reported a different mismatch. Because of various gas imbalances, closed experimental systems with animals did not last long. In such systems, water was not recycled, although in some experiments animal wastes were used to cultivate *Chlorella.*

By 1965, about 6 years after NASA's formation, a consensus was forming among NASA scientists and engineers about how to support humans in space for extended periods of time (Bioregenerative systems, NASA-SP-165, 1968). The view was that the most reliable (and most thoroughly understood) methods were based on physical chemical techniques, for example for CO_2 removal, for O_2 regeneration through electrolysis of water and for separation and storage of wastes. However, it was acknowledged that bioregenerative approaches showed some promise, and that efforts should be expended in investigating biological approaches to the problem.

The two major bioregenerative approaches considered were those employing photosynthesis by algae or chemosynthesis by "hydrogen bacteria." The use of algae assumed that the CO_2 produced by humans would be removed by the algae, which would incorporate its carbon into their structure. At the same time the organisms would capture light photons and use the energy to "split" water into H (as NADH2) and O (to form O_2). The O_2 would be released for human use, and the NADH2 would be used to chemically reduce the carbon (from CO_2) to cellular materials. Initially it was assumed humans could directly eat algae as food.

The bioregenerative systems based on chemosynthetic hydrogen bacteria assumed that water would be electrolyzed with electric current, generated by photocells, to H_2 and O_2 (sometimes described as an "artificial photosynthesis"). The O_2

would be made available for human respiration, and the H_2 would be fed (along with some of the O_2) to the hydrogen bacteria. The bacteria would consume the H_2 as an energy source, and remove CO_2 (exhaled by humans) which it would reduce with H_2 to form cell materials. Extensive studies on the growth of hydrogen bacteria under controlled conditions were conducted at the Battelle Memorial Institute by Foster and Litchfield (1967). Other studies, also supported by NASA, included metabolic investigations ranging from cell composition to the behavior of enzyme control systems (MacElroy *et al.*, 1969; Ballard and MacElroy, 1971).

While schemes for either photosynthesis or chemosynthesis would regenerate the O_2 needed by humans and remove the CO_2 that they produced, it was clear that these processes would be much more efficient as life support technologies if the cell materials produced could be used as human food. To this end a series of studies was conducted on the human food value of microorganisms.

It was found from animal feeding studies that green algae caused digestive disturbances in rats when the algae constituted only a small fraction (1% or less) of the edible mass. It was determined that the cell wall material was responsible for most if not all of the animal digestive disturbances. In contrast, blue-green algae (*Cyanobacteria*) were known to compose a significant fraction of the diet of several African tribes during certain times of the year. In reality, although both organisms were called "algae," green algae are much more closely related to plants, and *Cyanobacteria* are photosynthetic bacteria, and lack the cell walls of "true" algae.

Feeding studies of hydrogen bacteria were conducted at the University of California at Berkeley by D. Calloway's laboratory. The final report of her studies (Calloway and Margen, 1968) suggests that the species used in the studies caused digestive disturbances in humans that would probably preclude the organism's being used as a life support food source. Subsequent work on the classification of hydrogen bacteria indicates that of the two commonly used species *Hydrogenomonas eutropha* is in fact more closely related to the genus *Alkaligenes*, while *H. Facilis* is probably a member of the genus *Pseudomonas*.

The negative consequences to humans of the hydrogen bacteria feeding were widely noted. By the mid-1970s NASA's support for research on bioregenerative life support systems had dwindled to a few small contracts dealing with the photosynthesis of algae. In retrospect, the reaction was overly harsh, and it is likely that if extracts from *H. facilis* had been used instead of material from *H. eutropha*, less severe human reactions would have been observed. But, also in retrospect, the concept of conducting such human feeding studies, in the absence of extensive knowledge of the organisms involved, is amazing, even startling.

In 1976 a workshop, conducted under the aegis of the National Research Council (Washington, DC), was held at Snowmass, Colorado. The participants were invited to consider whether NASA should re-examine the issue of biological life support. While the minutes of the meeting were not published in the open literature, attendees included J. Spurlock (Georgia Tech University), P.D. Quattrone (Ames Research Center) and D. Popma (NASA Headquarters). The consensus of the meeting was that biological approaches appeared reasonable for long-term missions, and that a wide variety of organisms should be explored. The manager of the existing life support program (D. Popma) began to consider various engineering approaches to biological life support.

At approximately the same time, R.S. Young, the Program Manager of the NASA Exobiology Program, began to consider the role of ecological principals in exobiology, and in life support. He convened a group of scientists, primarily ecologists (D. Botkin, chair, Yale University; L. Slobodkin, SUNY Stonybrook; B. McGuire, University of Texas, Austin; H. Morowitz, Yale; and B. Moore, University of New Hampshire), who met for extended periods over two years (1978 and 1979). The "Botkin group" was to consider the role that ecology should play, as a discipline, in issues such as life on other planets, and regenerative life support systems. In summary, the group's comments about biological life support stressed that the interactions among organisms within a life support system should be a principal focus of research; however, they stopped short of projecting failure of such a system because of ecological interactions (Cooke, 1971).

From the Snowmass meeting emerged a concept of biological life support as an element of an ecological system, as well as a possible name for the research activity: the Controlled Environment Life Support System (CELSS). This concept was strengthened by the Botkin group's emphasis that ecological principles must be considered when life support system for use in space was developed.

These data and philosophies contributed to the background material and rationale for a NASA "new start"—essentially additional funding for the start-up of a new NASA program: the Controlled Ecological Life Support System (CELSS) program, which was initiated in 1978. As money became available for research, two NASA centers became involved in the activity: the Johnson Space Center, which focused on food processing and human diets in space, and the Ames Research Center, which was concerned with basic research in system controls.

The philosophy that Ames assumed in its approach to the CELSS program was projected in an internal publication that is available in the open literature: "Space Ecosynthesis: An Approach to the Design of Closed Ecosystems for Use in Space," MacElroy, R. D. and Averner, M. M., NASA Technical Memorandum 78491 (1978). There were three elements that became the bases for subsequent approaches:

1) that a regenerative space life support system would indeed be an ecosystem, but that it would resemble a terrestrial farm, rather than a typical isolated ecosystem;
2) that humans would be a major component of the ecosystem; and
3) that an ecological life support system would have a goal, unlike a natural ecosystem, namely the support and sustenance of its human beings.

Thus, like a perfectly operated farm, a space ecosystem would be stringently controlled as to the type and number of species supported, but unlike a natural ecosystem, it would be directed to support its farmers.

The conceptual approach adopted at Ames influenced the kinds of research that would be supported by the CELSS program. It also focused the competing influences addressing long term life support in space. The major directions in the early 1980s were set by the advocates of physical–chemical life support and bioregenerative life support. Within the bioregenerative life support camp were the ecologists, whose approach might be (possibly unfairly) summarized as supporting the long-term, natural evolutionary development of ecologically based life support systems, and the "practical" ecologists who supported the idea that by controlling environmental parameters, it would be possible to establish a stable, functioning, goal-oriented life support system using biological components. This latter approach, adopted by the CELSS research program assumed that, to the extent that the biological components were "ecologically compatible," less effort would be devoted to system control. Thus, the most "natural" systems would be the most efficient.

An innovative research approach that emerged from these considerations focused on issues of system control, and was embedded in theoretical, mathematically-based studies. The most far-ranging investigations were initiated by D. Auslander and R. Spear of the University of California at Berkeley. Auslander, Spear and their students, over the next several years, developed dynamic computer models that allowed the simulation of simple, closed-life-support-system stability. A particularly notable result, derived from the simulation, was that an apparently insignificant but persistent system fault could, over time (months to years), result in a major system failure. This result emphasized the importance of operating "balanced" biological systems in which the production and utilization of essential components is stringently controlled. Early experiments, in which the importance of "balancing" was not directly addressed, had begun in the late 1950s.

The first U.S. laboratory experiments on the characteristics of a balanced, closed system were conducted by Eley and Myers (1964). These studies were followed up on a larger scale when the private aerospace companies Boeing and General Dynamics constructed a regenerative life support facility (1966). Averner (1981) and Rummel and Volk (1987), working in MacElroy's laboratory, developed mathematical models of the process of microalgal cultivation. Studies conducted by Kok and Radmer extended our understanding of the basic photosynthetic processes used by algae.

However, research done with most closed systems yielded disappointing results, primarily because the systems used did not attain a stable, steady state. At the time it was assumed that this result confirmed the theoretical deductions of ecologists, and that system unreliability was due to the experimenters' inability to understand the intricacy of controlling biological regenerative systems which were based on monocultures of unicellular organisms, either photosynthetic (*Chlorella*) or chemosynthetic ones (*Hydrogenomonas*).

Misgivings were expressed that unicellular cultures would require complicated and energy-consuming control systems; that cultures could degenerate genetically over time; that the biochemical composition of unicellular cultures did not correspond to human nutritional requirements. A symptom of this mismatch was the assimilatory quotient, the ratio of O_2 (produced)/CO_2 (consumed) by the photosynthetic (or chemosynthetic) organisms, compared to the respiratory quotient—O_2 (consumed)/CO_2 (produced)—of the animals that ate the unicellular culture as food. The general conclusion was definitely pessimistic: use of such simple ecosystems as the basis of the life support system should be discontinued (Cooke, 1971).

In our opinion, though, such a resolute conclusion was too hasty (see Conclusion). It is evident that algal and bacterial cells contain too much protein and too little carbohydrate to be suitable, over the long-term, as the sole food in a human diet. It is also quite true that an oversimplified culture system containing only a unicellular monoculture is ecologically and physiologically unstable.

On the other hand, more thorough investigations of microalgae, such as those pursued in the Bios-3 project in Krasnoyarsk, clearly demonstrated that if microalgae are cultivated in a mixed culture, essentially in a naturally established relationship with bacteria rather than as axenic algal strains, the system becomes quite stable. Chapter 6, and Section 1 of Chapter 8, describe the Life Support System (LSS) based on the use of algae which was constructed on the basis of these principles. It proved to be stable and controllable in long-duration experiments. However, because American researchers became disappointed with microalgae and chemosynthetic bacteria during the 1960s and 1970s, NASA's attention was drawn to agricultural crops, specifically to higher plants. Within the Controlled Ecological Life Support System (CELSS) Program—and in follow-up activities, such as the Advanced Life Support Program (ALSP)—closed system cultivation occupied a central position among the major research directions.

NASA's mode of conducting research has been to identify research directions at the higher levels of the organization, and to fund programs and projects at those levels. Research money is made available competitively to researchers from academia, from industry and from NASA research centers. Essential, high-cost research equipment is often constructed first at NASA centers, and its subsequent use is shared with the external research community. As a new interdisciplinary activity, the CELSS program was initially planned jointly between NASA Headquarters and the principal NASA biological research centers. Once a set of approximate directions was established, the program was opened to the scrutiny of researchers in all areas of the United States.

The CELSS program planning activities were conducted based on inputs from the academic community and industry. The planning activities identified the research goals that should be pursued for future life support work. The principal working assumptions about potential bioregenerative life support systems were 1) that they would be used in space, possibly on orbit or during transit to another planet, and possibly on a planet's (Moon, Mars) surface; 2) that they would require relatively large amounts of power, either from the Sun or from nuclear sources; 3) that dedicated heat dissipation devices would have to be an integral part of the system design; and 4) that if a bioregenerative system were ever going to be considered for use in space, it would need to be both efficient in its use of mass, power, volume and human labor, and as reliable as non-biological systems.

Based on these assumptions, a group of research proposals was selected by NASA to address the practical issues of growing a higher plant food crop within a closed system. Initially, the principal studies were conducted by D. Raper at the North Carolina State University (soybeans), T. Tibbitts at the University of Wisconsin (wheat and potatoes), R. Huffaker at the University of California at Davis (nitrogen metabolism) and F. Salisbury at the Utah State University (wheat). The team of plant physiologists initially headed by Salisbury and later by Bugbee, selected a highly productive semi-dwarf wheat cultivar and developed a cultivation procedure for it (Bugbee and Salisbury, 1988; Bugbee *et al.*, 1996). The period from seed planting to grain harvest (maturity) was 65 days.

A cultivar produced later (Super Dwarf) had many of the same cultivation characteristics as the semi-dwarf, but a slightly more favorable ratio of edible to non-edible biomass (harvest index). In experiments jointly conducted by the Utah team, the Institute of Biomedical Problems (Moscow), and NASA Ames Research Center, the cultivar called Super-Dwarf wheat was evaluated under the conditions of real space flight aboard the *Mir* station and the Shuttle during 1995 and 1996/97 (Salisbury *et al.*, 1995).

The reason for interest in dwarf cultivars is their harvest index. The results obtained by the Utah group indicated that a plant growth area of 13 m^2 containing wheat could provide for the entire caloric requirements (but not all of the nutritional essentials) for one human, provided that each day a fraction of the crop (1/65) was harvested, and new seeds were planted. This same crop area could absorb the metabolic carbon dioxide produced by one human, and produce enough oxygen to allow the human to oxidize the calories contained in the wheat.

In fact, this crop growth scheme would produce oxygen in excess. The excess oxygen would be equivalent to the amount of photosynthesis required to produce the non-edible (by humans) portions of the wheat plants (stems, roots, leaves). Further, production of the non-edible biomass would require more carbon dioxide than generated by a human in the same period. This discrepancy would be resolved by oxidation of the non-edible biomass, either by a physical chemical process, or by a biological waste digestion system.

This same set of rationales is applicable to any cultivars of any single or group of species selected for food production in space, where the limitations of area, volume, electrical power, mass, and human labor are extreme. In each case the harvest index will have to be carefully evaluated, not only on the basis of the production of a crop in a defined period, but also based on other characteristics of a species. For example, as Tibbits (1994) points out, the harvest index of potatoes actually increases over time because after the first crop of tubers is harvested, a second crop develops on the same roots, stems and leaves that the plant has already constructed. Other cultivation procedures can also significantly affect the efficiency of a cultivar in using scarce supportive resources. As an example, a highly productive procedure for cultivating potato under continuous lighting has been developed by Tibbits and collaborators at the University of Wisconsin (Tibbits and Alford, 1982; Tibbits and Cao, 1994; Wheeler and Tibbits, 1986, 1987).

CELSS Program Focus and Diversification

By 1985 the major thrusts of the CELSS program were 1) to find the environmental conditions that would maximize edible food yield per unit of volume occupied, per unit of mass launched into orbit (support machinery required), per unit of energy required for light (converted to mass of solar panels required to be launched), and per unit of human labor; 2) to develop models that would describe the functions, operational dynamics and stability of closed, regenerative systems that would support human life; 3) to understand the process involved in converting harvested crop materials into human food; and 4) to explore processes for converting non-edible plant material and human wastes into resources (gaseous and solid, organic and mineral) for continued plant growth.

To a significant extent, the major thrusts of the program between 1980 and 1990 were influenced by studies that were conducted under contract with the Boeing Company in Seattle, Washington. These studies, initially headed by R. Olson of Boeing, were called a "Transportation Analysis" and considered the places in space (specifically, in low Earth orbit, at geosynchronous orbit, on the Moon's surface, and on Mars) in which a bioregenerative life support system could be used. Surprisingly, the planned Space Station *Freedom* was considered to be the NASA mission that would most benefit from an operation of the bioregenerative life support system. The approach was to develop design concepts for life support systems capable of supporting a crew through the use of a recycling system and relying on the growth of higher plants (e.g. wheat, rice, oats, soy, etc.), estimate the requirements for the food growth and recycling machinery (mass, power, volume, labor requirements), and compare the costs of launching and operating such a system in comparison to the costs of periodically launching the supplies required by a non-regenerative life support system. The developers of the initial study, E. Gustan and C. Vinopal (1986), subsequently collaborated on a follow-up study along with R. Olson and T. Oleson, to more completely define concepts for a bioregenerative system.

By the mid- to late-1980s, the principal research and technology development thrusts of the CELSS program were identified. In brief summary they were:

1) to conduct research as if a mission using the CELSS system had been approved;
2) to continuously refine system requirements to meet foreseeable mission constraints (mass, power volume, human labor in space);
3) to maximize higher plant productivity of food based on the efficient use of power and volume;
4) to understand and exploit the linkage between environment control and plant productivity;
5) to increase the extent of materials recycling by maximally utilizing reduced carbon (potential food);
6) to efficiently oxidize (to CO_2) materials that could not be further utilized;
7) to develop rapid and efficient methods for water purification to provide for human drinking water;
8) to develop the tools necessary to conduct plant productivity experiments in the space environment;
9) to enlarge the size of the various functional units (e.g. plant growth, waste processing, etc.) to human process human-sized flows;

10) to link the various production units (e.g. plant growth with waste processing with water purification, etc.) to evaluate stable operation;
11) to make use of future long term human flights employing conventional life support systems by supplementing them with facilities for the growth of consumable salad vegetables;
12) to maintain cognizance of past and current activities associated with the development of closed recycling life support systems.

The NASA CELSS program developed research and development activities in response to the identification of these thrusts. The principal task areas, keyed to the thrusts outlined above, were:

I. and
II. various studies of plant productivity that employed parameters of the growth environment (day and night temperature, light levels, light cycles, humidity, air speed, etc.) to maximize food production rates per unit of volume; studies of algal growth and purification; studies of growth of blue-green algae (cyanobacteria);
III. studies of further use of non-edible plant materials;
IV. studies of waste oxidation by physical chemical means;
V. development of efficient methods of water purification, including recycling water from urine, gray water recovery etc.;
VI. development of plant growth habitats for use in the space environment—development of the CELSS Test Facility for potential use on Space Station;
VII. development of the Biomass Production Chamber at the Kennedy Space Center to study the growth of plant crops large enough to support humans (with food, oxygen, carbon dioxide removal, fresh water production);
VIII. development of a "human-sized" system consisting of large-scale plant growth chambers, human habitats, and physicochemical recycling equipment at the Johnson Spaceflight Center;
IX. development of concepts for a device for growing suplemental food for space crews, the so-called "Salad Machine";
X. evaluate, study and collaborate when appropriate with the projects developing the Bios-3 facility (Krasnoyarsk), the Biosphere 2 facility Arizona, USA) and the Institute for Environmental Sciences facility (Rokkasho-mura, Japan).

Five of these tasks (VI (The CELSS Test Facility), VII (The Biomass Production Chamber), VIII, IX and X) will be addressed in some detail below.

The CELSS Test Facility

The purpose of the CELSS Test Facility (CTF) project was to develop concepts for research equipment that could be used on the Space Station in order to gather specific data about the growth of higher, food-producing plants in the space environment. The intent of the equipment was to identify and stringently maintain a set of environmental variables, within an enclosed volume, that would allow and ensure the rapid and efficient growth of food plants in space.

The project's approach was to request that the plant growth research community generate the physical requirements for growing several kinds of plants, e.g. wheat, lettuce, soybean, rice. Requirements included such factors as temperatures, light intensity, nutrient delivery, gas composition, air flow, etc. The requirements were established by an empanelled Science Working Group.

After the requirements were established, engineers and designers working on the CTF project explored ways to translate the requirements into hardware. This involved developing ideas for sensors, control systems, packaging information, calculating power and volume demands, cooling schemes and designing information collection and distribution systems. Part of the design included fitting the design to the Space Station laboratory, and ensuring a match between the capabilities of Space Station (e.g. power, cooling systems, wastewater collection, etc.) with the demands of the plant growth device.

The initial CTF design was developed by C. Straight and R. MacElroy in 1988 and was further developed under the direction of M. Kliss, B. Borchers and C. Blackwell (Straight and MacElroy, 1990).

The Biomass Production Chamber

The Breadboard Project was initiated at the Kennedy Space Center in 1985, principally by W. Knott and J. Sager. Later they were joined by plant scientists R. Prince and R. Wheeler. A description of the chamber and some of the work conducted with it has been prepared by Dr. Wheeler.

NASA's Kennedy Space Center has been conducting bioregenerative life support testing since 1985 in support of NASA's Controlled Ecological Life Support System (CELSS) and Advanced Life Support (ALS) programs. The centerpiece to this testing has been a large, closed plant growth chamber called the Biomass Production Chamber, or BPC. The BPC provides 20 m^2 of plant growing area in a closed 113 m^3 volume. Lighting is provided by 96 400-W high pressure sodium lamps, although 400-W metal halide lamps were used on three occasions. For all studies, plants have been grown using a recirculating nutrient film technique. To date, five tests with wheat, five tests with potato, four tests with soybean, five tests with lettuce, two tests with tomato, and one test with rice have been conducted. One potato test ran continuously through 418 days (four plantings), and a mixed crop test with both wheat and potatoes ran continuously for 339 days. These studies provide large (20 m^2) samples of crop production data for ALS analysis, along with direct measurements of stand photosynthesis, respiration, transpiration, and production of volatile organic compounds (e.g. ethylene).

Other research at the Kennedy Space Center Breadboard Project has explored biological waste treatment and resource recovery. To date, this testing has focused on treating the inedible biomass to 1) recover minerals for subsequent plantings and to 2) look at alternative food production options (e.g. fungus, yeast, and fish culture). General approaches for processing the inedible biomass have utilized liquid stirred tank reactors (either aerobic or anaerobic), or solid composting techniques. More recent tests have studied the potential for adding liquid waste streams, such as gray water, containing soap or urine (waste streams high in sodium), directly to plant growth systems.

Results to date suggest that biological systems can be resilient and predictable life support components, provided adequate environmental control is maintained. However, further testing is needed to determine long-term reliability and performance of multi-crop systems (Prince and Knott, 1989; Wheeler *et al.*, 1991; Wheeler *et al.*, 1993).

The Advanced Life Support System Test Bed

This project, which was known as BIO-Plex at earlier stages of its development, was initiated at the Johnson Space Center principally by D. Henninger, T. Tri and D. Barta in the early 1990s. The system is the largest of the NASA life support test systems, and the first in the US to involve humans in a system based on technology that is both biologically and physicochemically grounded. In many ways, the system is the culmination of NASA life support system development approaches, and incorporates data developed in university and commercial laboratories and at the Kennedy and Ames NASA Centers. Figure 3.1 gives an idea of the three phases of the experiment with a crew. The principal references to this work include the following: Early Human Testing Initiative Phase I Final Report, 1996; Lunar–Mars Life Support Test Project, Phase II Final Report, 1997; Lunar–Mars Life Support Test Project Phase III Final Report, 1998; Barta and Henninger, 1996; Behrend and Henninger, 1998; Tri, 1999; Rummel *et al.*, 1998; Barta and Henninger, 1994; Vodovotz, 1998.

Advanced Life Support System at Johnson Space Center

The Advanced Life support System Test Bed (ALSSTB) involves the use of crops to perform life support functions under the restrictions of optimizing volume, mass, energy, and labor and with the goal of nearly complete closure of processes for regenerative approaches to Advanced Life Support, and the closure percent or resupply required can vary with design. The purpose of the ALSTB Program at the NASA Johnson Space Center is to:

- Validate regenerative life support technologies (e.g. air revitalization, liquid and solid waste recycling, active thermal control) through long term testing of integrated biological and physicochemical life support systems with human test subjects.
- Advance the technology readiness level of regenerative life support components and thermal system components.
- Identify terrestrial applications of life support technologies.

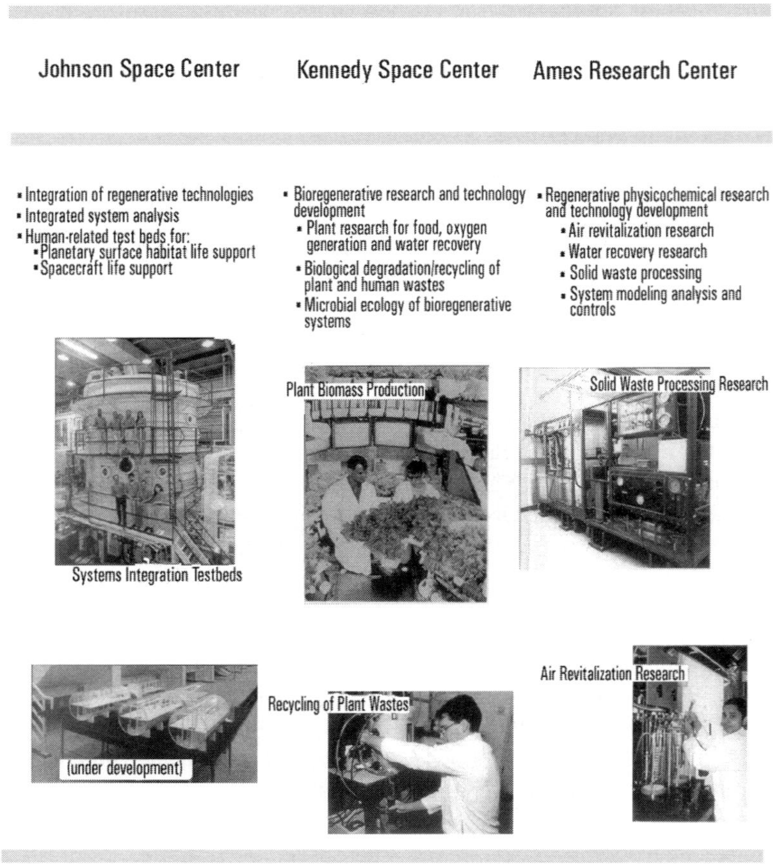

Figure 3.1 NASA research centers. (From NASA Advanced Life Support Program. NASA Headquarters Office of Life and Microgravity Sciences and Applications. http:/www.hq.nasa.gov/office/olmsa/.)

To fulfill this purpose, a coordinated set of activities to perform the necessary research, technology development, integration, and verification of regenerative life support systems to provide safe, reliable, and self-sufficient human life support systems has been established. As part of these coordinated activities, a series of four advanced life support systems tests with human test subjects were conducted (Figure 3.2).

The Lunar–Mars Life Support Test Project (LMLSTP) Phase I Test was performed in August of 1995 (Edeen and Barta, 1996). The purpose was to obtain engineering and scientific data to demonstrate the ability of a crop of wheat to provide air revitalization for a human test subject for a 15-day period. The test also characterized crop growth and test bed performance of wheat, from seed to harvest, in the closed, controlled atmosphere of the growth chamber section of the Variable Pressure Growth Chamber (VPGC) using hydroponics and high-intensity light. The test chamber was divided into two sections, the plant growth chamber, and the airlock that was used as the human habitation chamber.

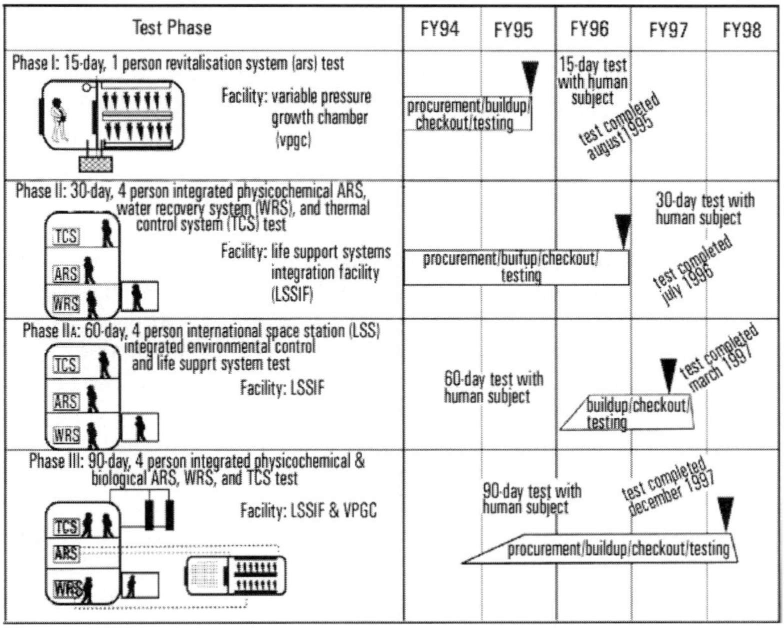

Figure 3.2 Johnson Space Center—Houston, Texas.
LMLSTP 90-Day Phase III Test. Quick Look Presentation.
Crew and Thermal System Division. Marybeth Edeen, 2/11/98.
Lunar–Mars Life Support Test Project Schedule

The objectives of this test were successfully met demonstrating that an 11.2 m^2 crop of wheat could continuously provide the CO_2 removal and O_2 production functions for the air revitalization needs of a single human test subject for 15 days. Also, the ability to control the photosynthetic rate of the crop in different ways was demonstrated. Results of this testing suggest that plants can be controlled to provide a specific desired performance comparable to a physicochemical system. These control systems have been demonstrated to operate autonomously, which will be necessary to maintain a future long-term human presence in space. Also, the results demonstrate the robustness of plant systems as part of a human life support system.

Over the short duration of this test, the populations of microorganisms in the airlock were shown to increase. This is consistent with Shuttle and previous closed chamber tests. No microorganisms were identified which would be of concern to human or plant health at the levels measured. Because of the limited duration of the test, it is not known if the microbial populations had reached a steady state level. Trace gas contaminants generated during the test were those expected from human, plant, and facility systems. The unexpected trends in ethylene and carbon monoxide levels warrant further investigation.

LMLSTP Phase II was a 30-day test with 4 crew members beginning on June 12, 1996. The primary test objective was to develop and test an integrated human life support system capable of sustaining a crew of four for 30 days in a closed chamber. The secondary objective was to 1) provide a regenerative air revitalization subsystem capable of removing carbon dioxide from the internal atmosphere of a sealed chamber, recovering oxygen from the carbon dioxide, and controlling trace gas contaminants for a crew of four for 30 days; 2) provide a regenerative water recovery subsystem capable of recovering potable water from hygiene water (shower, hand wash, laundry), urine, and humidity condensate for a crew of four for 30 days; and finally, 3) evaluate an active thermal control subsystem capable of acquiring heat from the chamber interior, transporting the heat to the exterior, and simulate the capability of rejecting the heat in a lunar day environment.

LMLSTP Phase IIA began on January 13, 1997 and was the third human test to validate regenerative life support technologies. This test used hardware representative of the International Space Station. As in Phase II, this was an integrated test recycling the air and water required for a crew of four. This test was scheduled for 60 days. Results from this test will be combined with results from NASA/Marshall Space Flight Center tests for evaluation and comparison of advanced life support system technologies. The test was successfully completed on March 14, 1997.

Both Phase II experiments were aimed to test physicochemical systems of water and air recycling. Both were successful. A more detailed description is out of the scope of the book devoted to biological systems.

The overall objective of the Phase III test was to conduct a 90-day regenerative life support system test with four human test subjects demonstrating an integrated biological and physicochemical life support system. In support of this objective, the test demonstrated the ability to produce potable water using a combination of biological and physicochemical systems, and demonstrated the ability to maintain a shirt-sleeve environment using an active thermal control system. Additionally, the test obtained the data necessary to perform a mass balance on the system, perform an energy balance on the system and determine the specific energy of all subsystems.

A peculiarity of Phase III was integration of biological and physicochemical regenerative processes.

The key component of the biological regeneration was a crop of wheat (Barta and Henderson, 1998; Brasseaux *et al.*, 1998).

The Lunar–Mars Life Support Systems Test Project's Phase III Test utilized the Variable Pressure Growth Chamber to contribute to the air revitalization and food requirements of a crew of four for a period of 91 days. USU-Apogee wheat was planted and harvested using a staged approach to provide more uniform levels of air revitalization and a staggered production of grain. The wheat crop provided an average of 1.1 person-equivalents per day of carbon dioxide removal for air revitalization over the 91-day human test. It was found that staged cropping required more intensive management of the nutrient solution than single batch cropping. The test confirmed that recycled nutrients, recovered from inedible plant biomass, can be effectively used for crop growth. Staging multiple crops in the same hydroponic system requires careful management of the elemental composition of the nutrient solution.

During the Phase III test, the air was revitalized using a combination of physicochemical systems and biological systems (Brasseaux *et al.*, 1998). The physicochemical systems consisted of a four-bed molecular sieve carbon dioxide removal subsystem, a carbon dioxide reduction subsystem, and oxygen generation subsystem, and a trace contaminant control subsystem. These systems provided approximately 75% of the air revitalization. The remainder of the air revitalization was provided by a crop of wheat. The test results demonstrated that the Phase III physicochemical technologies were capable of maintaining an acceptable atmosphere for a crew of four for 90 days in a closed environment. Also, it was demonstrated that physicochemical and biological systems can effectively be integrated to provide the air revitalization function.

It should be noted, though, that the conclusion made by the authors of the experiment about compatibility of biological and physicochemical systems for air revitalization is only true with respect to quality. In terms of quantity, such duplication of the same function must lead to a number of undue difficulties and imbalance of gas flows.

A system for recovering nutrients from inedible plant biomass had to be added to the life support system with introduction of wheat, which, unlike lettuce, contains inedible parts (Strayer *et al.*, 1998).

During Phase III this process was tested at a distance. The system for bioprocessing of plant residues was at the Kennedy Space Center. The wheat straw was brought there and, upon processing, returned to the Johnson Space Center (JSC) to be used in the nutrient medium. Thus, oxygen used to oxidize organic substances of wheat residues was not part of gas exchange balance in the closed system at the Kennedy Space Center.

The Breadboard Scale Aerobic Bioreactor was used to bioprocess inedible wheat crop residues to provide recycled nutrients to support crop growth in the JSC as part of the 91 day JSC-Lunar–Mars Life Support Test Project Phase III. Filtered broth was amended with nutrients and transported to JSC for integration into the VPGC wheat growth component. Biodegradation of JSC wheat residues was a constant 45% during steady state bioreactor operation.

To increase the closure, a system of Waste Incineration for Resource Recovery in Bioregenerative Life Support Systems was devised (Fisher *et al.*, 1998; Patterson *et al.*, 1996; Bubenheim and Wingnarajah, 1995). This system was used during Phase III.

Over the last three years, the University of Utah, NASA Ames Research Center, and Reaction Engineering International have been developing an incineration system for the regeneration of components in waste materials for long-term life support systems. The system includes a fluidized bed combustor and a catalytic flue gas clean up system. An experimental version of the incinerator was built at the University of Utah. The incinerator was tested and modified at the Ames Research Center and then operated during the Phase III human testing during 1997. The tests to date—chemical analysis and tests with growing plants—indicate that the cleaned flue gas can be used as a carbon dioxide source for plant growth without causing harm to the plants. The ash that was produced from biomass and feces was appropriate for reuse in making hydroponics solution for plant growth. The condensate water is high in metals, and this is probably due to degradation of the incinerator and the catalyst system. This may be controllable by improvements in the incinerator. The biomass and feces feed systems were technical challenges that were only partially met. Both the biomass feed system and the feces feed system are far from being the reliable systems that would be needed for space application (Fisher *et al.*, 1998).

Note that the same problems faced us when we tried to incinerate solid waste material (straw) during our experiments in the Bios-3 in 1970–80. We also left them to future researchers.

An integrated water recovery system was operated for 91 days in support of the Lunar–Mars Life Support Test Project Phase III test. The system combined both biological and physicochemical processes to treat a combined wastewater stream consisting of waste hygiene water, urine, and humidity condensate. Biological processes were used for primary degradation of organic material as well as for nitrification of ammonium in the wastewater. Physicochemical systems removed inorganic salts from the water and provided post-treatment. The integrated system provided potable water to the crew throughout the test (Pickering and Edeen, 1998).

Marybeth Edeen and Karen Pickering made a wonderful summary of the instructive results obtained during Phase III (Edeen and Pickering, 1998).

"The Lunar–Mars Life Support Test Project Phase III test was the final test in a series of tests conducted to evaluate regenerative life support systems performance over increasingly longer durations. The Phase III test broke new ground for the U.S. Space Program by being the first test to look at integration of biological and physicochemical systems for air, water and solid waste recovery for a crew of four for 91 days. Microbial bioreactors were used as the first step in the water recovery system. This biologically based system continuously recovered 100% of the water used by the crew consistent with NASA's strict potable standards. The air revitalization system was a combination of physicochemical hardware and wheat plants which worked together to remove and reduce the crew's metabolically produced carbon dioxide and provide oxygen. In addition, for the first time, the crew's fecal matter was used as a source of carbon to form carbon dioxide in an incineration system. The carbon dioxide was then used to support the plants for a portion of the test. After harvesting of the wheat, it was provided to the crew in the form of flour to use in baking bread. Overall, the test successfully demonstrated that biological systems can be integrated as part of a regenerative life support system. The use of plants to provide air revitalization while providing food for the crew was successfully demonstrated with both the VPGC and the GARDEN. The use of microbes to purify the wastewater and the development of an integrated WRS that recovers all of the influent wastewater and treatment of combined waste stream of urine and hygiene water was successfully demonstrated. As a first attempt at the processing of human solid waste, the incinerator showed that the concept of recovering carbon from the waste and converting it into CO_2 for use by plants is feasible."

It is essential not only to report progress but also to analyze the difficulties that emerged during the experiment, to detect defects, and to state new tasks.

M. Edeen and K. Pickering wrote in the same paper:

"However, further work needs to be done in many areas. Nutrient management of staged crops needs to be understood to prevent plant stress. Control systems must be developed to respond to events when the plants are maturing faster or slower than expected and harvests need to occur differently than predicted. While incineration of feces proved successful, there is significant room for improvement in the design of the incinerator. The feed system for slurried feces was adequate but not user friendly. The overall system was very large and power intensive and must be made smaller before it can reasonably fit in any space vehicle. The controls need to be improved to prevent operation in uncontrolled conditions. The physicochemical air revitalization system needs work in the area of component reliability. Prediction of faults and better recognition of faults and their causes would aid overall system performance. The Water Recovery System suffers from two major technical problems. Conversion of raw food products to edible material was shown to be critical for using plants for human consumption. While the lettuce was eaten as is, the problems with the wheat indicate that this is perhaps the tip of the iceberg in the development of food processing systems. Additionally, while the incinerator was successful for human waste, the real load is going to be inedible plant biomass and solid waste treatment systems need to be directed in this area. While the microbial based water treatment system worked flawlessly during the test, the limits of these biologically based systems need to be determined. The limits of concentrations that can be treated and the operating envelope of these systems needs to be defined through testing. Finally, integrated control systems that take into account the overall operation of the whole life support system and make adjustments as necessary without human intervention would have a huge payoff in crew and ground personnel time."

A future test complex referred to as ALSSIT (Advanced Life Support Systems Integration Test Bed (formerly known as BIO-Plex) will be the basis for future long-duration human missions on lunar and planetary surfaces. The overall objective of the ALSSIT Project is to support large-scale, long-duration testing of integrated, high fidelity, biological and physicochemical regenerative life support systems with human test subjects under closed, controlled conditions. Human accommodations will be provided in the habitat to provide for the needs of four crew members and up to eight crew members during 48 h crew changeovers.

ALSSIT will be comprised of a series of interconnected chambers with a sealed internal environment outfitted with a system of internally distributed utilities capable of supporting a test crew of four for periods exceeding one year. The full configuration calls for a habitat chamber, a life support systems chamber, two biomass production chambers, and a laboratory chamber, all of which will be linked by an inter-connecting tunnel with access through an airlock. The multichamber test complex will be monitored and controlled from a nearby control center. The life support system will perform air revitalization, waste recovery, solid waste processing, thermal management, food production, and integrated command and control functions.

In order to accomplish this goal ALSSIT will be capable of supporting crew size and test duration scenarios representative of future extraterrestrial surface systems. Since the ALSSIT is intended to be built over the course of several years, its physical construction and general testing capabilities have been divided into two distinct stages of operation. The initial stage of operation will accommodate life-support-system testing in which approximately 45% of the crew diet will be produced and 55% will be resupplied. The secondary stage of operation will accommodate testing of integrated systems that produce 90% of the crew diet. ALSSIT will serve as the focal point for other disciplines to conduct research and to develop supporting technologies, techniques, and procedures pertinent to future planetary missions via cooperative and collaborative experimentation and testing. A 240-day human test is planned for 2004.

Phase III of the experiment at the Johnson Space Center has been completed, and now science has results of two experiments with the largely closed balanced material cycling inside manmade ecosystems.

The distance of half of the Earth and nearly a quarter of a century lie between our experiments with Bios-3 and the experiment at the Johnson Space Center. The American test complex BIO-Plex, which was constructed in 20 years after the construction of the Russian Bios-3, is incomparably better equipped with analytical devices as well as systems for monitoring and computer data processing. Nevertheless, they are quite similar in their philosophy, biotechnological methods, and principal results.

The reader will be able to compare the systems and results as well as to judge their advantages and disadvantages after reading Chapter 9, describing experiments in Bios-3.

However, we will take the liberty of coming up with one conclusion here. Since the world scientific community has at its disposal only two experimental systems that can be used to conduct such experiments (one or two more manmade ecosystems can be probably constructed in Japan and Europe in future), such complicated and costly experiments should be performed following coordinated, maybe, mutually reviewed programs.

Different versions of largely closed manmade ecosystems are available or will be available in the nearest future to every team within the scientific community. Considering their importance not only for future space flights but also as a scientific instrument for the investigation of sustainability of closed ecosystems as models of the Earth's biosphere under the conditions of the imminent technogenic ecological crisis, it seems natural to join forces, for instance, to establish an international club uniting possessors of manmade closed ecosystems.

Biosphere 2

A special case is Biosphere 2, an ambitious project devised in 1980–90 and realized in the USA by a team of enthusiasts supervised by J. Allen and M. Nelson. It was generously sponsored by the Texas oil multi-millionaire Ed Bass. The project attracted attention not only of the scientific community but also, and to a greater extent, of the general public. The mass media devoted numerous discussions to the project, mostly expressing a negative attitude. The magnitude of the project, ambitious objectives, and contradictory results deserve special consideration.

Biosphere 2 can be regarded as a triumph of American engineering genius. An impressive structure was erected in the Arizona desert: the hermetically sealed construction built on two acres of land housed typical biotopes of the Earth—miniature desert, steppe, tropical forest, ocean, field and farm, and habitat for 8 persons. It was assumed that photosynthesis under the transparent hermetic roof would bring the artificial "biosphere" to the state of dynamic equilibrium and support the life of all the organisms that it included.

Eight enthusiasts sacrificed two years of their lives to Biosphere 2. The authors of the project intuitively expected that in the system even roughly portraying the Earth's biosphere the main property of the biosphere would be automatically reproduced: the balanced material cycle driven by solar energy. The system, however, behaved contrary to intuition—no equilibrium had been established, oxygen concentration of the atmosphere dropped, and oxygen had to be pumped into Biosphere 2 from the outside, to save people and to keep the experiment going for the designed time, two years. The initiators of the experiment were subjected to destructive criticism in the press, and the results were described as a failure. Apparently, the criticism was heated with the enormous amount of money spent on this enterprise. It is true that biologists could find a more effective (but maybe not so spectacular!) application for this money. (It would be unjust not to mention that initiators of the project did not take any money from state or private funds intended for the advancement of science. It was merely the good will of a very rich person, who could, if he would, lose this same money in Las Vegas.) There is a more important thing to consider. In our opinion, it would be a real waste if this experiment were simply cursed and forgotten. The results of the experiment suggest two very significant conclusions.

The first is a scientific conclusion: isolation of an arbitrarily chosen part of the Earth biosphere, complex in structure and rich in diversity of species, even outwardly resembling the biosphere, cannot automatically close its material cycle. Dynamic equilibrium of material exchange in the system cannot set in spontaneously, as cycling, the fundamental property of the biosphere, making it potentially immortal. It means that designers of the portrait analog of the biosphere did not take into account mechanisms maintaining steady-state dynamic equilibrium of material exchange in the biosphere. These principal mechanisms, ways of their evolutionary origination, and limits of sustainability still remain unknown. These mechanisms, rather than the present-day portrait of the biosphere, must be preserved and protected against destruction. "Failure" of Biosphere 2 is objective progress along this line.

The second conclusion is rather social than scientific. Two hundred million dollars invested in the two acres failed to condition the atmosphere above them, while above millions of acres of land and sea over the whole planet we have it free and forever, unless we break the reproduction mechanism.

Concluding the description of Biosphere 2, we should admit that this experience was not a scientific experiment in a strict sense. However, the history of science can provide examples of situations when the negative result, pointing to incomplete knowledge, gave a stronger impetus to development than the positive result, closing the issue.

3.2 IN RUSSIA

First of all, it should be noted again that V.I. Vernadsky and K.E. Tsiolkovsky were the conceptual predecessors of experimental work on creating closed ecological systems in Russia.

V.I. Vernadsky devised and elaborated the concept of the biosphere as an essentially closed material cycle on a planetary scale (1926). The branch of ecology studying biogeochemical processes on Earth, which has vigorously progressed in the second half of the twentieth century, is a current stage of the development of V.I. Vernadsky's fundamental ideas.

When we deal with the problem of creating artificial closed life support systems, it is essential to remember the conclusion that follows from Vernadsky's basic idea: it is feasible to sustain life indefinitely long only in a system with a closed material cycle.

K.E. Tsiolkovsky was a forerunner of Russian cosmonautics. In his book published in 1895 and 1926, he revealed astonishing sagacity, defining the principal problems of the future activity of humanity in space—from the mechanics of rocket engines to human life support in a long-duration spaceflight. He foresaw the use of plants to regenerate human environment and produce food for humans on board a spacecraft. In the first decade of the century it was vision that read like fiction, but as soon as half a century later, a moment in terms of human history, the first human went to space.

Experimental work on creating biological life support systems on the basis of a closed material cycle started in the Soviet Union in the 1960s. Yu. Gagarin's flight and successful manned flights that soon followed it provided a powerful impetus to the development of this work. Construction of a lunar base, missions to Mars and other nearby planets seemed a matter of only several decades away. S.P. Korolev was extremely enthusiastic about it. In 1965, shortly before he prematurely passed away, he told the author (I.G.): "I must take people to other planets, and I have only 10 years ahead of me. (Actually it was a few months!) I have to hurry!" He provided powerful intellectual and material support to the progressing work on creating biological life support systems.

The Institute of Biomedical Problems (IBMP), founded in Moscow, took on the team of physicians and biologists contributing to the first spaceflights with animals and then Soviet cosmonauts aboard. The Institute included a specialized department that worked out the biotechnological foundations for biological life support systems. The founders of this line of investigation in the Institute in the 1960s were O.G. Gazenko, A.M. Genin, Ye.Ya. Shepelev, and G.I. Meleshko.

Ye.Ya. Shepelev took up the development of the general biological concept of closed ecological ecosystems and their space applications (1963, 1966). He also put forward a number of ideas about nontrivial factors of human habitation.

In the beginning, the attention was focused on microalgae. By that time, the first experiments using microalgae as atmosphere regenerators had been carried out in the USA (see Section 3.1). Japanese scientists popularized the potentialities of microalgae as food producers (Nakamura, 1961, 1963, 1963a). In the Soviet Union the potentials of microalgae attracted the attention of G.G. Vainberg and A.A. Nichiporovich. V.Ye. Semenenko, M.G. Vladimirova and colleagues, Nichiporovich's co-workers, performed a series of physiological and biochemical investigations of microalgae and reached a number of biotechnological decisions on the intensive mass cultivation of microalgae (1967). G.A. Rubenchik and V.A. Kordyum of the Institute of Microbiology, Ukrainian Academy of Sciences, introduced into the culture a thermophilic strain of *Chlorella vulgaris*, which was further used by Soviet researchers in Moscow and Krasnoyarsk in most experiments on human life support (1969). Note that intensive continuous cultivation for over 30 years did not lead to any degradation of the strain. Apparently, it is the continuity of the reproduction and removal of part of the cells from the cultivator, the analog of the natural process of stabilizing selection, that prevents the culture from genetic degeneration.

This stage of work could be considered complete when the first experimental installation was constructed in the IBMP, which allowed direct gas exchange, and later water exchange, between a human and microalgal culture. During 1964–65 five experiments were conducted, in which 3 cultivators 15 l in volume each, containing culture of density 10–12 g/l dry matter, continuously provided requirements of the human, placed in the sealed cabin 5 m^3 in volume, for atmosphere and water regeneration during 29–32 days. In the course of the experiments, a CO_2 imbalance, amounting to 5–17% per day, was recorded, a result of the difference between the human respiratory quotient and the assimilatory quotient of microalgae. In some experiments *Chlorella* biomass constituted up to 10% of the weight of the test subject's ration. The "human–microalgae" system also exhibited stabilizing properties for carbon monoxide, methane concentrations, and quantity of microorganisms in the atmosphere. The gas closure of the system reached 90%. It is obvious that to close the system further is impossible without closing the trophic cycle. It is also obvious, however, that initial hopes of extensive use of microalgae as food were too optimistic. The biochemical composition of algal biomass, containing over 60% of proteins and nucleic acids, does not conform to human nutritional requirements, not less than 60% of which must be supplied by carbohydrates. That is why in the experiments that followed the system included higher plants as a traditional source of food for humans. A greenhouse designed under the supervision of V.P. Dadykin was part of the Ground Experimental Complex (the so-called GEC); experiments conducted in it lasted up to a year (1968, 1968a). Environment in the GEC was principally regenerated by physicochemical processes, and the greenhouse was a source of fresh vegetable food. The greenhouse, 15 m^2 in area, daily produced 482 g of dry biomass, including 86 g of edible biomass (54 g of wheat and 32 g of vegetables). Together with *Chlorella* biomass, the sources within the system provided 26% of the human requirement in terms of mass and 19% in terms of calorific value. There is a reserve to close this system more—transformation of unused plant biomass to food. As can be seen from the values given above, only 18% of the produced biomass was used for food.

Later, in the 1970–80s, Ye.Ya. Shepelev and G.I. Meleshko headed investigations aimed at selecting animals to be included in the human life support as producers of animal protein. The possible candidates were quails, fish (Levinskikh and Sychev, 1989), and mollusks. For the first time animals were tested in real space flights (Meleshko *et al.*, 1991). A major achievement of the 1990s was cultivation of wheat in the phytotron "Svet" ("Light") aboard the *Mir* space station. Experiments were performed in the framework of the joint Russian–US program, using an ultra-dwarf wheat cultivar selected by F. Salisbury, who took part in these investigations (Salisbury *et al.*, 1995).

Simultaneously, work on creating closed ecological systems, life support systems among them, was progressing in Krasnoyarsk, at the Department of Biophysics of the Institute of Physics, which later developed into the Institute of Biophysics of the Siberian Branch of the Russian Academy of Sciences. The works published before 1988 under the heading of the Institute of Physics and afterward under the heading of the Institute of Biophysics are the works of the same team of researchers.

Medical–physiological aspects of the investigation—human response to living in a biological life support system and the safety of test subjects—were studied in collaboration with the IBMP, which set up a laboratory in Krasnoyarsk for the period of experiments with humans as test subjects.

Results of this work done at the Institute of Physics, since 1989 at the Institute of Biophysics, are discussed in greater detail in the chapters to follow.

In the first half of the 1960s at the Institute of Physics in Krasnoyarsk, researchers formed a team with the objective of designing manmade ecosystems of different complexity and studying specific qualities of such systems stemming from the condition that they must be closed. The practical incentive to this work was the creation of a prototype of the biological LSS that would accompany humans beyond the Earth's biosphere.

This many-sided problem required the integrated effort of biologists, physicists, mathematicians, designers, and physicians. During the 30 years of joint effort the work has gone through the following stages. Between 1961 and 1965 researchers developed the biotechnology of intensive continuous microalgal culture as the key link for atmosphere and water regeneration in a biological LSS. During 1964–65, experiments with the two-link Bios-1 "human–microalgae" system were carried out (Kirensky *et al.*, 1967, 1967a). By 1969, a procedure of conveyer cultivation of higher plants was elaborated, using a subirrigation technique and hydroponics (Lisovsky and Shilenko, 1971). The combined Bios-2 system, including microalgae, higher plants, and humans, allowed experiments lasting up to 90 days. The experiments showed the feasibility of entirely balanced gas–water exchange and steady-state regeneration of vegetable food for humans, with the use of human wastes by plants. It is significant that entire agreement between the assimilatory quotient of the plant link and the human respiratory quotient proved to be attainable by the selection of appropriate crop species. These results served as a basis for creating the Bios-3 experimental complex, experiments in which were started in 1971. Their results were published in detail in 1975 and later in three monographs and dozens of papers; most of them are listed in References.

The fundamental distinction of the complex is independent control: the processes all are controlled and maintained by the inhabitants of the system, and all the requisite information is obtained within the system, as it actually must be aboard a spacecraft or at a planetary station. The vegetable fraction of the crew's diet can be entirely regenerated inside the system (Lisovsky *et al.*, 1969; Gitelson *et al.*, 1976, 1989). During 1974–1985, the Bios-3 complex was used to conduct experiments on human life support in biological LSS, lasting up to half a year. The results were communicated to the scientific community in many publications. They are summed up in three monographs published in Russia (Gitelson *et al.*, 1975; Zamknutaya sistema, ed. Lisovsky, 1979; Gitelson *et al.*, 1981) and have been regularly reported at congresses of IAF and COSPAR since 1967 and in an American scientific journal by Salisbury, Gitelson, and Lisovsky (1997).

3.3 IN EUROPE

As far as is known, the scientific community in European countries did not initially intend to address such a universal problem as the creation of a complete biological human life support system.

However, elements of the system were developed thoroughly and fundamentally, which is traditional for European university culture.

Progress in the investigation of individual elements of the system led to the next step—construction of the model system of a closed material cycle by putting together separate links elaborated in several European countries—France, Belgium and Germany as well as in Canada. Spain was chosen as the assembly site: Universidad Autonoma de Barcelona. This is a good example of effective international co-operation in solving a problem that is of importance to humanity as a whole. Figure 3.3 shows the general view of Melissa Pilot Plant laboratory (MELISSA, 1998).

The European Space Agency (ESA/ESTEC) organized work on the MELISSA project (Micro-Ecological Life Support Alternative), which can be regarded as an analog to the US program.

The project is "intended as a tool to gain understanding of the behavior of artificial ecosystems, and for the development of the technology for future biological life support systems for long-term manned space missions, e.g. a lunar base or a mission to Mars."

The international project has united CNRS/IBP Gif sur Yvette/Orsay, University of Clermont Ferrand, ADERSA from France; University of Ghent and VITO Mol, Belgium; Universidad Autonoma de Barcelona, Spain; and University of Guelph, Canada.

In the framework of the project five compartments of an aquatic ecosystem are being developed, from the anaerobic fermenter to the photosynthetic link made up of microalgae and higher plants.

Figure 3.3 General view of the MELISSA pilot plant laboratory. (From MELISSA Final Report for 1997, Activity memorandum of Understanding ECT/FG/MMM/97.012. Edited by Ch. Lasseur, April 1998, p. 92.)

The concept of MELISSA is presented in Figure 3.4.

The general goal of the project has been stated as the closing of the material cycle to regenerate food by recycling human exometabolites.

The starting compartment of MELISSA is a Liquefying Compartment, designed for biodegradation of human fecal material and other waste generated by the crew. Degradation fails to exceed the average of 40% either under thermophilic or superthermophilic conditions. Interestingly, autochtonous bacterial consortium, which has formed evolutionarily and is present in native feces, proved to be a better transformer than any of the tested strains of microorganisms of the genera *Clostridium* or *Coprothermobacter*. The mathematical model of the process also turned out to be of little use, apparently due to the complexity of this multifactor process. The experimental part of the work was performed at the University of Ghent, Belgium. The mathematical model, including the nitrification process, was developed at the University of Clermont Ferrand, France.

The microalgal link is represented in the MELISSA project by the microalga *Spirulina*, investigated by C. Tamponnet, R. Binot, C. Lasser, and C. Savage (1991), and J. Cornet, A. Marty and C. Dussap at the University of Clermont Ferrand, France (1997; MELISSA, 1998, 1999). An important innovation was to introduce into the culture *Rhodospirillum rubrum* as an absorbent for volatile fatty acids (MELISSA, 1998).

As it is obviously impossible to close the food chain without regenerating traditional vegetable food, the MELISSA project included a higher-plant link. Experiments with higher plants performed at the University of Guelph, Canada by M. Dixon and G. Cloutier (MELISSA, 1998) confirmed earlier results of similar investigations of higher plants carried out in the Soviet Union (Lisovsky, 1972) and the USA (Bugbee and Salisbury, 1988).

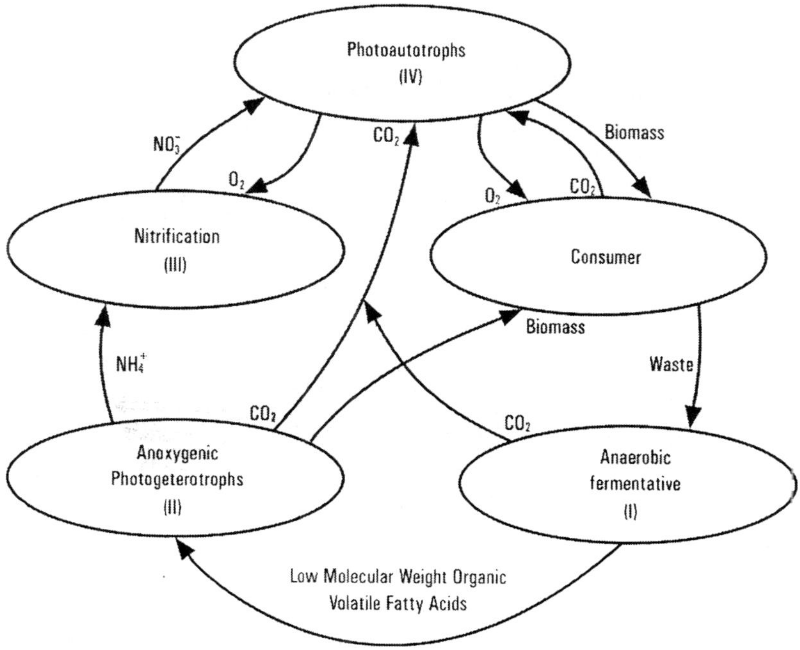

Figure 3.4 Concept of MELISSA (from the same source as Figure 3.3).

The above-mentioned results made possible an important step on the road to creating a prototype of a biological LSS within the MELISSA project. At Universidad Autonoma de Barcelona, Spain (1988), a Pilot Plant was established (MELISSA, 1988). Complete closure is planned for 2003.

So as to investigate the effect of factors of the space environment on the life support ecological system, including the effect of gravity reduced to that on the Moon, a two-link "photoautotroph–heterotroph" system is being prepared for space experiments.

Thus, the ESA project MELISSA in its current state is an important step in elaborating biological life support systems. The new elements are the nitrification compartment (MELISSA, 1998) and the compartment of the absorption of volatile organic acids. A detailed analysis of the efficiency of the compartment for the mineralization of feces suggests a conclusion that the anaerobic method, as well as the aerobic approach we have investigated before (Posadskaya, 1976), is not efficient enough or too slow for an LSS. Hence it follows that a physicochemical link for the mineralization of waste products is essential for an LSS. In the MELISSA project much attention has been concentrated on the development of algorithms of control and process modeling. The proposed models are not quite adequate yet to describe such complicated processes, but this avenue of investigation is reasonable and necessary (MELISSA 1998).

The essentially closed ecosystem C.E.B.A.S., which was the first to be launched into space aboard the Shuttle in 1998, was designed at the University of

Bohum, Germany, under the supervision of V. Bluem (Bluem, 1998). The mathematical model of the system was computed at the Institute of Biophysics, Krasnoyarsk.

The aquatic ecosystem includes fish, aquatic invertebrates, and aquatic plants. The experiment showed that it could function under microgravity conditions.

The second flight of this system was successfully made in 1999. It must become one of the elements intensively used in biological experiments aboard the international space station (Bluem, 1999, personal communication).

3.4 IN JAPAN

The history of work on closed ecosystems in Japan is rather brief but bright.

Though potentialities of microalgae were first reported by Japanese researchers, during the 1970–80s no investigations on the biological LSS of any significance were carried on in Japan. In the 1990s, however, the project Closed Ecological Experimental Facility was supported by the state, and the Institute of Environmental Sciences was founded.

We can say with satisfaction that in the 1990s a great stride toward creating artificial closed ecological systems has been made by Japanese researchers, supported by exceptionally generous material aid from the state.

It makes sense that Japanese society, with a very high population density and insular psychology, was able to comprehend the crucial importance of biospheric problems for civilization. Awareness of the problem brought about a definite action.

In a period of a few years, the project CEEF—Closed Ecology Experiment Facilities—has been devised and basically implemented. For this purpose in December 1990 the Institute for Environmental Sciences (IES) was founded in Rokkasho, a small township in prefecture Aomori on the Island of Honshu.

CEEF consists of three subsystems for the cultivation of plants: Closed Plantation Experiment Facility, Closed Animal Breeding and Habitat Experiment Facility, and Closed Geo-Hydrosphere Experiment Facility (Figure 3.5). Plantation Modules and Material Processing Equipment are shown in Figure 3.6, and the CEEF schematic in Figure 3.7. "The material circulated in CEEF is strictly controlled in a closed system by air-conditioners and material processing subsystems (Figure 3.8). Only energy and information are exchanged with the outside. Each facility can be independently operated or cooperated with another facility (see Figure 3.7). [The subsystems of] CEEF are a unique tool for the environmental sciences and other fields of research such as test beds for life support systems for human and Mars base application, the global change problem and furthering the solutions for a zero-emission society." (Outline of CEEF, Institute of Environmental Sciences, Japan, 1988.)

An important incentive to CEEF creation is the topical problem of thorough investigation of the migration of radioactive elements by the metabolic pathways in ecosystems. Another objective is to model global change, specifically the ecological consequences of global warming. Thus, the closed ecological system is increasingly perceived not only as a means to support human life in a hostile environment—in space—but primarily as a tool for the experimental investigation of the problems of the Earth's biosphere. The complexity and cost of constructions that are designed for experiments are comparable to the complexity and cost of installations for physical experiments, such as elementary particle accelerators. An example can be provided by the Japanese system CEEF.

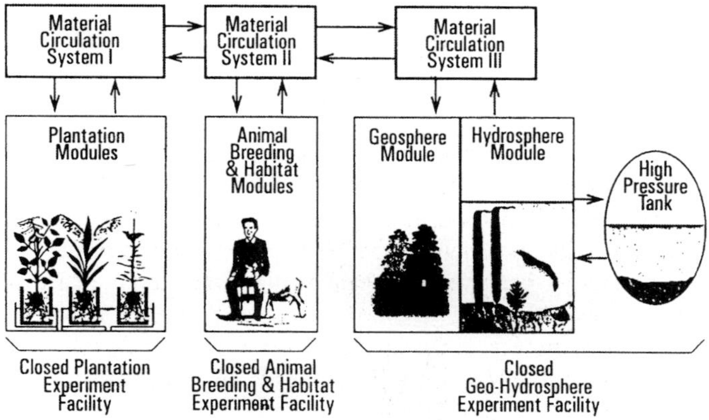

Figure 3.5 Closed ecology experimental facility (CEEF). (From Closed Ecology Experiment Facilities (CEEF), Institute for Environmental Science, Booklet, 1998.)

Special emphasis was placed on designing a physicochemical subsystem to form a closed loop of the material circulation of biological processes via the mineralization of wastes and end products to return the elements to the biological cycling—Artificial Material Processing Equipment.

The largest industrial companies of Japan—Mitsui, Hitachi, Kawasaki, and others—took part in manufacturing equipment and elaborating technology. When the construction of CEEF is completed, it will be highly instrumental in studying the processes forming a closed biospheric material cycle. CEEF and its subsystems are described in detail in publications of the Institute for Environmental Sciences (Nitta, 1998).

Comparison of the complex Bios-3, operating in Siberia, and the system CEEF allows the conclusion that they complement each other. CEEF can break apart the closed material cycle to obtain numerous components, which can be analyzed individually. Bios-3 has made it possible to investigate the properties of the ecosystem that emerge when it gets closed, and that are not peculiar to any of its elements taken separately.

Interestingly, the integrated system proves to be far simpler owing to cyclic processes occurring in it and autoregulatory feedbacks. As an example of a complication caused by the separation of functions in CEEF, we can mention that to avoid disruption of metabolic closure, the personnel servicing the system inside is provided with specially designed suits, similar to spacesuits, so that their respiration does not mix with the system gas exchange. In Bios-3, however, the crew services and controls the system entirely, their respiration being part of the system metabolism. However, to study particular processes, it is inconvenient, and sometimes absolutely impossible, to use the integral system. For this purpose it would be expedient to turn to the Japanese "analytical" system separated into subsystems. Results of the first experiments in CEEF are to be expected with interest.

Figure 3.6 Plantation modules with artificial lighting (CEEF). (From Closed Ecology Experiment Facilities (CEEF), Institute for Environmental Science, Booklet, 1998.)

Taking into consideration that the USA has nearly closed the work on CELSS and that the economic situation in Russia does not allow proceeding currently with the work on the LSS in Krasnoyarsk, where the ground full-scale prototype of a biological human LSS had been tested for many years, MELISSA, ALSSIT, and the Japanese project acquire particular significance.

Thus, now in the world there are only seven experimental ecosystems that can be used to conduct investigations with the material cycling closed to a greater or lesser extent. They are: the Ground Experimental Complex at the Institute of Biomedical Problems in Moscow; Bios-3 in Krasnoyarsk; BIO-Plex at the Johnson Space Research Center in Texas, USA; Biosphere 2 in Arizona, USA; the C.E.B.A.S. aquatic ecosystem at the University of Bohum in Germany; the CEEF complex at the Institute of Environmental Sciences in Japan; and the Pilot Plant that is being constructed in the framework of the European program MELISSA at Universidad Autonoma de Barcelona, Spain.

All these unique facilities have a common fundamental goal—to model the biosphere—and a common practical objective—to create closed human life support systems.

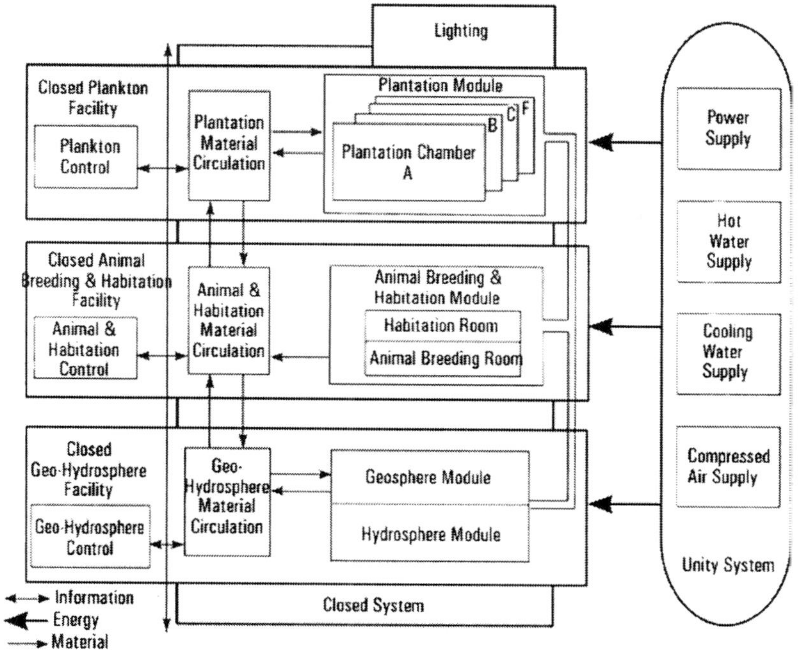

Figure 3.7 System configuration of Closed Ecology Experiment Facilities (CEEF). (From Closed Ecology Experiment Facilities (CEEF), Institute for Environmental Science, Booklet, 1998.)

Figure 3.8 Artificial material processing equipment (CEEF). (From Closed Ecology Experiment Facilities (CEEF), Institute for Environmental Science, Booklet, 1998.)

It is very complicated and costly both to construct these facilities and to use them for experiments. Since this work is important for humanity as a whole, it would make sense to coordinate the experiments at the preparatory stages and while analyzing the results, following the example of atomic physicists cooperating in using the few nuclear-particle accelerators existing in the world.

* * *

As the reader can conclude from our brief review, the creation of manmade closed ecosystems, which started in the 1950s with only a few projects, has by now developed into a rapidly advancing field of research.

The importance of this work for humanity as a whole, identifying our purpose, as well as the complexity and high cost of experiments—all make further international cooperation in this field of knowledge imperative.

CHAPTER 4

The Human Component in a Closed Life Support System

The single, existent, all-encompassing closed ecosystem on this earth—the biosphere—does not have a purpose, at least one that is known to humankind. In contrast to the biosphere, manmade closed ecosystems are intended to fulfill specific purposes.

The purpose of the closed ecosystem that is intended for human life support is the restoration of an optimal ecological habitat for the humans that are included in it by virtue of their material exchange.[1] The goal of the closed ecosystem should also be to compensate for the disturbances that human external metabolic activity introduces into it. These goals determine system structure and design requirements. For this reason we call the human link the reference element of such an ecosystem in the sense that that system is built as a metabolic counterbalance to humans and is reciprocal with respect to the humans living in it. That is, in the conditions of minimal closure, the system should consume all that the human organism excretes and restore all that it consumes. All other links in the ecosystem perform functions determined by the reference element—humans.

Therefore this chapter is devoted to describing human external material exchange, because it is on the basis of this information that such a system can be considered and designed, and this information is needed as a basis for the remaining chapters of this book. It is on the basis of the knowledge presented in this chapter that experimental ecosystems for human life support were planned and created.

Human material exchange and energy have been studied intensively, starting with the second half of the 19th century, as soon as physical, chemical, and physiological methods began to develop and render this study possible, although it is most likely true that our concepts about human external material exchange were solidly established only toward the middle of the 20th century. The development of aviational, space, and underwater medicine spurred research conducted in the second half of the 20th century, which confirmed classical ideas and introduced greater clarification and accuracy. When creating ecosystems for human life support, we based our experimental work on the vast amount of data and normative documents written about human external exchange and nourishment that were published in the 1950s and 1960s.

---

[1] Human metabolism can be subdivided into internal metabolism—a totality of biochemical processes in the cells and tissues of the organism—and external metabolism—material exchange between the human and his/her surroundings, involving consumption of elements from the environment and excretion of the metabolic end products. Here we consider the external metabolism of humans.

HUMAN ENERGY

Human life support is enabled by energy drawn from nutritive substances. The level of energy consumption for a human within a closed ecosystem must be determined in order to establish the energy content of system nutritive rations and to calculate the power needed to run adequate thermostabilization devices in the system.

All human energy exchange is conventionally divided into two categories: the minimal energy expenditures to maintain vital functions and the supplemental energy necessary to enable organism activity. The first category, called basal metabolism, corresponds to the level of energy expenditure in the state of complete rest. There is no standard value for the basic exchange. It depends on the influence of various internal and external factors, which determine the state of the organism. The basic energy exchange of a healthy person has been rather well documented. It does not fluctuate more than ±10% from its normal level.

The second category of human energy exchange undergoes a much greater fluctuation. The main influence on its value is the intensity of muscle activity. The human energy exchange depends in an essential way on the effects of various factors in the external environment, and before all others on air temperature. Other factors include the character of the food eaten, its specific dynamic effect; emotions, and mental activity.

This is by no means a full list of the factors that determine the level of human energy exchange, but it demonstrates the complicated nature of the interrelations between the organism state in the widest sense and the intensity of its exchange.

However, the practical task of defining the average energy expenditure level for the organism is significantly simplified by the absence of any sharp, prolonged deviations from the established stereotype in a healthy person. This is conditioned by the fact that during the course of a short amount of time energy expenditure fluctuations are leveled out. This leveling out enables establishing energy exchange norms for people based on their physical exertion. Thus, for instance, energy requirements for men weighing 70 kg, whose activity does not include great physical exertion, are around 7000–7700 kJ/day. Data published during recent years do not differ from these values as shown in Table 4.1 (Eckart, 1996).

Table 4.1 Daily energy budget for an adult male (70 kg).

	Energy consumption (kJ/day)
General budget	7 140
Resting	8 400
Leisure	9 660
Working	10,080–20,160

Methods of direct or indirect calorimetry must be used for a more exact determination of energy expenditures. The most reliable results can be obtained during repetitive measurements made while the subject is restricted to one chamber; during such experiments real human living conditions can be recreated. Caloric content (kilojoules) of the rations can be controlled to correspond to the weight dynamics of the human body, while introducing the necessary corrections in the rations. In this way, the body weight of the tester of the biologic life support system can be successfully maintained at a constant level for a long time (up to 6 months of continuous experimentation) with weight fluctuations of not more than ±830 g.

RESPIRATORY QUOTIENT

A source of energy for a heterotrophic organism is the oxidation of the rich energy of the chemical bonds of organic compounds found in nutritive substances. Energy is also supplied by oxidizing simpler organic substances with lesser energy stores. The energy produced does not depend on the quantity and composition of intermediate oxidation products, but is determined by the structure of the initial substance and the final oxidation products in accordance with Hess' law. For each nutritive substance, the quantity of oxidation products (in part, of CO_2) completely corresponds to the specified quantity of O_2 absorbed from the atmosphere. In this way, there exists a stoichiometrical ratio between the released energy in the organism, the carbon dioxide, and the absorbed oxygen. The energy expenditures of the organism can therefore be indirectly determined by the amount of oxygen assimilated.

The data of many researchers indicate that a young adult male on average absorbs 550–600 nl of oxygen a day, manufacturing 500 nl of carbon dioxide from this. Gas exchange intensity during the course of a day is not even. During 8 h of sleep, a person consumes about 2 times less oxygen than during a waking period of the same length. This circumstance can lead to fluctuations in the concentrations of O_2 and CO_2 in the closed system atmosphere. The amplitude of such fluctuations will depend on the correlation between the intensity and duration of human physical exertion and between the efficiency of the device that levels off these fluctuations. The fluctuation amplitude also depends on the volume of gas in the system environment.

The value that characterizes the ratio of the excreted carbon dioxide volume to the absorbed oxygen volume is known as the respiratory quotient. The respiratory quotient is exceptionally important in balancing human gas exchange with the process of regenerating the atmosphere. The respiratory quotient can change within a rather large range: from 0.7 (when only fats are being oxidized) to 1.0 (when only carbohydrates are being oxidized). The majority of authors consider 0.82–0.89 to be the average respiratory quotient during the consumption of a normal, mixed diet. The average daily respiratory quotient is equal to 0.89–0.90 when a person is being fed a ration having a biochemical composition that corresponds to physiological norms (Okladnikov and Kasaeva, 1969; Okladnikov in Eksperimentalnyye ecologicheskiye sistemy…, Chapter 7, 1975). The level of energy expenditures corresponds to organism requirements and does not depend on the quantity of food, if the specific dynamic effect of the food is discounted. Therefore, the excess or lack of food can change only body mass and the intensity with which carbohydrates are transformed into fats, and vice versa. In these cases, the respiratory quotient may not reflect the ratio of oxidized nutritive substances.

In this way, the qualitative and quantitative composition of the nutritive ration can substantially change the gas exchange indicators. This must be taken into account when creating a closed ecological system for human habitation.

FOOD REQUIREMENTS

Besides supplying the organism with the necessary energy, nutritive substances fulfill another as important function—the biosynthetic function. A full value food ration should not only satisfy an organism's energy requirements, but should contain all the substances necessary for biosynthetic purposes in the optimal quantities and ratios.

The existent norms are but a particular point in the optimal zone of possible values for the nutrient type quantities that make up the ration (see Tables 4.2 and 4.3). This zone is sufficiently broad. In the opposing case when food intake is not monitored and when it is practically impossible to establish or observe any kind of precise norm, the health of human participants would often be upset. If it is possible to establish one nutritional norm for different people that would answer the physiological requirements of them all, then this implies that there is quite a wide range of admissible variations of the diet composition for each person, and that within this range ration quantities can vary with respect to proteins, fats, and carbohydrates without changing the completeness of the rations as a food source.

Table 4.2 Physiological nutritional norms for adults. The adults are placed in 4 groups according to the labor intensity of their daily activities.

Nutrient type	I	II	III	IV
Protein	109	122	141	163
Animal protein	67	72	82	94
Carbohydrates	433	491	558	631
Fats	106	116	134	153
Animal fats	91	95	108	121
Calories	3208	3592	4112	4678
Calories from animal foods	1211	1287	1449	1641

Dietary flexibility is confirmed by a series of studies of the food intake of various populations of many regions of the former Soviet Union that experience varying climactic conditions (Priputina *et al.*, 1964). It turns out that nutritional ratios in the rations of those people under observation differed from the average accepted norm, and even so, the health condition of these people was completely satisfactory.

Table 4.3 Recommended daily nutrient intake for cosmonauts and astronauts.

Nutrient type	Russian cosmonauts	American astronauts
Energy (kJ)	13,400	9,600–12,950
Protein (g/kg body weight)	1.5	0.8
Fat (g/kg body weight)	1.4	1.3
Carbohydrates (g/kg body weight)	4.5	4.8
Phosphorus (g)	1.7	0.8
Sodium (g)	4.5	3.5
Iron (mg)	50	18
Calcium (g)	–	0.8
Magnesium (g)	–	0.35
Potassium (g)	–	2.7

The digestive system possesses the capability of making broad adaptations to differing dietary conditions and food quality, and this is an important element in the normal functioning of the system. The body adapts very quickly to a new food regime. This adaptation is enabled by quantitative changes in the secretion of digestive juices and by a change in their fermentation activity (Ugolev, 1958). A fluctuation in the ration composition within the bounds of this physiological zone can be assumed as not only allowable, but useful for maintaining the vigor of the digestive system, especially for such an omnivorous species as humans. Only when these bounds are violated—for instance, when it is necessary to adapt to a ration that is primarily composed of only one of the nutrient types—great stress is put on the body's fermentation systems, and the material exchange processes within the organism are frustrated.

Juxtaposing the recommendations made by several researchers, a daily food ration can be considered complete as long as it contains sufficient energy and as long as the quantities of basic nutrient types in the ration are within the following bounds (for men with a body weight of 70 kg): protein, 100–105 g; fats, 50–150 g; carbohydrates, 300–600 g. Analogous data were gathered on Russian cosmonauts and American astronauts as listed in Table 4.3 (Eckart, 1996).

Knowledge of nutritive bounds and the allowable variations in the ratio of proteins, fats, and carbohydrates is necessary when manufacturing the foods intended for consumption in closed ecosystems. This knowledge is also necessary for balancing the gas exchange between the humans and the link in the closed system that regenerates the atmosphere.

Human energy requirements can be satisfied in accordance with the principle of isodynamics. In this way, 1 g of oxidized carbohydrates is equivalent to 1 g of oxidized protein and to 0.45 g of oxidized fats. However in reality, the principle of isodynamics can be considered limited because, besides being an energy source, the nutrients perform a structural function, and they are not interchangeable in this function.

Now human requirements for essential nutrients will be discussed, that is, for those nutrients that are not synthesized by the human organism, in terms of the ratio of proteins, carbohydrates, and fats that should be used to constitute the ration. The daily requirement for protein is 0.8–1.2 g/kg of adult male body mass. The essential amino acids are leucine, isoleucine, lysine, methionine, phenylalanine, threonine, valine, and tryptophane. Human requirements for various essential amino acids are listed in Table 4.4. The concentrated source for essential amino acids is animal proteins. This explains their essential quality.[2]

The human requirement for carbohydrates is 5–7 g/kg of adult male body weight. Accordingly, carbohydrates constitute a large portion of the ration—from 60–70% of the ration's mass.

Table 4.4 Daily amino acid requirements for males and females.

Amino acid	Male (mg/day)	Female (mg/day)
Histidine	700	450
Isoleucine	1100	620
Leucine	800	500
Methionine		
without cystine	1100	550
including 810 mg cystine	200	180
Phenylalanine		
without tyrosine	1100	
including 1100 mg tyrosine	200	180
Threonine	500	300
Tryptophane	250	160
Valine	800	800
Lysine	800	800

Fats (lipids) are a considerable source of energy for humans thanks to their high caloric content. An adult human male requires about 1 g fat per kg of body weight per day. But the function of fats is not exhausted by their role as an energy-provider. Fats participate in biosynthetic structural processes. Many fatty acids that are necessary for biosynthesis cannot be synthesized by the human body. The requirement for essential fatty acids is about 10–15 g a day, and for the most part these fatty acids usually come from animal sources, but, again, vegetarians get them from plant food.

[2] The same essential amino acids can be obtained from vegetable food, but larger amounts of it will be required. Many vegetarians can serve the proof of this possibility.

Vitamins are included in the list of essential nutrients. Daily requirements for vitamins are listed in Table 4.5.

To maintain human life support, minerals must be supplied to the human body. Minerals play a substantial role in many body functions, among them: maintaining equilibrium, stabilizing the internal sphere of the body, ensuring the basic organs are kept in a normal state.

Table 4.5 Daily vitamin requirements and sources.

Vitamin	Food containing the vitamin	Daily requirement (mg)
A	Green vegetables, carrot, fruits, milk, liver	0.8–1.2
D	Liver, animal fat	0.005–0.01
E	Vegetable fat, grains	10–15
K	Green vegetables, liver	0.07–1.5
B_1	Liver, grains, yeast	1.0–15
B_2	Liver, milk, yeast	1.5–2.0
B_6	Green vegetables, yeast, liver, grains	1.8–2.0
B_{12}	Liver, eggs, milk	0.005
Niacin	Liver, yeast, milk	15–20
Folic acid	Green leafy vegetables	0.4
Pantothenic acid	Liver, eggs, yeast	8
H (biotin)	Liver, eggs, yeast	0.1–0.3
C	Citrus fruits, potato, green leafy vegetables, red pepper	75

Recommendations by many researchers on daily ration requirements for the basic minerals are presented in Table 4.6. The wide variations in recommended amounts are readily apparent. The establishment of a mineral "norm," especially a minimal mineral norm, meets with definite difficulties. These difficulties arise because: minerals in the body can be repeatedly reused in metabolic processes; practically all minerals are stored in the body to some extent; and the body can adapt to a considerable lack or surplus of minerals. Besides this, human metabolism is practically never limited by these minerals since the basic quantity of these minerals enters the body with food intake, and since a decrease of minerals in the ration inevitably leads to a reduction of the physiological value of other food nutrients.

For a closed biological system where unicellular algae and higher plants are used as the regenerative link, the part sodium has to play is of the utmost interest since it is not used by most plants.

The average daily human requirement for NaCl is about 12 g. However, many authors insist that this quantity is excessive and indicate that it is not justified by any physiological requirements since the body is able to quickly and precisely eliminate sodium depending on the amount taken in. It is possible to decrease the quantity of salt intake to 4 g/day, and to even 3 g/day, under normal human living conditions. So there is a real possibility that salt intake can be safely decreased. However, when food is less salted than usual, the “taste” of this food changes, and this must be taken into account since this “taste” may have been acquired from early childhood. Low-salted foods can call forth negative emotions and can even go so far as causing the subject to refuse eating altogether.

Table 4.6 Daily mineral requirements (g).

Author	P	Mg	S	K	Na	Ca
Kaplansky, 1938	1.0–1.2	0.7	–	2.4	4.0–5.0	0.6–0.7
Nikolaev, 1948	–	–	1.2	–	–	–
Shtenberg *et al.*, 1959, 1961	1.5	0.5–0.7	1.2–1.3	2.0	4.0–5.0	0.8
Clark, 1958	1.5	0.5	0.8	1.0	1.0	1.0
Pokrovsky, 1964	1.0–1.5	0.3–0.5	–	2.5–5.0	4.0–6.0	0.8–1.0
Kraut, 1966	0.84–1.12	–	–	–	–	0.98–1.19
Rappoport, 1966	1.0	0.3	–	2.0–3.0	4.0–6.0	0.8
Dubrovina *et al.*, 1967	–	–	–	–	–	0.6–0.7
Range total	0.85–1.50	0.3–0.7	0.8–1.3	2.0–5.0	4.0–6.0	0.6–1.2

Minerals are eliminated from the body through feces, urine, and to a lesser degree, sweat. Long-term experiments conducted in our laboratory with the participation of various people demonstrated that from 70–97% of P, S, K, and Na are eliminated with urine, but that only 41% of Mg and Ca are eliminated with urine. The remaining quantity of each mineral is eliminated in the feces. It is interesting that the human organism excretes minerals in a very irregular way over time. Daily fluctuations in eliminations often significantly exceed the fluctuation level that adheres for minerals consumed with food intake (Bazanova, 1969); that is, the precision with which internal exchange is regulated is not very great.

The daily human requirement for trace elements is even less studied. Since there are few studies on this topic, it is adequate to take the daily human requirement for these substances to be those introduced by the work of A.A. Pokrovsky: iron, 15 mg; zinc, 10–15 mg; manganese, 5–10 mg; chrome, 2–2.5 mg; copper, 2 mg; cobalt, 0.1–0.2 mg; molybdenum, 0.5 mg; selenium, 0.5 mg. The recommendations of other researchers do not vary significantly from these values (see Table 4.7).

Table 4.7 Daily trace mineral requirements and stores (Pokrovsky, 1964).

Trace element	Stored within the body	Daily requirement (mg)
Fe	4–5	10–12
Zn	2–3	15
Cu	0.10–0.15	1–5
Mn	0.01–0.03	2–5
Mb	0.001	0.2–0.5
J	0.01–0.02	0.1–02
Co	0.01	< 1
Cr	0.006	0.02
F	3	0.5–1.0

WATER EXCHANGE REQUIREMENT

The majority of researchers have defined the daily human requirement for water in comfortable climactic conditions and under moderate physical activity to be 2200–2500 ml/day. Water enters the body with food and in the form of drinks; about 300 g of water is formed in the body during the oxidation of nutrients. This is the so-called metabolic water.

Water is eliminated in the urine (1200–1500 ml/day), the feces (100–200 ml/day), and evaporation from skin and mucous membranes (750–1000 ml/day).

These intake and elimination values for water are average and can significantly change under various circumstances. For instance, physical work leads to an increase in lung ventilation, and consequently, to an increase in the water lost through the lungs. The value of this loss is proportional to the intensity of the work and can surpass the water volume discharged during respiration when the subject is at rest by 2 to 5 times. Even more significant can be the loss of water through the skin. The thermoregulatory role of perspiration evaporating from the skin significantly increases during physical work or when the ambient temperature is high. The quantity of perspiration can reach 1.5–2 kg/h. These losses must be compensated to preserve homeostasis in the human body.

The topic of minimal water loss from the body is of special interest. The minimal amount of water that can be lost in a day is about 1400 ml. Of this, 900 ml is lost through perspiration and 500 ml is lost in the urine, and this constitutes the minimal amount of water necessary to maintain thermoregulation and elimination of all metabolites.

External metabolism requires yet one more point of water consumption: the satisfaction of sanitary and housekeeping needs. It is true that in the literature there are very scanty data about the necessary amount of water for the procedures of personal hygiene in the narrow sense (washing of the hands, face, and body). However, if personal hygiene is considered to be inseparable from maintaining the hygienic condition of underwear and bed linens, clothing, shoes, and all other objects surrounding people, then it must be noted that these kinds of hygienic procedures are at their earliest stages of development with respect to the closed ecosystem. Defining methods of personal hygiene in the broad sense under the conditions of isolated human existence meets with well-known difficulties. This is because answers to such questions as "How?" "With what?" and "With what regularity?" must be answered taking into account the cleaning and disinfecting substances, as well as water, all of which are available in limited quantities and qualities.

Many years of work experience with a biological life support system allows the definition of the minimal necessary quantity of water for sanitary–hygienic purposes as being 6.5 l/day on average for one person: 3 l for washing hands, face, and body; 0.5 l for processing food products; 1.5 l for washing dishes and kitchen implements; and about 1.5 l for washing underwear and linens. It stands to reason that these values can have meaning only as a guide. But it seems that substantially decreasing these water consumption values is not expedient.

Besides this, the normal functioning of the human body requires maintaining the chemical composition and physical characteristics of the surrounding environs within specified limits. These parameters are more closely examined in Chapter 9 on the basis of the experimental system for human life support—Bios.

CHAPTER 5

Possible Components of a Closed Human Life Support Ecosystem

5.1 VARIOUS METHODS FOR BIOLOGICALLY REGENERATING THE HUMAN HABITAT

Humans are the heterotrophic agents with respect to mass transfer in a closed system. As such, they oxidize the ready organic substances. It is only natural that their metabolic antipode in a biological system should be autotrophic agents. These autotrophs synthesize organic substances from human oxidation by-products while at the same time forming molecular oxygen.

The primary synthesis of organic substances from inorganic in the biosphere is realized almost exclusively by the phototrophs when they use sunlight energy in photosynthesis. The role of chemotrophs as compared with that of phototrophs is vanishingly small.

Higher and lower plants can act as autotrophic agents in a closed system where light is the primary source of energy. Phototrophic bacteria are scarcely suitable as autotrophs since they 1) are anaerobic, 2) require specific hydrogen donors (H_2S), and 3) do not release oxygen during photosynthesis.

In mid latitudes, during the course of a single day the strength of photosynthetically active radiation (PAR) found in the sunlight radiation stream that is used by plants varies from 0 during night hours up to 400–450 W/m^2 of continuous illumination during the clear-skied noon hours. During the growing period, PAR is equivalent, on average, to 100–110 W/m^2 of continuous illumination. In orbit and beyond the bounds of Earth's atmosphere, PAR intensity is significantly greater, coming up to about 600 W/m^2. Even when orbiting Mars, PAR intensity is sufficiently high, measuring about 260 W/m^2. Under orbital conditions, and on Earth, photosynthesis is a promising method for creating organic substances in life support systems.

Using direct sunlight proves to be impossible or impractical (for instance, in subterranean ecosystems, in submarines, and in orbits or on planets distant from the Sun) and when various sources of electric power and artificial light sources are applied in its stead, the advantages of using phototrophs for biological synthesis are far from being undisputed. In these conditions, it might be more propitious to use chemotrophic organisms on the basis of their use in artificial systems, although there are no analogs of such usage in the natural system. For instance, hydrogen-oxidizing bacteria might replace the phototrophs.

Heterotrophic organisms cannot act as metabolic antipodes to the heterotroph—the human. Therefore the participation of other heterotrophs in a human system is not as necessary as the participation of autotrophs, and heterotrophic participation can be reduced to the performance of auxiliary functions. First and foremost, these functions are the decomposition and mineralization of those human and autotrophic by-products that cannot be directly used by these organisms themselves—for instance, organic substances found in liquid and solid human wastes, cellulose, and autotroph lignins. Various microorganisms that function as independent agents in the system or as symbionts for other agents in the system are the best suited for this role.

The synthesis of new organic substances that are suitable or necessary for other agents in the system—and above all, humans—accompanies the decomposition of organic substances. Lower organisms (for instance, fungi, infusoria, etc.) as well as various higher heterotrophs (fish, birds, mammals, etc.) can be used as such heterotrophic synthesizers.

Thus, various species of microorganisms, plants, and animals from the most diverse systematic groups can be proposed to fulfill the various roles necessary to sustain the human life support system.

It is impossible and perhaps pointless to conduct an empirical investigation of all the plant species, animals, and microorganisms in nature that could potentially be included as components of a biological system since there are about half a million possible candidates. A preliminary selection of potential components can be based on various evaluation criteria.

The level of a species' participation in naturally formed ecological material exchange cycles that include humans can serve as one of the basic selection criteria.

However, a more solid criterion is to select participants not so much by the traditional character of their use in human activity, as by their compatibility with the human and other agents in the system. That is, closed system participants should be chosen based on their real ability to produce exchange products that are qualitatively acceptable for the satisfaction of human requirements and the requirements of other agents in the system, and by their ability to utilize human metabolites and those of other agents in the system. This approach makes imperative a broad preliminary search for suitable candidates among the many species of plants, animals, and microorganisms that may have hardly been used by humans up to this point. The benefit of such a search is obvious. The biosynthesis products of one species can scarcely be counted on as a sufficiently complete supplier of human requirements. Parametric considerations should play a great role in adapting these species to humans and to each other as agents in the biological system. Parametric changes can be realized through cultivation and genetic regimes, and by controlling the qualitative direction of biosynthesis. However, satisfying the real and various human requirements (especially for food) can apparently be accomplished only by creating a human habitat containing multiple species that serve as regenerating agents.

Control of the quantitative side of the agents' metabolic processes is of great significance if the system is to function in a self-sustaining way. In a manmade ecological system, only a determined control structure can ensure agreement between the rates and direction of various exchange processes. From this point of view, the choice of potential system components is made based on the degree to which these components can be controlled or governed. The most easily controlled are unicellular organisms that lack the supercellular determined control system that has been developed in complex multicellular organisms. Even so, some degree of external control is possible with respect to higher organisms, up to and including humans.

The integral energy efficiency of processes transpiring in a species is another basic species selection criterion that can be used to determine whether a species is fit as a closed-system component. The efficiency coefficient for phototrophic, chemotrophic, and heterotrophic synthesis determines the effectiveness of energy and mass transfer, and as a result, these determine the weight, volume, energy, and other closed system characteristics that are limitable by the technical means employed in system creation. A high energy efficiency coefficient with respect to the biosynthetic processes of autotrophs and heterotrophic synthesizers and a null energy efficiency coefficient with respect to heterotrophic decomposers can serve as a positive indication favoring the selection of a particular species.

The above-designated basic evaluation criteria for species selection can be strongly modified and supplemented by numerous, more particular criteria that depend on the type and purpose of the closed manmade system to be created. Thus, providing for one or another particular closure—among others, closure with respect to gas, with respect to gas and water, or with respect to gas, water and the vegetable portion of the ration—substantially changes the compatibility requirements of system participants. The way the system functions under various conditions—whether in weightlessness or in gravity, whether using natural or artificial energy sources, etc.—will require a series of supplementary selection criteria and an evaluation of components intended for use as biological agents in the closed system.

Some authors consider it possible to reduce various selection criteria for system components to one overriding criterion. Usually mass characteristics are proposed as the overriding criterion for closed human life support systems meant for use in space. Recently, an attempt has been made to examine closed system reliability as the basic evaluation criterion when assembling both technical and biological system components (see Chapter 9). Adopting a basic, overriding criterion applicable to such biological qualities as energy consumption, compatibility, self-reproduction, and so on, is doubtless alluring.

However, it is impossible not to reckon with the fact that, for instance, besides using mass characteristics to evaluate the potential of a biological component, the mass characteristics must also be used to judge technical components. The degree to which engineering methods can cope with the design of various technical components as dictated by the overriding evaluation criterion decisively influences the formulation of that overriding evaluation criterion. In reality, when using either mass criteria or reliability criteria, it is necessary to give the greater consideration to engineering capabilities with respect to technical components rather than to the capabilities of potential biological components. Despite the significance of such an overriding evaluation criterion, it should not be viewed as the sole and decisive factor in choosing potential agents/actors. Besides this, any overriding evaluation criteria should be deemed fundamental only when they are based on sufficiently precise and experimentally established characteristics gathered under conditions in a real closed system with respect to the capabilities of various biological components.

Photosynthesis: Algae

Aqueous phototrophs, algae, play a significant role in the photosynthetic restoration of organic substances in the biosphere. The annual production of organic substances accorded to algal photosynthetic activity is evaluated at 55 billion tons of the total 164 billion tons of all organic substances that are synthesized in the biosphere as a whole (Woodwell, 1970).

Unicellular algae—diatoms, green algae, blue-green algae, and other algae—are the main producers of primary organic substances in the seas and oceans.

The great diversity of their biological and ecological traits, the biochemical composition of their cells, and the high potential multiplication rate yield a rich selection of species that can be used as potential participants in closed manmade systems.

When creating a system that is closed only with respect to water and gas, the biochemical composition of the algal mass is not critical. The main species selection criteria for autotrophs in this case would be the degree to which they can be controlled and the degree to which they are energy efficient. From this point of view, the following algae are not very promising: brown and red algae, multicellular algae, unicellular yellow-green and golden algae with pigmentation systems that are adapted to acceptable illumination with respect to the spectral composition, and that have complicated methods of sexual reproduction.

Diatomaceous algae, despite the fact that they are essential in the process of creating organic substances in the biosphere, are not very desirable because of their specific skeletal structures, which contain a large quantity (20–40% of biomass weight) of silicon, which is of no worth in a closed system, and because of the difficulties in cultivating them.

The most attractive candidates are the various species of green and blue-green unicellular algae due to their consummate autotrophism, their simple and asexual method of reproduction, and their broad ecological plasticity.

As early as the 1960s, various algal species and strains were cultivated under artificial conditions while maintaining basic growth parameters at an optimal level, and this demonstrated that the upper productivity limits under identical illuminations for these species and strains were quite close (Vladimirova, *et al.*, 1966; Lisovsky *et al.*, 1966).

This allows selecting species and strains based more on how they react to the technological character of the cultivation process than on the biological supremacy of one species or strain over the other. Such technological considerations include: temperature gradient resistance, resistance to mechanical effects experienced during the culture mixing process, ability to stay in suspension and not cling to cultivator working surfaces, simplicity of separating the biomass from the suspension, and so on. No less essential criteria for algal species and strain selection are their liquid/gas phase compatibility with humans, their "self-compatibility"—that is their resistance to the metabolites they themselves produce and that accumulate in culture liquid.

Experience in algal cultivation has led to the understanding of the diverse methods needed to accommodate the various species and strains. For instance, all *Chlorella* and *Scenedesmus* species tested have dense cell envelopes, freely withstand intensive culture mixing by centrifuge pumps and separators, and their cells are not destroyed as they are deposited in a centrifuge at high speeds. Other algae, like *Asterominas*, *Platymonas*, and all algae that have large cells with thin cell envelopes, cannot withstand this kind of cultivation technology; their cells are destroyed, and when subjected to slow mixing speeds, although they stay in suspension, they quickly settle on cultivator working surfaces.

For many algae, for instance, for *Dunaliella*, *Asteromonas*, and other such algae that demonstrate a high productivity in culture, high salt concentrations are required in the medium (Masyuk, 1966; Milko, 1963; Voskresenskii and Yurina, 1965; Yurina, 1966). This makes them less acceptable in closed system conditions than, for instance, *Chlorella* and other fresh water forms that are sufficiently resistant to some salinization of the medium (Trukhin, 1976; Rerberg and Vorobyeva, 1967; Lisovsky and Sypnevskaya, 1969).

There are great differences not only between genera and species, but even between algal strains within one species with respect to acceptable levels of human liquid metabolite concentrations in the medium surrounding the cells. For instance, according to data procured by Rerberg and Vorobyeva (1964), only 2 of the 9 *Chlorella vulgaris* strains studied display a high tolerance to these metabolites, and only 2 of the 4 forms of *Scenedesmus* studied do the same. To a great degree, tolerance to human metabolites in the cell medium and the ability to utilize them in vital processes is characteristic of nonsterile algal cultures, which are in fact algal–bacterial cenoses (Rerberg and Vorobyeva, 1964). It is possible that the greater or lesser tolerance to human metabolites for various algae is connected with the way they differ in their compatibility ratings with respect to the bacterial components of such a cenosis.

The resistance of algae, or more precisely, of the algal–bacterial cenosis, to the accumulation of those metabolites that they produce in the culture medium under repeated reuse was studied for *Chlorella* (Kuzmina and Kovrov, 1967; Tauts, 1966; Meleshko *et al.*, 1967; Kurapova, 1969). It turned out that *Chlorella* was very resistant to such metabolite accumulation.

Prolonged and intensive *Chlorella* cultivation under artificial light has demonstrated that there is no seasonal rhythm to its growth (Feoktistova, 1965; Vladimirova *et al.*, 1966).

It is already apparent from the brief enumeration of the differences in various algae that the unicellular algae *Chlorella* possess the largest number of positive features that better qualify them for use in closed systems. However, it is difficult to say if only *Chlorella* possesses this array of features, or if this conclusion was reached because *Chlorella* is the best alga studied. Nevertheless, given the current state of knowledge, several *Chlorella* species and strains are the most promising candidates as phototrophic agents when creating and conducting experiments on human life support systems that are closed only with respect to gas and water. For this reason, it was precisely *Chlorella* that caught the eye of numerous researchers who were investigating theoretical variants of closed systems (Myers, 1954; Stern, 1966; Nakamura, 1961; Nichiporovich, 1963; Krauss, 1964; and others), who were developing systems with animals (Golueke and Oswald, 1963; Masachito, 1964; Savkin *et al.*, 1970), and who were developing systems with humans (Kirensky *et al.*, 1967, 1967a; Gitelson *et al.*, 1970). Together with this, a number of researchers have examined several other unicellular green (*Scenedesmus*, *Chlamydomonas*) and blue-green (*Anabaena*, *Anacystic*, *Microcystis*) algae as alternatives to *Chlorella* in the role of the gas–water exchange link in a closed system (Nakamura, 1963; Vladimirova *et al.*, 1966; Galkina *et al.*, 1967; Semenenko *et al.*, 1969; Bolsunovsky and Zhavoronkov, 1996).

In the task of choosing between algae, it is more complicated to find a way to close the system with respect to food. If an algal culture, for instance *Chlorella*, can provide for the daily human requirement for oxygen, then it also produces 400–500 g of biomass in that same period. Numerous studies of *Chlorella* biochemical composition have demonstrated that a daily harvest of *Chlorella* contains all amino acids essential for human nourishment in large quantities, as well as a full complement of vitamins, a sufficient quantity of lipids, and a series of ashy elements. There is an insufficient amount of carbohydrates in this harvest with respect to human needs, however (Combs, 1952; Schieler *et al.*, 1953; Aach, 1955; Sisakyan *et al.*, 1962).

Attempts to use large quantities of *Chlorella* and *Scenedesmus* as a direct food for animals and humans have not yielded positive results. Together with a low digestibility, a series of deleterious effects upon the functioning of the animal and human bodies were observed (Powell *et al.*, 1961; Krauss, 1962; Bychkov *et al.*, 1967). It was supposed that using algae as a food would be much more possible if individual food substances were extracted from the biomass of unicellular algae, or if toxic or allergic substances were eliminated from the biomass (Bychkov *et al.*, 1967). Experiments were conducted in which rats were fed protein that had been extracted from *Chlorella*, and these experiments demonstrated that this protein was of full biological value (Klyushkina and Fofanov, 1967; Klyushkina *et al.*, 1967). But the extraction technology is still rather complicated and requires further development.

The possible food value of blue-green algae *Spirulina* has been advocated (Abakumova and Fofanov, 1972; Dillon and Phan, 1993; and others). However, biochemical data about this alga are for the time being limited, and direct experiments with its use as a food in large quantities have not yet been conducted. At present, the most positive approach is to favor a search for algal forms capable of satisfying human nutritional requirements to a larger extent than *Chlorella*. Barring special work on selective breeding directed at bettering unicellular algal food value, it can also be supposed that even among thousands of species it is unlikely that one will be found capable of satisfying essential human food requirements by itself.

But the way in which algae satisfy human food requirements is only one side of the picture. The other side is whether algae can directly, or by means of a decomposing heterotroph, satisfy their own requirements for biogenic elements from human excretions. For instance, many disparities appear when the biogenic elements present in human excretions are compared with biogenic elements required by *Chlorella. Chlorella* needs a larger quantity of nitrogen, potassium, and sulfur than is found in human excrements. At the same time, *Chlorella* cells do not use all the calcium, sodium, and chlorine that are contained in human wastes (Bazanova and Kovrov, 1968; Bazanova, 1969).

Summarizing the above, and considering the present state of research on algae, the conclusion can be drawn that unicellular green algae, primarily *Chlorella vulgaris* and perhaps some forms of blue-green algae (*Spirulina*), were and remain the most fitting model objects of the known algae for experimental development of the problem of creating closed ecological systems.

If unicellular algae have never in essence been used as human food, with the exception of *Spirulina*, others, like *Laminaria*, *Ulva*, *Alaria*, and *Porfira* are rather well known as foods and are specially cultivated in many areas (Yamakawa, 1953; Barashkov, 1963). But people who do eat these algae use them more as salad ingredients than as a main staple. As such, the edibility of these algae does not at all prove their suitability as the main component in a human diet. Many of their biochemical traits—their particular types of carbohydrates and amino acids, their high ashiness, and so on—do not allow considering them for the role of the leading autotrophic agent designated to regenerate food. In addition, these algae have complicated cycles of sexual reproduction, they require specific light and saline regimes, and can scarcely be fitting for intensive cultivation using limited volumes of medium.

Other multicellular algae are so little studied from the biochemical point of view, and all the more so from the nutritive point of view, that it is still too early to speak of them as the agent or link designated to regenerate human food in a closed system. With regard to regenerating the atmosphere and water, all multicellular algae without doubt defer to unicellular algae because of the complicated technology necessary to cultivate them that are connected with the peculiarities of their biology.

Higher water plants, and mainly duckweed, have been discussed in a series of works as possible autotrophic links in closed systems (Nichiporovich, 1963). The value of duckweed is in its simple method of vegetative multiplication, its complete edibleness for several animals, for instance, for ducks, and in its supposed high productivity. However, various attempts at cultivating it have for the time being not confirmed its high productivity. When cultivating several different species of duckweed using natural light in naturally occurring ponds, as well as in artificial nutritive media, the harvest calculated in terms of dry biomass usually came out to 7–12 g/m^2 per day (Landolt, 1957; Schulz, 1962; Sukhoverkhov, 1964; Taubaev *et al.*, 1971; Muzafarov *et al.*, 1971). Only for the duckweed variety wolfr cultivated under laboratory conditions a harvest of up to 25 g/m^2 per day was observed (Nakarmura, 1961).

Many authors make note of the great food value of duckweed for fish, water birds, pigs, and other domestic animals (Nikolaeva, 1956; Schultz, 1962; Galkina, 1964; Muzafarov *et al.*, 1968; Taubaev *et al.*, 1971; Abdullaev, 1971; and others).

Duckweed is generally used as a food supplement for mammals, but not as the main staple. Attempts to use duckweed for human food are unknown.

The above discussion does not yet allow speaking of any advantage that higher water plant cultures have over higher land plants as the phototrophic link in a closed system, and that includes, first and foremost, traditionally grown crops.

Photosynthesis: Higher Land Plants

Higher phototrophic land organisms can perform, just as algae do, in the capacity of a polyfunctional link in the closed system. In principle, both algae and land organisms are capable of regenerating the atmosphere, water, and food by processing human gaseous and liquid excretions, and mineralizing solid excretions. Because the perspective for using unicellular algae and other lower plants as food regenerators is still far from clear, for those closed systems that require regeneration of full value human food along with water and atmosphere regeneration, the first and most probable pretenders for the role of the polyfunctional link are higher plants (Tsiolkovsky, 1964; Dadykin, 1968a; and many others).

Many criteria must be considered all at once when selecting higher-plant species for inclusion in a closed system.

V.P. Dadykin (1968) named the following plant selection criteria for life support systems: high productivity, ability to maximally satisfy human requirements for total mass and biochemical composition, mutual biological compatibility with humans, and also the absence of any airborne excretions that are harmful to humans. Additional criteria include the requirement that the plant's optimal temperature and cultivation humidity coincide with the optimal human temperature and humidity values. The technological simplicity with which the harvested plant can be prepared as food, and the variety of dishes that can be prepared from the plant, are also important criteria.

V.G. Chuchkin and coauthors (1975) tied culture selection to the ease with which polysaccharides are produced, to the possibility of preparing diverse dishes, and to receiving a balanced ratio between the food substances in the supplied food rations and the produced plant matter. Together with this, it must be stressed that including higher plants in a life support system is only reasonable under conditions where the mass characteristics of those links are less than the whole system of supplied food rations. According to calculations performed by these authors, necessary mass limitations for systems with a functional period of 5 years can be enabled if 1) a photosynthesis performance coefficient for the plants is not less than 7–8%, 2) the edible fraction of the total biomass is not less than 70%, 3) the average light period length is 22 h/day, and 4) PAR is not less than 150 W/m^2.

The noted values, even given their conventional nature, can be thought of as a concretization of the general criterion of "high productivity" that was named so by V.P. Dadykin. The remaining criteria formulated in the mentioned works are not mutually exclusive and might fully be employed as orienting factors in the selection of a higher plant culture for use in life support systems.

For systems meant to remain closed for many years at a time of autonomous functioning, in our opinion (Gitelson *et al.*, 1975), it would make sense to introduce criteria governing the plant reproduction. Cultivating plants only on the basis of stored seeds is less dependable (the seeds age, they can perish by chance in emergency situations, and so on) than reproducing the full life cycle with a yield of new seeds (tubers, cuttings, etc.). And so, there is an additional criterion for plants that employ seed regeneration: self-pollination or apomixis. This is because cross-pollination requires special agents to transport the pollen dust, and this lowers the reliability of seed and fruit setting. Not only this, but those wind-pollinated cultures cast their pollen into the air in great quantities, and since the pollen of many plant species is a strong allergen, they end up polluting the closed system. Many authors (Milov and Balakireva, 1975; Nikishanova, 1977; Hoff *et al.*, 1982; Tibbits and Alford, 1982; Hill, 1984; and others) have more exactly defined and expanded their enumeration of selection criteria, boosting the number of criteria up to 20 or more. It is clearly axiomatic that no single species of the 4 million species of angiosperms taken individually can fulfill all of the enumerated criteria simultaneously.

Many authors in the main have proposed, and still are proposing, all possible vegetable cultures—potatoes and sweet potatoes included—for the higher plant link (Golueke and Oswald, 1963; Nilovskaya and Bokovaya, 1967; Dadykin, 1968; Lebedeva *et al.*, 1969; Milov and Balakireva, 1975; Tibbits and Alford, 1982; Hill, 1984; Wheeler and Tibbits, 1986, 1987). The main arguments in favor of these cultures are: 1) a high yield of food production per a small fraction of wasted plant matter (i.e. a high harvest index), 2) they are rich in carbohydrates and vitamins, and 3) they require minimal processing before they can be used as food. But there are other factors that must be weighed when considering such cultures. These cultures have 1) a low calorie content per unit mass and per unit volume of the edible portion, 2) the potato–vegetable diet is too monotonous, 3) many of these plants require photoperiods that would lead to an enlargement of the needed cultivation area, and so on.

The greater part of humanity satisfies its basic requirements for vegetative food by its consumption of grain cultures. The following facts should attract the attention of researchers working on the problem of creating life support systems and in selecting higher plants for those systems. Grain products are highly digestible. Many grain cultures are self-pollinating and therefore produce a small quantity of airborne pollen as opposed to the high quantity produced by some vegetable cultures. And grain cultures can grow under continuous and intense illumination (except for rice). Even considering these positive factors, grain cultures have often been rejected as the possible phototrophic link in a life support system because of their low harvest index and the technological difficulties encountered in processing the harvested product (Dadykin, 1968; Dadykin and Nikishanova, 1969). Apparently, the negative reaction to introducing grain cultures in a life support system due to their supposed low productivity was above all the result of the little amount of study that had been afforded the question of their potential as a photosynthesizing system.

It is well known that grain cultures sown in open fields make poor use of sunlight energy, and that this is because the growth conditions and the development of these plants in the field are often far from optimal. Research conducted on cereal grains in the 1960s through the 1980s, and primarily on wheat, under optimized conditions of artificial illumination demonstrated that the total productivity of cereal grains did not defer to other highly productive cultures, such as beets, carrots, etc. (Moshkov, 1966; Lisovsky and Shilenko, 1971). Moreover, it became clear that a compacted (crowded) wheat cenosis can deliver a substantial productivity increase under high irradiation intensities that exceed the intensity produced by a half day of sunlight by 1.5–2 times (Polonsky *et al.*, 1977; Polonsky and Lisovsky, 1980; Bugbee and Salisbury, 1988). This opens wide perspectives for further intensification of processes for regenerating the atmosphere, water, and part of traditional human food while at the same time reducing the area and volume occupied by plants in a closed life support system. In this way, there is sufficient basis to consider cereal grains as worthy candidates for the role of the leading components in future closed systems of prolonged human life support, and consequently, in experimental models of such systems.

In a closed system, cereal grains, vegetables, and tuberous plants are capable of restoring, in addition to the expected oxygen and water: vegetable proteins, the main carbohydrate mass, and most of the essential vitamins. However they do not regenerate vegetable fats with essential fatty acids in sufficient quantities. Therefore, those plant species that produce significant quantities of nutritive fats can be desirable as potential components in a closed ecosystem. Many cultures that are used to produce vegetable oils in agricultural production at first seem to be of little promise for use in closed life support systems due to large plant size and low harvest index. These undesirable qualities are noted in olives, sunflowers, cotton, sesame, rape, and others. Others are of little promise because they are potentially toxic—poppies and mustard, for instance. Therefore, the greatest interest among researchers has been excited in the main by such traditional species as soy and peanuts from the legume family and a long forgotten culture from the sedge family—chufa (Milov and Balakireva, 1975; Lisovsky *et al.*, 1979).

Soy and peanuts yield fat, and at the same time, a large quantity of vegetable protein. In its bulb, chufa contains (as dry mass) around 20% fat and up to 50% starch. Because of these facts, these cultures are today the most probable candidates for inclusion in the composition of the photosynthetic link of a closed human life support system that is designed to regenerate the entire complement of vegetable food. From the above discussion it is clear that the task of regenerating the full complement of vegetable food intended for human consumption cannot be decided by the choice of one species of photosynthesizing plant. Humans require a complement of proteins, carbohydrates, fats, vitamins, and other substances that can realistically be provided only by creating a higher plant link composed of many species. If the system were to employ many higher plants, such selection criteria used for individual species as high productivity, high energy efficiency, and a few others, can be relegated to the background, so that the more important selection criteria would then become their ability to satisfy isolated human nutritional requirements, the fact that they could not be replaced by long-lasting or conserved foods, etc.

Chemosynthesis

Several microorganisms—sulfur bacteria, iron bacteria, nitrogen-fixing bacteria, hydrogenating bacteria—are capable of synthesizing organic substances on the basis of the energy they receive from oxidation of reduced or not fully oxidized compounds. For instance, hydrogenating bacteria grow using the energy produced when molecular hydrogen is oxidized up to water. They do this by using molecular oxygen and reducing the carbon in carbon dioxide according to the following general equation:

$$6H_2 + 2O_2 + CO_2 \rightarrow CH_2O + 5H_2O.$$

Applying electric energy for electrolysis of water enables procuring the necessary quantity of oxygen and reduced hydrogen:

$$6H_2O \rightarrow 6H_2 + 3O_2,$$

and during this process oxygen that is superfluous to the bacterial reactions is released, and this oxygen can then support human respiration. During human respiration, oxidation of organic substances employs the oxygen superfluous for hydrogenating bacteria and it produces the carbon dioxide that is so necessary for them, as well as metabolic water:

$$CH_2O + O_2 \rightarrow CO_2 + H_2O.$$

The totality of these processes can be viewed in general detail as a balanced exchange of atmosphere, water, and even food, if it is supposed that the biomass of hydrogenating bacteria is suitable to perform a nutritive role.

As early as the 1960s, there were ideas about employing hydrogenating bacteria as a component in a bioregenerative system meant for human habitation (Mattson, 1966; Imshenetsky, 1967; and others). Technologies were developed for continuous cultivation of these bacteria, as well as methods to use human liquid wastes as a nitrogen source. Biochemical analyses of the biomass were conducted, as well as experiments to evaluate the suitability of the bacterial biomass as animal feed, and so on (Bongers and Kok, 1964; Galloway and Kumar, 1969; Kesler *et al.*, 1983; and others). The positive results in a series of experiments on feeding animals with the biosynthesis products of hydrogenating bacteria unfortunately were not confirmed in experiments on human subjects (Waslien and Galloway, 1969).

The possibility of creating bioregenerative systems with hydrogenating bacteria have been discussed many times at the International Astronauts Congress (Athens, Greece, 1995; Baku, USSR, 1973; and others). Some of the qualities that were touted in favor of hydrogenating bacteria were 1) high energy efficiency, 2) ability to use electric, instead of photosynthetic, energy, and this is important because then it is not necessary to transform electric energy into light energy, 3) a developed technology of continuous and intensive cultivation, 4) the possibility of directly employing human wastes, and so on. The gas and water exchange function of hydrogenating bacteria in a closed system was the first thing examined. However, there were problems that dampened the enthusiasm of researchers, including our researchers, with respect to the possibility of the experimental realization of the system "human–hydrogenating bacteria," including: 1) the safety features on the bacterial cultivator, which worked using components of detonating gas (O_2 and H_2), 2) the technological complexities of organizing the gas exchange between humans and the dangerously explosive bacterial cultivators, 3) the unsolved questions of how the biomass itself can be used. To date, such systems have not been created. Therefore, in the chapters of this book describing experiments conducted with closed systems, those systems that employ hydrogenating bacteria and other chemosynthetic organisms are not examined.

In principle, other chemosynthetic bacteria are very promising for use in a closed cycle. Such chemosynthetic bacteria procure their energy from ferrous oxide, sulfur, and other elements. They present no danger of explosion inherent in processes associated with hydrogenating bacteria, and the technological process of their cultivation by means of direct electrical restoration of the energy substrate has already been developed (Kovrov *et al.*, 1967). However, these bacteria have not been researched as components in a closed system.

Heterotrophic Biosynthesis and Decomposition

When the biological material exchange is closed in an ideal way, the heterotrophic link should decompose all compounds synthesized by the autotrophic link in the system to simple initial substances, and all energy that is released during this decomposition should be converted to heat and then withdrawn from the system entirely.

In a closed human life support system, the humans themselves perform the role of the main heterotroph. The rest of the system material exchange characteristics are tailored to complement human material exchange requirements. However, there is no known autotroph that can fully utilize all the gaseous, liquid, and solid human wastes and at the same time produce biomass that is suitable as human food. Moreover, there is no basis for expecting that such a pair of species-antitypes can be found in nature such that their metabolisms would be fully and mutually compensated for by one another. Evolution, apparently, never traveled in the direction of creating such isolated ecosystems; on the contrary, each species interacts like an open system with respect to material exchange in the whole biosphere and strives to occupy as maximal place as possible. From this, it follows that it is necessary to create a multispecies ecosystem for closed systems aimed at supporting all human material exchange requirements. This ecosystem should use both autotrophs and heterotrophs to reduce the system efficiency coefficient.

There are three functions performed by the heterotrophs in the system. The most important is transforming an elemental biomass which is qualitatively unfit or excessive for direct human consumption into substances that do not exist or that are lacking in phototrophic biosynthetic products in order to produce a full value human nourishment (animal proteins and fats). This heterotrophic function is unique, and so far autotrophs and chemical synthesis cannot replace it. The two other functions are the decomposition of human metabolites that are not assimilated by plants and the decomposition of autotrophic biosynthesis products that are not needed by humans. It would be optimal to find one link that could perform all three of these functions, but this possibility is not very probable. It is more probable that these functions be divided between two agents—between the agent that transforms biomass into human food and the agent that mineralizes the metabolites of all other system links.

The agents that most completely perform the first role could be domestic animals that produce milk, meat, and eggs from green plant biomass. Satisfying human desires for animal protein, they at the same time introduce a substantial complication into the system. Several tasks arise in this connection: 1) to provide a full value ration for animals, 2) to create for them an adequate habitat within closed-system conditions, and 3) to return their excretions and inedible parts to the material exchange.

These tasks are not much less complicated than those associated with providing life support for the humans themselves. Besides this, general bioenergetic considerations dictate that introducing an interim trophic level between humans and the agents that produce the primary biomass would reduce the human energy usage factor a minimum of 10–15 times for that energy captured in photosynthesis and directed along this trophic canal.

As early as the 1960s the first theoretical and experimental research was conducted on the keeping and breeding of domestic animals in a closed system—goats, rabbits, ducks, chickens (Abakumova *et al.*, 1965; Voronin and Polivoda, 1967).

They demonstrated that there was a real chance to automate and realize an intensive yield of animal products from chickens, ducks, rabbits, and goats in a relatively small area.

The loss in system energy efficiency when domestic animals are introduced corresponded to that predicted by calculations. Thus, according to the data procured in works by I.A. Abakumova, and her coauthors to satisfy the requirements of one cosmonaut for meat it is necessary to raise 30 ducks of various ages, or 37 chickens. The oxygen required to sustain these fowl is almost twice that needed to sustain the cosmonaut, and the quantity of carbon dioxide excreted by them also grows correspondingly.

Moreover, the problem of a balanced animal diet and the consumption of their metabolites has not yet been solved.

Today's grasp of this issue allows concluding that animals may be introduced into a closed system as a heterotrophic link at the price of significantly complicating the system, enlarging its dimensions, and reducing its energy efficiency by several degrees.

Therefore it is natural that researchers strive to find another more efficient method to satisfy human requirements for animal food. Cultivating herbivorous fish, land mollusks, insects (locusts), lower crabs, protozoa, yeast, fungi, and finally animal cell cultures or tissue is proposed.

As far as fish are concerned, the density of their living habitat poses a significant difficulty that surpasses the population density of land animals by almost a thousand times. The cultures of mollusks, insects, and crabs have not been examined enough to state anything definite about their value, and what is more important, to make judgments about the obstacles that are usually not apparent during a surface acquaintance, but that are sketched out only on a deeper knowledge of the subject. The main task would be to find an adequate food ration for these animals, and this might prove to be more difficult than providing the human ration. It must be added however that a system for cultivating fish in space has been developed (Bluem, 1998), and this affords an opportunity to receive experimental data for calculating various parameters connected with the creation of biological life support systems.

When it comes to protozoa, yeast, and fungi, intensive cultures of which have been sufficiently developed, it remains to be proven that their proteins completely replace traditional animal protein as a human food. Only after positively solving this fundamental problem can further technological improvement of their cultures acquire any significance.

Cultures of animal cells and tissue attract attention for a promising yield of animal protein under conditions of an intensive, controlled process in compact reactors. However, the complexity of the nutritive medium needed for the tissue of a multicellular organism, the cells of which are auxotrophic in many components and which must receive a wide range of substances from blood, makes this process unrealizable on a grand scale for the time being. The problem of procuring a culture medium for the tissue from the primary biomass is especially complicated in a closed system. The task of maintaining a sterile environment for the tissue and governing the differentiation and proliferation of the tissue in the culture is only somewhat less difficult. For this reason, at present, we have not yet seen serious bases from which to discuss the possibility of using animal tissue cultures in life support systems.

The most promising path to ensure an agreement between the metabolism of the links that regenerate the ecosystem habitat and the human metabolism is genetically engineering a shift in the direction of the material exchange for phototrophic plants cultivated in the system to the synthesis of products needed by humans: amino acids, peptides, proteins, fatty acids, and other substances traditionally provided by animal products. But a few more decades are needed before practical application of such a project in closed systems can prove possible.

The second and third functions of heterotrophs in a closed system are the decomposition of metabolites and deadlock products: solid human wastes and inedible plant parts—the stems, the roots, the leaves, and mainly the cellulose and lignins.

There are two possible paths for the destruction of organic substances: aerobic and anaerobic. The latter is more universal because more substances are accessible to this method, but it is less intensive and gives off a wide range of gaseous and liquid products that require further utilization. Thus, for instance, the methane given off in the anaerobic process can be used by methane-oxidizing bacteria, but this means the introduction of yet one more link. And the biosynthetic product of these methane-oxidizing bacteria does not now have an addressee in the closed system. A complete technological scheme for processing cellulose and lignins by means of the anaerobic process and returning consumable end products into the material exchange has not yet been developed.

The aerobic method for processing human waste under closed system conditions is better studied (Rerberg *et al.*, 1968). Aeration tank performance under conditions of a closed experimental system can be judged by the data of real experiments described in Chapters 8 and 9. When there is intensive aeration and a balanced habitat, a relatively quick and deep decomposition of metabolite masses and deadlock products can be achieved. The so-called active sludge in the biomass system could be a strong argument in favor of mineralization. It would be a great gain to use this biomass as a replacement for animal protein, and there are biochemical precedents for doing this, but there are no direct physiological experiments to support this replacement.

A promising approach to the biological mineralization of inedible plant parts (straw, roots, tops, etc.) might prove to be using higher fungi to process them, for instance, oyster fungus (the *Pleurotus* species). As I.M. Pankova *et al.* (1985) have demonstrated in their research, when an appropriate technology is employed the fungi oxidize up to 50% of inedible plant biomass. During this process, the fungi form edible, fleshy bodies (up to 10% of the mass of the dry mass of straw). The remains not used by the fungi are mainly cellulose and lignins, as well as fungal spawn. These remains, according to N.S. Manukovsky *et al.* (1996) can be subjected to the California worm for deep destruction, and then it can be used as a nutritive substrate for the plants.

The above section has laid out several means for using heterotrophs for the biological decomposition of wastes in a closed ecological system. An evaluation of the true merits and deficiencies of each of these demands further experimental investigation.

5.2 PHYSICOCHEMICAL PROCESSES NECESSARY FOR MAINTAINING SYSTEM CLOSURE

It is difficult to assume that consumers will be found for each of the substances formed in the life support processes of the various organism species inhabiting the system, and that these consumers can completely use these substances. When the topic is higher organisms, such a situation is imaginable for a completely closed ecosystem only as an apparently inaccessible ideal. Substances that are not required by the organisms in the system or substances that are not consumed quickly enough in its biological processes inevitably inhibit prolonged functioning of the system for two reasons: 1) a portion of the substances fall out of the material cycle, and 2) such deadlock or slowly-consumed products pollute the human habitat or the habitat of other organisms populating the system. The destruction of the system can be elicited even by an insignificant quantity of accumulated substances (for instance, oxides of carbon, ethylene, and other substances in the system atmosphere).

Apparently, prolonged existence of manmade ecological systems is impossible without a special link that has as its task returning substances to the material cycle that defy biological destruction.

The simplest method to resolve this task is the mineralization of organic substances to higher oxides of the mineral components, carbon, and water. In a number of cases, when the quantity of accumulated substances are negligible, but even this accumulation is inadmissible, it is possible not only to apply oxidization, but to use chemical or physical combining and removing these substances from contact with the organism's vital medium. Such a method can be used to remove toxic volatiles from the system atmosphere and from the drinking water prepared for the crew, although in both cases a biological oxidation process could have been applied.

The problem of deadlock products in the ecological system can be broken into two. The first of these is purification from the small quantities of substances that pollute the atmosphere and water. This purification can be accomplished by sorption of these substances, by concentrating them with the last oxidation products or without them. The second problem is the oxidation of one or another of the quantities of organic substances not used in the system with the aim to return fundamental components found in any biomass—carbon, oxygen, and hydrogen—to the material exchange process.

It must be noted that there is no sharp boundary between these problems. Purification of system atmosphere from volatiles can be successfully realized with the help of catalytic oxidation (Gusarov *et al.*, 1967, 1973; Sinyak *et al.*, 1971) in a platinum–palladium catalyst or by means of volatile sorption over activated charcoal (Kirensky *et al.*, 1967).

The problem of water purification is solved in an analogous way—either by catalytic oxidation in superheated steam mixed with air, or by volatile sorption over activated coal and ion-exchanging resins (Ballod *et al.*, 1967), as well as by evaporation during a decrease in the barometric pressure or lyophilization (Moiseev *et al.*, 1967; Gusarov *et al.*, 1967; Sinyak *et al.*, 1972). The method of electroosmosis was researched with this aim in mind (Seryapin *et al.*, 1966). One of the first reviews of methods for procuring drinking water from untraditional sources were done by S.V. Chizhov and Yu.E. Sinyak (1973).

The choice of existing methods to purify water and atmosphere from pollutants of various types is rich enough and enables considering this problem as principally resolved. In each concrete instance, taking the quality and quantity of pollutants as the starting point, the optimal purification method can be chosen.

Substances can fall out of the material exchange cycle when sorption methods are used to purify the water or atmosphere in the system. When the absorbent is not further used after the purification process, sorption can be employed in systems that are meant to function for short periods only.

The need for physicochemical processes that oxidize some portion of system organic substances arises when the material exchange is closed not only with respect to air and water, but when it is closed with respect to nutritive substances as well. Inedible biomass is produced at the same time as edible biomass when a system is completely closed with respect to food. Inedible biomass products include roots, stems, plant leaves, and food wastes. It might prove necessary to oxidize portion of the inedible biomass not subject to heterotrophic decomposition. If part of the food comes from food stores not produced in the system, then it becomes necessary to remove the quantity of elements introduced by the food stores to preserve the material exchange. The best option would be to remove the inedible biomass in exchange for these food stores elements. If the quantity of inedible biomass yielded by the system is larger than the quantity of food stores introduced into the system from without, then the surplus must be oxidized and returned to the material exchange. Solid human waste is also included in the general idea of inedible biomass, although it makes up a small part of the general inedible biomass.

The problem of oxidizing inedible biomass by physicochemical means is not as pressing as the problem of purifying the system atmosphere and water. The need for physicochemical oxidation arises only when 1) the system is essentially closed with respect to human food exchange; that is, if food is provided only by in-system components, and not by food stores provided from without, and 2) when the inedible biomass does not undergo any heterotrophic decomposition, otherwise known as "biological oxidation."

Physicochemical methods have been developed to oxidize inedible organic substances. One of these methods is "wet combustion"—oxidation of solutions and suspensions dissolved in water under high temperatures and pressures (Agre *et al.*, 1966; Gusarov *et al.*, 1967; Drigo *et al.*, 1967). Another method is the thermal combustion of fine and large biological particles of differing water content in various types of furnaces—for instance, in shaft furnaces and cyclone furnaces (Balin *et al.*, 1967).

These oxidation methods are highly compatible with the end products excreted by biological links in the system. These are first and foremost the CO_2 and H_2O consumed by autotrophs in the system. During wet combustion, the mineral components remain in solution and can be used by the plants. During thermal combustion, a part of the ash components can be dissolved in water or in the acids that can be procured from the airborne nitrogen and sulfur oxides that form as a result of the biomass burning. The solution yielded from this process can be used as a component in the nutritive medium consumed by the autotrophs.

The most preferred method at present is thermal combustion because it does not require fine grinding of the substrate, or high pressures, and is more economical energy-wise than wet combustion, especially when oxidizing significant quantities of such substances as straw, plant tops, and kitchen waste.

In concluding this examination of the possible ways to structure a closed human ecosystem, it can be ascertained that obligatory participants in such a system should be a metabolic counterbalance to the human in the form of an autotrophic link. This link might include unicellular algae, higher plants, or chemotrophic bacteria. Autotrophs and the human, the heterotroph, by their nature can form the basis of a closed material exchange. There are unresolved problems in such a system: supplying humans with animal protein and returning the products of autotrophic biosynthesis to the material exchange. However, this system can fully provide a closed gas and water exchange and to a significant extent can close the trophic chain, and with this supplying humans the vegetative portion of their food ration while consuming their metabolites. The role of the decomposing agent can be fulfilled by other heterotrophs and physicochemical processes.

Therefore it might be expedient to realize the "human–autotrophic organisms" ecosystem and then, if this proves successful, to try to widen the sweep of the closure in the next system by means of complicating its structure by the introduction of links that enable decomposing surplus and deadlock substances. It was precisely in such a way that research on closed ecosystems for human habitation has developed. This research is described in the upcoming chapters.

CHAPTER 6

Principles of Controlled Continuous Cultivation of Microalgae

In Chapter 5, we examined the potential for using the photosynthesis of green plants as the main method for regenerating the human habitat in a closed ecological system.

In order to realize this potential, a technological process had to be developed that would ensure that photosynthesis in the system would transpire with the same continuity and sustainability as does the human metabolism. In one design of this process, microalgae were chosen as the opposing counterpart to the human metabolism.

Unicellular algae are especially suitable for this type of closed system for the following reasons:

1. Almost the whole biomass of algal cells is an apparatus that promotes photosynthesis and the biosynthetic processes that are associated with it. There is no inert or dead biomass in it at all.
2. Potential growth rates and multiplication rates are extremely high. Under optimal conditions, cell division occurs every few hours.
3. The efficiency of light energy usage in photosynthesis reaches more than 10% under experimental conditions; theoretically, it can reach as high as 19%.
4. Human excretions can serve as a source of biogenic elements for the microalgae. Microalgae can procure carbon dioxide directly from the air.
5. Microalgae metabolism is extremely plastic and controllable, and therefore the protein content of their biomass can be maintained anywhere within the range of 20 to 70%.
6. The microalgal suspension is a suitable subject for technological processes. The cells can be easily mixed, submitted to a centrifuge, and sparged, yet they suffer no practical damage.
7. Autoselection in a continuous microalgal culture provides a high level of protection for the microalgae and ensures that culture sustainability experiences no harm.
8. Under optimal growth conditions, microalgae do not require special antibacterial measures and can be cultivated in nonsterile media.

All of the above properties make microalgae an extremely attractive candidate for one of the main regenerative tools in a biological life support system. A good deal of effort was expended in finding a process for cultivating algae that would safeguard the balance of material exchange as well as the equilibrium of the system as a whole. To realize the full benefit of the above-listed properties, this technological process had to render microalgal cultivation sufficiently intensive, sustainable, and controllable.

Having these goals in mind, two of the authors of this book (I.I. Gitelson and G.M. Lisovsky) carried out an experimental complex in which *Chlorella vulgaris* microalgae were studied. *Chlorella vulgaris* is the thermophilic strain originated by I.M. Rubenchik and V.A. Kordum of the Microbiology Institute of the Ukrainian Academy of Sciences. I.A. Terskov and I.I. Gitelson directed all experiments conducted in this complex. B.G. Kovrov, A.A. Shtol, E.S. Melnikov, A.S. Budanov, and V.S. Filimonov developed cultivator design and devised the technology used in the cultivation process. F.Ya. Sidko, V.N. Belyanin, and N.S. Eroshin researched microalgae optical properties and the culture's light regime. V.A. Batov studied the temperature regime. M.I. Bazanova, G.I. Sadikova, and R.I. Kuzmina researched chemical nourishment of the culture. Yu.N. Okladnikov and G.E. Kasaeva worked on gas nourishment. M.S. Rerberg, I.M. Pankova, L.S. Tirranen, L.A. Somova, and M. Posadskaya investigated the microflora.

This chapter is based mainly on materials presented in their numerous publications summarized in our three books published in Russian (Gitelson *et al.*, 1975; Zamknutaya sistema, ed. Lisovsky, 1979; Gitelson *et al.*, 1981).

6.1 CONTINUOUS CULTIVATION OF MICROALGAE

Two groups of factors define the growth rate for unicellular organisms: external and internal. There are two internal factors: 1) the maximum decoding and duplicating rates inherent in the genetic code specific to the particular species and housed in the cell's genome, and 2) the maximum growth rate of the slowest enzymes that participate in the cellular self-reproduction cycle.

External factors in the culture medium are accessibility of: 1) nutritive substances necessary for growth, 2) energy, and 3) water. Other factors can stimulate or inhibit growth.

As a rule, the potential growth rate of unicellular organisms as defined by their genetic code repeatedly exceeds the average rate with which this same genetic code acts in nature. In an experiment, the researcher can create optimal conditions that do not limit cellular growth rate, thereby eliciting the genetically-coded maximum rate of self-reproduction. In one such experiment made on the bioluminescent marine bacterium *Photobacterium leiognathi*, cells duplicated at the rate of 8 to 10 min (Gitelson *et al.*, 1973). If all the descendents of just one cell were to grow at such at rate, in only 16 h their total biomass would exceed the volume of the Earth, and in a single day the biomass would expand beyond the bounds of the solar system. This calculation helps us to appreciate the enormous potential inherent in the cell's genetic code with respect to biosynthesis rates and the growth rate. Of course, in real life, nothing of the kind ever happens for more than short periods of time due to external factors that prohibit unlimited growth.

However, such a combination of favorable, or nonlimiting, factors for a particular species can rarely and momentarily occur in nature. For instance, there are times when an outburst of algal multiplication begins, the result being such a grandiose phenomenon as an oceanic "red plankton blossom." But the outburst of growth is always short-lived and finishes with "red tide." The cells quickly exhaust one of the factors necessary for their growth and they begin to die off in masses. In the natural environment, the growth of cellular populations is always limited by the deficit of one or another necessary factor. This deficient factor is termed as *limiting.*

When the cells chance into an optimal, that is, a non-limiting medium, they grow and multiply until they exhaust one of the factors necessary for their growth, and as they do this, they themselves render the medium limiting. Why does nature endow the cell with such a high potential multiplication rate if this rate can be realized only by chance and for short durations? Evidently, this potential is a powerful survival tool for the species under competitive conditions. When favorable conditions for unlimited growth accidentally occur, those species that can grow the fastest win out because they can more fully utilize the auspicious moment that exists until the medium once again becomes limiting.

This is the way the process works in nature. A question arises. Cannot artificial conditions be created that maintain such external factors at an optimal level continuously and thereby allow the cell's internal mechanism to reach its potential and promote growth at the maximal rate generation after generation while still yielding a useful product?

It seems that if a non-limiting medium were simply created, then this maximal growth could be sustained. The only problem is that the cells themselves will destroy this optimal medium in the course of their growth. They will consume those substances that are essential for growth and excrete other substances that inhibit growth. Consequently, enabling maximal cell growth consists not only in creating an optimal medium for the cells, but in constantly restoring that medium by eliminating the very changes that the growing cells themselves introduce into it. If this is done at a rate commensurate with that by which cell growth alters the medium, then a "constant medium" in a state of dynamic equilibrium is the happy result.

This medium will be as constant and as flowing as a river. Since the medium is optimal, the cells will grow at the maximum rate possible. And since the medium is constantly optimal, cell growth rate will be constantly maximal. Herein lies the principle of *continuous cultivation.*

The principle of continuous cultivation can be more easily understood by studying Figure 6.1 and extrapolating from it. This figure schematically illustrates the typical dependence of cell population growth on time.

In the *lag phase* of this figure, the sown cells adapt to the medium and begin to grow at an accelerating rate. In the *exponential phase*, they reach the maximal growth rate. After the exponential phase, one of the factors becomes exhausted and therefore limiting, at which time growth rate falls off to a level where division rate and death rate are equivalent. Now the total number of cells remains constant. This last phase is called the *steady state* phase. This scenario is typical for what transpires in nature. These same phases are experienced by cell populations known to exist within batch cultures.

Now imagine that at a certain point in the exponential growth phase, cell mass increase is constantly removed. At the same time, substances are constantly returned to the culture medium at exactly the same rate that the cells consumed them during the course of their growth, and those substances that they have excreted into the medium are eliminated. The resulting culture will have a constant number of cells surrounded by a constant and optimal medium, and the cells will grow at the maximum rate allowable for the given conditions. This is a continuous densitostat or turbidostat flowing culture, and it is illustrated by the saw-toothed curve 4 in Figure 6.1.

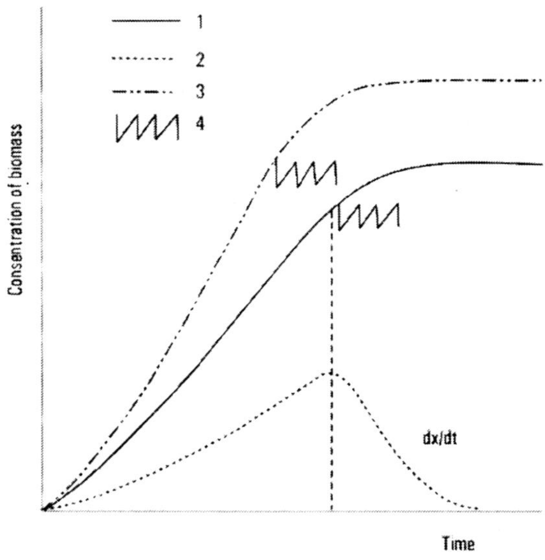

Figure 6.1 Stabilization of culture biomass concentration at the growth peak by removal of part of the suspension and addition of an equivalent part of nutrient solution for two different cultivation regimes. 1—culture growth under specified conditions; 2—biomass increase rate; 3—culture growth at more optimal combination of parameters; 4—stabilization of culture density at the optimal level.

A continuous culture can be established at any point in the exponential growth phase, but it is most effective to do so at the uppermost point of this phase before it transitions into the steady state phase. This means that the most effective point for establishing a continuous culture is at that point where a maximally high cell concentration occurs and where the cells are growing with maximum speed. Each cultivator volume unit will yield the greatest production under these conditions.

There is another less obvious but nevertheless important quality of continuous flowing cultures. The flow protects against both a decrease in the growth rate and against the accumulation of slow-growing mutants or cells that have been damaged. This important trait is further examined in Section 6.7 because it is so vital to the development and evaluation of bioregenerative life support systems.

Continuous cultivation had been experimentally implemented as early as the 1920s, and later found a wider application in the microbiology industry.

6.2 MATHEMATICAL MODEL OF ALGAL CULTURE GROWTH

This section contains a mathematical model for algal cell growth and multiplication based on and confirmed by experimental data. Calculations premised on this model are used to design algal cultivators as well as the technological processes of cultivation used in them.

Mathematical modeling of cell population dynamics has a richly developed history. Cultures comprised of unicellular organisms were studied as a particular population type and became an object of mathematical modeling. The classic equations describing the dynamics of microorganism population counts in a culture were formulated by Monod and then by Novik and Szilard (Monod, 1950; Novick and Szilard, 1950). These equations served as the springboard for a whole series of works in this field.

Technical microbiology generally bases its studies on cultures of heterotrophic microorganisms. Progress in this field, by its very nature, demanded that the mathematical models of these cultures be developed in an accelerated way. The Monod–Ierusalimsky (1965) model is the most widespread model typifying the heterotrophic continuous culture. It has become the classic model in this field. It is expressed by the following equations:

$$\frac{dx}{dt} = \mu \cdot x - Dx\,; \tag{6.1}$$

$$\frac{dS}{dt} = D \cdot (S_0 - S) - \mu \cdot \alpha \cdot x\,; \tag{6.2}$$

$$\frac{dP}{dt} = -\beta \cdot \mu \cdot x - DP\,; \tag{6.3}$$

$$\mu = \mu_m \frac{K_p \cdot S}{(K_s + S + \frac{S^2}{K_m}) \cdot (K_p + P)}, \tag{6.4}$$

where x is culture biomass concentration; μ is specific cell growth rate; D is culture dilution; S is the concentration of the medium component limiting the growth rate; S_0 is this substrate concentration of the medium entering the cultivator; α is the coefficient representing the amount of substrate consumed in synthesizing a biomass unit; P is culture metabolite concentration; β is the coefficient representing metabolite excretions produced during the synthesis of a unit of biomass; and K_s, K_m, and K_p are constants.

There is no biological specificity inherent in Equations (6.1), (6.2), and (6.3) of this model, and they can describe both heterotrophic and autotrophic microorganism cultures, as well as a large class of chemical technology processes that accompany the continuous synthesis of a certain product X from raw material S while generating by-product P.

Equation (6.4) of this model reflects the way in which the specific cell growth rate quantity depends on medium conditions. In particular, this equation expresses two dependencies: 1) growth rate dependence on the particular component concentration in the medium that limits or inhibits cell growth and 2) growth rate dependence on the concentration of inhibiting metabolites in the medium. That is, this equation describes the specific reaction of the cells to a combination of parameters in the medium.

Microalgae, however, are photoautotrophs. Photoautotrophic growth is distinguished by the fact that it can be limited by light as well as by substances dissolved in the medium.

Since culture illumination consumes great amounts of light, it is characteristic that light be the limiting factor. In fact, to ensure that the light factor is maximally exploited, light must be the main limiting factor. Although mineral components can also limit microalgal growth, it is easy to maintain a constant and abundant supply of these components within in the medium.

In real microalgal cultivation conditions, besides light, the parameter that most often limits cell growth rate is carbon dioxide as a source of carbon. For this reason, we examine in detail microalgal growth rate dependence on light and on carbon dioxide.

Any consideration of cultivation productivity for unicellular microalgae must be based first and foremost on microalgae dependence on light intensity as a limiting factor.

Unicellular Microalgal Growth Dependence on Illumination

B. Kovrov and his coauthors (1967) undertook to design a growth model for a unicellular photoautotrophic culture. In this attempt, they relied exclusively on experimental data and dependencies.

A continuous culture of the thermophilic strain *Chlorella vulgaris* was the object of study in these experiments. The *Chlorella* cells were grown in optically dense, flat layers. The experimental material has been systematized, and this has enabled its presentation in the form of exceedingly simple dependencies (cf. Figure 6.2).

The y-axis in Figure 6.2 represents quantity μ/β, where μ is specific cell growth rate and β is the relative content of chlorophyll in the cell culture at the moment of the cell growth rate measurement. Quantity μ/β then characterizes the photosynthetic effect of the cell's chlorophyll.

The x-axis represents light energy dosage absorbed per unit time by the photosynthetic cell apparatus—that is, it is the quantity of light energy absorbed by a unit of pigment mass per unit of time:

$$\mathbf{D} = \frac{E_{\text{absorb}}}{\beta \cdot \bar{x}} = \frac{E_0}{\beta \cdot \bar{x}} \cdot (1 - e^{-1\beta\bar{x}}), \text{ W/kg}, \tag{6.5}$$

where E_0 is the illumination of the culture surface, expressed as W/m^2; x is the cell biomass surface concentration, expressed as kilograms of dry cell mass per square meter of illuminated culture surface; l is the specific coefficient of white light absorption by chlorophyll and is equal to $4.6 \cdot 10^3$ m^2/kg. This equation and its constituent parts were calculated based on the data gathered by N.S. Eroshin (Eroshin *et al.*, 1964) and F.Ya. Sidko (Sidko *et al.*, 1967) and should be considered together with the information below, which describes the optical properties of microalgae.

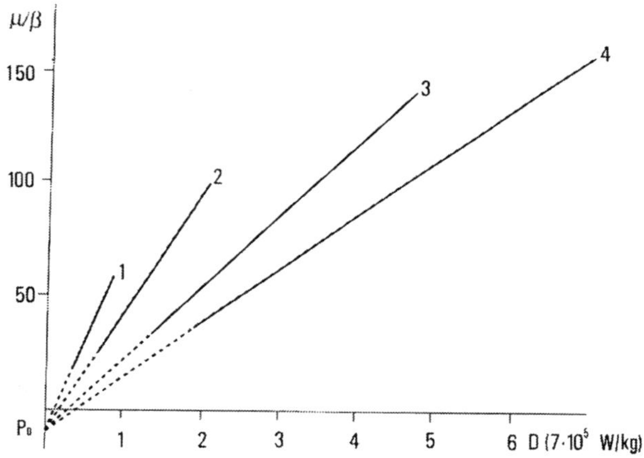

Figure 6.2 Dependence of cellular chlorophyll photosynthesis intensity μ/β on the PAR dosage per unit of time (**D**) which is absorbed by chlorophyll under various flat layer surfaces illuminations. 1—E_0 = 44.5 W/m^2; 2—E_0 = 102 W/m^2; 3—E_0 = 250W/m^2; 4—E_0 = 361 W/m^2.

Equation (6.5) was formulated under the assumption that light absorption in a microalgal cell suspension occurs in accordance with the Lambert–Beer law. In actuality, the properties of curved attenuated light in the suspension make its formulaic representation differ in exponential degree from the members of Equation (6.5), although this difference becomes essential only at a significant optical depth where light is attenuated by tens of times. Therefore any error in calculation of **D** based on the Lambert–Beer law does not exceed errors made in experimental measurement of the quantities x, l, β, and μ.

Experimental data in the region of coordinate values under consideration yield a family of straight lines. These lines can be extended (dotted lines in Figure 6.2) down into the region of small light energy dosage values until they intersect on the negative half of the y-axis.

Each of the straight lines in this figure illustrates photosynthetic dependence μ/β on the light energy dosage absorbed per unit time when the culture's surface is illuminated by constant white light. Changing the light energy dosage absorbed per unit time while the culture is illuminated in a constant way can be experimentally achieved by changing the biomass surface concentration. All experimental data for various illumination intensities were gathered using optically dense culture layers where light transmission is less than 3% and can therefore be disregarded.

The following equation describes the family of straight lines in Figure 6.2 and consequently defines cell growth rate in the culture:

$$\mu/\beta = \frac{K \cdot D}{E_0 + q} - P_0, \tag{6.6}$$

where experimental constants K and q are defined as $K = (0.192\ \text{kg/m}^2 \cdot \text{days})$ and $q = 243\ \text{W/m}^2$; P_0 is the section of the y-axis from 0 to the intersection point for the dotted lines in Figure 6.2. This section is negative, and the biological interpretation is that it represents the diminishment of the microalgal biomass in the absence of light due to the fact that culture respiration is not compensated by photosynthesis in the dark.

Since the relation of growth rate to pigmentation content is traced along the y-axis, it then follows that:

$$P_0 = \frac{\gamma}{\beta}, \tag{6.7}$$

where γ is the respiration factor and is expressed in the same units as specific growth rate μ (d^{-1}). According to measurements, γ equals $0.32(d^{-1})$.

Substituting the values for P_0 and **D** from Equations (6.5) and (6.7) into Equation (6.6) the following more explicit cell growth rate equation results:

$$\mu = \frac{KE_0}{\bar{x} \cdot (E_0 = q)} \cdot \left(1 - e^{-1\beta\bar{x}}\right) - \gamma. \tag{6.8}$$

Note that Equation (6.8) is not applicable for layers with an infinitely small biomass concentration ($x \rightarrow 0$) since in reality a cell layer cannot be thinner than one cell. The growth rate dependency on illumination for a cell monolayer as described by Equation (6.8) is illustrated by curve 2 of Figure 6.3. Layer width for this calculation is taken as 5 mm. An experimental curve representing cell growth rate in a diluted culture is also plotted in Figure 6.3, curve 1. Curve 1 is constructed according to V.N. Belyanin's experimental data (1964) taking into account the decrease in cell chlorophyll content that obtains when illumination is increased, but not taking into consideration cell geometric parameters. Curves 1 and 2 are of similar character and they coincide near their points of origin.

Culture productivity, that is, the cell biomass increase per unit time per unit of illuminated surface area, is constituted as follows:

$$\mu = \frac{KE_0}{\bar{x}(E_0 + q)}\left(1 - e^{-1\beta\bar{x}}\right) - \gamma\bar{x}, \text{kg/m}^2 \cdot \text{days}. \tag{6.9}$$

The coefficient $(1 - e^{-1\beta\bar{x}})$ represents layer light absorption, and for optically dense or fully-absorptive layers this coefficient is close to unity. For such cultures, the productivity equation takes on the form:

$$P = K\frac{E_0}{E_0 + q} - \gamma\bar{x}. \tag{6.10}$$

Interestingly enough, the first member of this equation is an hyperbole analogous to the famous Michaelis–Menten equation. In our equation, illumination E_0 figures instead of a substrate, and constant q is analogous to the constant in the Michaelis–Menten equation, being numerically equal to the illumination value at which productivity is half of maximum. Maximal productivity is introduced by the coefficient $k = (0.192 \text{ kg/m}^2 \cdot \text{day})$.

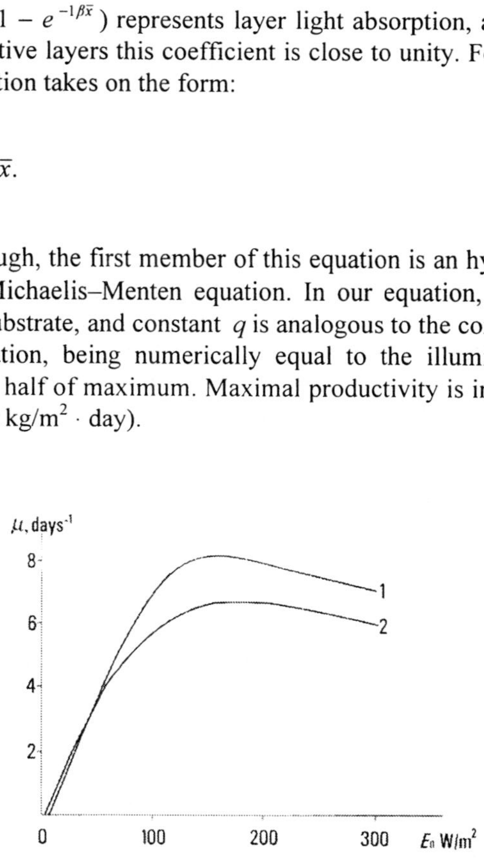

Figure 6.3 Cell growth rate in an optically thin layer of algal culture. 1—the experimental dependency according to V.N. Belyanin's data; 2—the theoretical curve yielded by Equation 6.8 for a cell monolayer.

Although the authors did not study this question, the fact that these dependencies take on an identical form in both Equation (6.10) and in the Michaelis–Menten equation possibly attests to the formal similarity that also adheres in the two systems with respect to cell mechanisms for assimilating quanta and molecules.

Employing Equation (6.9), the efficiency coefficient for photosynthesis can be formulated as follows:

$$\eta = \frac{QP}{TE_0\left(1-e^{-1\beta\bar{x}}\right)} = \frac{Q}{TE_0\left(1-e^{-1\beta\bar{x}}\right)} \times \left[\frac{KE_0}{E_0+q}\left(1-e^{-1\beta\bar{x}}\right) - \gamma\bar{x}\right], \tag{6.11}$$

where $T = 8.64 \cdot 10^4$ and is the normalizing multiplier when the unit of time is taken to be 1 day (1 day = $8.64 \cdot 10^4$ s). Q is the specific calorie content of the microalgal biomass.

The specific calorie content of microalgal biomass is experimentally defined to be 5435 cal/g. When microalgae are cultivated in a fully mineralized medium, all of this energy is received from light. When cultivation is achieved using urea as a source of nitrogen, then the calorie content of the urea assimilated by the cells during biomass synthesis must be subtracted from the above figure. When urea consumption is 170 mg/g and its calorie content is approximately 2700 cal/g, assimilated light energy comes to $Q = 5435 - 460 = 4975$ cal/g. This number should be used in calculating the efficiency coefficient for microalgal photosynthesis when using urea as a source of nitrogen in cultivation.

From Equation (6.9) the optimal cell biomass concentration that will support the maximum culture productivity for a given illumination value can be easily defined.

Taking $d\eta/dt = 0$, we derive

$$\bar{x}_{\text{opt}} = \frac{1}{1\beta}\ln\frac{E_0K1\beta}{\gamma(E_0+q)}. \tag{6.12}$$

Then the dependence of productivity on illumination (substituting (6.12) into (6.9)) will be

$$P_{\text{opt}} = K\frac{E_0}{E_0+q} - \frac{\gamma}{1\beta}\left(1-\ln\frac{E_0K1\beta}{\gamma(E_0+q)}\right), \tag{6.13}$$

and the performance coefficient for photosynthesis is now

$$\eta = \frac{Q}{TE_0\left(1-\frac{\gamma(E_0+q)}{E_0K1\beta}\right)} \times \left[\frac{KE_0}{E_0+q} - \frac{\gamma}{1\beta}\left(1-\ln\frac{E_0K1\beta}{\gamma(E_0+q)}\right)\right]. \tag{6.14}$$

Dependency graphs for Equations (6.12), (6.13), and (6.14) are illustrated in Figure 6.4. Equation (6.14) does not permit calculating the maximally possible photosynthetic efficiency coefficient since this equation is targeted for conditions that make for maximum productivity, and these two things are not the same. So we can see that maximum productivity is achieved when the culture is optically dense and practically fully-absorbent, whereas the maximum performance coefficient for photosynthesis can be reached only when layer width is minimal, that is, when there is a cell monolayer. Figure 6.4 contains a graph of the photosynthetic efficiency coefficient for a layer where x = 1.5 g/m^2, and this approximately corresponds to that of a cell monolayer. From Figure 6.4 it is clear that the photosynthetic efficiency coefficient for a cell monolayer is higher than for the optimal layer, although the monolayer strongly defers to the culture's optimal layer with respect to productivity. Optimal surface concentration value x for the microalgal biomass under various illumination intensities (shown by curve 2 in Figure 6.4) allows us to calculate the amount of light penetrating through the culture's optimal layer. The value of this quantity varies from 2.5 to 6% of E_0 for different illumination intensities. Together with this, comultiplier $(1 - e^{-1\beta\bar{x}})$ in productivity Equation (6.9) is always close to unity. The second member of this same equation describes culture respiration and contains comultiplier x; this second member is small compared to the first member. So we can say that when layers are optically dense, then microalgal culture productivity is first and foremost a function of illumination and depends only weakly on the biomass concentration.

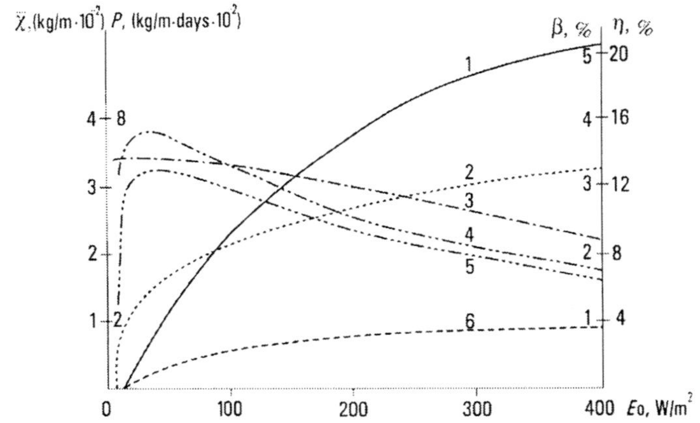

Figure 6.4 Parameter dependence on surface illumination: 1—the productivity of optimal culture layer P_{opt}; 2—biomass optimal surface concentration $\bar{x}_{opt}$; 3—the chlorophyll concentration in algal cells; 4—the photosynthesis performance coefficient for a cell monolayer; 5—the photosynthesis performance coefficient for optimal layer η; 6—cell monolayer productivity.

The dependency graphs in Figure 6.5 illustrate culture productivity for the microalgal biomass surface concentration based on exposure to various illumination intensities. These curves have 1) a maximum that is commensurate with

Equation (6.12); 2) a steep decline in the region of low concentrations where the layer becomes not fully-absorptive; and 3) a gradual linear decrease in productivity in the region of high concentrations that is elicited by culture respiration. Figure 6.5 shows that in the neighborhood of x_{opt}, there is a region where biomass changes by a factor of two: from $1.75 \cdot 10^{-2}$ to $3.5 \cdot 10^{-2}$ kg/m^2, whereas culture productivity in this region changes by not more than 5%. This quantity is of the same order as experimental errors made in the determination of culture productivity.

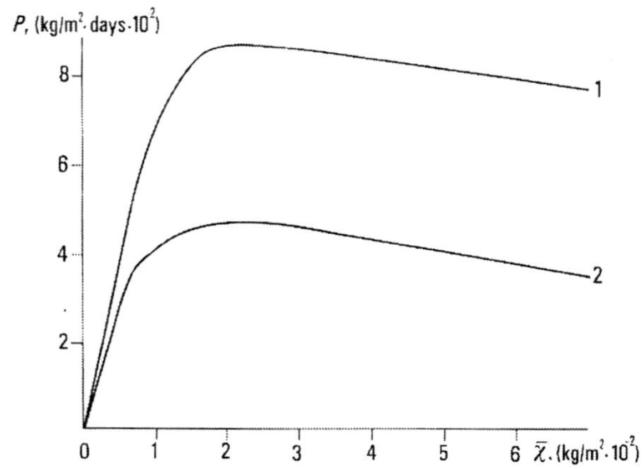

Figure 6.5 Layer productivity dependence on biomass surface concentration when illumination is 250 W/m^2 (1) and 100 W/m^2 (2).

This means that there is a wide zone in which productivity of the cultivation process is insensitive to changes in biomass, and this zone is of crucial importance when inventing ways to automatically regulate biomass concentration.

When Equations (6.8) and (6.9) are set equal to 0, the compensating point of photosynthesis becomes clear:

$$E_{\text{comp}} = q\frac{\gamma x}{K\left(1-e^{-1\beta\bar{x}}\right)-\gamma\bar{x}}, \text{ W/m}^2. \tag{6.15}$$

Equation (6.15) clearly shows that the value representing the compensating point of photosynthesis is not constant, but is a function of biomass surface concentration. It turns out that the compensating point of photosynthesis for a cell monolayer is 3 W/m^2 where $x = 1.5 \cdot 10^{-3}$ kg/m^2, and for a fully absorptive cell layer where $x = 3 \cdot 10^{-2}$ kg/m^2 the compensating point has grown to 13 W/m^2.

All of the above dependencies were procured for a culture layer illuminated from one side. When the transition was made to two-sided illumination, experiments revealed that the culture responds in an additive way. The productivity of such a layer is described as follows:

$$P_2 = K\left(1 - e^{-1\beta\bar{x}}\right)\left(\frac{E_{01}}{E_{01} + q} + \frac{E_{02}}{E_{02} + q}\right) - \lambda\bar{x}, \tag{6.16}$$

where E_{01} and E_{02} are illumination intensities from the first and second sides of the layer respectively; P_2 is the productivity of layer unit area in response to both light streams.

When E_{01} and E_{02} are equal,

$$P_2 = 2K\frac{E_0}{E_0 + q} - \gamma\bar{x}\,. \tag{6.17}$$

Equation 6.17 clearly shows that productivity for a layer illuminated from two sides more than doubles the productivity yielded when only one side is illuminated. Accordingly, such a layer also has a higher performance coefficient for photosynthesis. So the maximal photosynthetic efficiency coefficient for a one-sided optimal layer is equal to 13% (curve 5 in Figure 6.4), whereas the same layer has a maximal photosynthetic efficiency coefficient of 14.4% for two-sided illumination. When illumination intensities are high, this difference is not as significant.

When performing practical calculations, it helps to use the productivity equation for a unit of illuminated surface exposed to two-sided illumination when the culture is arranged in flat parallel layers:

$$P_2' = K\frac{E_0}{E_0 + q}\left(1 - e^{-1\beta\bar{x}}\right) - \frac{1}{2}\gamma\bar{x}. \tag{6.18}$$

The culture productivity equation can be transformed for layers configured in another more complex geometric form. For instance, according to Equations (6.6) and (6.9), a cylinder that is uniformly illuminated within and that is packed with an algal culture can be described as follows:

$$P_c = \mu\bar{x} = \frac{K\beta D}{E_0 + q} - \lambda\bar{x}. \tag{6.19}$$

The light energy dosage absorbed per unit time can be defined as follows in this case. Consider that all the light energy radiated by a unit arc length of the cylinder's circumference is absorbed inside the sector that corresponds to the arc. This is true because culture homogeneity and the identical luminous emittance of all points on the cylinder's surface ensures that an energy exchange between any selected sector and its neighboring one is equivalent (cf. Figure 6.6).

The light absorption coefficient on the path that is equal to the diameter is $(1 - e^{-1\beta x 2R})$ where x is the relationship of biomass volume concentration to surface concentration yielding $x = 2x/R$, where R is the cylinder radius.

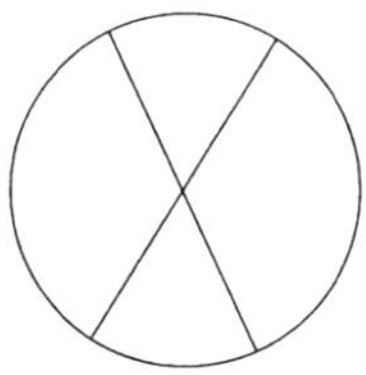

Figure 6.6 A cylinder evenly illuminated within and packed with algal culture.

Now the light energy dosage absorbed per unit time is

$$\mathbf{D} = \frac{E_0}{\frac{R}{2}x\beta}\left(1 - e^{41\beta x 2R}\right), \qquad (6.20)$$

and the productivity of a unit surface area for a cylinder packed with culture is

$$P_c = K\frac{E_0}{E_0 + q}1 - e - 41\beta\overline{x} - \gamma\overline{x}. \qquad (6.21)$$

Using the already described method to find the optimal biomass and substituting it into Equation (6.21), the following equation for the cylinder can be derived:

$$P_{\text{c opt}} = K\frac{E_0}{E_0 + q} - \frac{\gamma}{41\beta}\left(1 + \ln\frac{4K1\beta E_0}{\gamma(E_0 + q)}\right). \qquad (6.22)$$

The analogous method for a sphere packed with culture yields

$$p_{\text{sph opt}} = K\frac{E_0}{E_0 + q} - \frac{\gamma}{61\beta}\left(1 + \ln\frac{6K1\beta E_0}{\gamma(E_0 + q)}\right). \qquad (6.23)$$

When the optimal productivity equation for a culture in a dense layer (6.14), in a cylinder (6.22) and in a sphere (6.23) are compared, it can be seen that the first member of these equations is identical, whereas the second member is the sum of losses due to culture respiration and light penetration, and therefore depends on the capacity of the geometric form packed with culture.

All equations examined above describe so-called "apparent photosynthesis," that is, true photosynthesis adjusted for culture respiration. However, it is possible to evaluate photosynthesis proper within the framework of the present model. If the member of Equation (6.11) that describes respiration is set equal to 0, $\gamma\bar{x} = 0$, then the photosynthetic efficiency coefficient will be

$$\eta = \frac{QK}{1(E_0 + q)}, \tag{6.24}$$

that is, the photosynthetic efficiency coefficient depends on biomass and is inversely proportional to illumination. The maximum photosynthetic efficiency coefficient as $E_0 \to 0$ and the above-defined values obtain is a constant equal to 19.1%. This constant does not go beyond the bounds of known theoretical predictions.

Special attention must be paid to parameter β, which enters into all model equations. β—the photosynthetic pigmentation content in algal cell biomass—is an exceedingly complicated function of culture illumination and biomass surface concentration x. Until now, β had no satisfactory mathematical description. This dependence had already been introduced for a fully absorptive culture layer (see curve 4, Figure 6.4). Using this information, the calculations for all the above equations of the model where β was a constant value can be corrected. However the necessity to correct productivity falls off if the calculations are made for fully absorptive culture layers, that is, when $e^{-1\beta x} = 0$. As Equation (6.10) shows, parameter β does not enter into the calculation in this case. Neither is it necessary to keep track of the changes in β under low illumination intensities where $E_0 <$ 100 W/m^2 since β changes little in this region. A calculation of changes in β are necessary when computing growth rate and culture productivity under high illumination intensities (300–400 W/m^2) and when the layers are of low optical density.

The productivity equations introduced above for algal cultures arranged in dense and parallel layers were applied to performing the calculations necessary to construct algal cultivators for a target productivity. These calculations are presented in Section 6.4. The lengthy use of cultivators built according to these calculations confirmed the veracity of these calculations and equations. The basic parameter value region for which the veracity of calculations was experimentally confirmed is: surface culture illumination set to 25–400 W/m^2, and biomass surface concentration set to $1.5 \cdot 10^{-3}$ to 10^{-1} kg/m^2.

A Thin-layered Algal Culture

From basic Equation (6.6) describing algal cell growth dependence on illumination, it is clear that growth is directly proportional to the light energy dosage absorbed per unit time and is inversely proportional to culture illumination. Growth acceleration (or an increase in the photosynthetic efficiency coefficient) is possible when light energy dosage per unit time has been augmented without increasing culture surface illumination. This is easily achieved under experimental conditions when algal layer optical thickness is decreased or, which is the same thing, when biomass concentration x is decreased. In this case, the cell layer will turn out not to be fully-absorptive. Even so, light loss can be easily avoided if we place a cell layer under the first thin layer and ensure that this layer does not cohere or commingle with the first layer, and then place a third layer under this in a similar manner, and so on, until full light absorption is obtained.

The value of dividing the culture into layers that do not commingle is easily demonstrated by a numerical example. Let the culture layer be illuminated from one side with an intensity of 300 W/m^2. Now let optimal biomass concentration be set equal to $3 \cdot 10^{-2}$ kg/m^2 (as shown in Figure 6.4), let productivity be set to ($9.43 \cdot 10^{-2}$ kg/m$^2 \cdot$ day), and let the photosynthetic efficiency coefficient be set to 8.2%. Divide this layer into two identical layers that do not commingle. Now the illumination of the first layer remains at 300–400 W/m^2, whereas that of the second layer is: $E_{02} = E_{01} \cdot e^{-1\beta\bar{x}} = 48.5$ W/m^2. According to Equation (6.9), the productivity of these layers is P_1 = (0.0872 W/m$^2 \cdot$ day) and P_2 = (0.0228 W/m$^2 \cdot$ day), which totals to 0.110 W/m^2, that is, 16% higher than the productivity defined for the whole layer before it was divided. Correspondingly, the photosynthetic efficiency coefficient increases from 8.2% to 9.5% when the layer is split.

The extreme case of culture multilayering occurs when the culture is divided into an infinite number of infinitely thin layers. For any culture monolayer of thickness $d\bar{x}$ that is situated at some depth $\bar{x}$ and is illuminated with an intensity of E_i, the light energy dosage per unit time absorbed by chlorophyll is:

$$\lim_{\bar{x}\to 0} D = \lim_{\bar{x}\to 0} \frac{E_i}{\beta\bar{x}}\left(1 - e^{-1\beta\bar{x}}\right) = 1E_i. \tag{6.25}$$

According to Equations (6.6) and (6.7), cell growth rate in this layer will be

$$\mu_i = K\frac{D\beta}{E_i + q} - \gamma = \frac{E_i}{E_i + q} - \gamma, \tag{6.26}$$

and productivity of the monolayer is

$$P_i = \mu_i d\bar{x}\left(K1\beta\frac{E_i}{E_i + q} - \gamma\right)d\bar{x}. \tag{6.27}$$

Surface illumination of the isolated monolayer is

$$E_i = E_0 e^{-1\beta\bar{x}}. \tag{6.28}$$

Substituting this illumination value into Eq. (6.27) and integrating productivity over layer thickness x, we obtain the following equation for culture productivity when the culture consists of an infinite number of infinitely thin layers that do not commingle:

$$\begin{aligned} P &= \int_0^{\bar{x}}\left(K1\beta\frac{E_0 e^{-1\beta\bar{x}}}{E_0 e^{-1\beta\bar{x}} + q}\right)d\bar{x} = \\ &= (K1\beta - \gamma)\bar{x} - K\ln\frac{E_0 + qe1\beta\bar{x}}{E_0 + q}. \end{aligned} \tag{6.29}$$

When $dP/dx = 0$, optimal algal biomass concentration for such a layer is:

$$\bar{x}_{\text{opt}} = \frac{1}{1\beta}\ln\frac{E_0(K1\beta - \gamma)}{\gamma q}. \tag{6.30}$$

The case examined here is of an algal culture with immobilized cells. Technical difficulties inhibit the creation of such a culture. Its closest analogue might be a culture consisting of a finite number of layers where each layer is a cell monolayer.

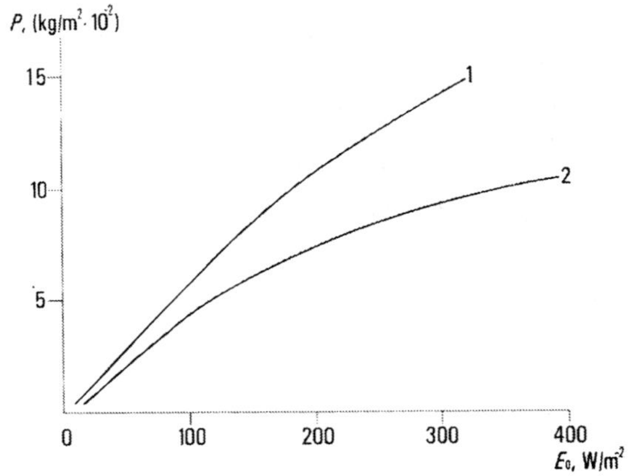

Figure 6.7 Dependence of productivity on surface illumination for an algal culture layer with immobilized cells (1) and one with a commingled layer (2). Biomass surface concentration in both instances is $x = 3 \cdot 10^{-2}$ kg/m^2.

The layers can be divided by transparent dividers, and the cells can freely commingle within the layer while being constantly illuminated. Now the productivity equation assumes the form

$$P = \frac{K}{x}\left(1 - e^{-1\beta\bar{x}}\right)\left[\bar{x} - \frac{1}{1\beta}\ln\frac{E_0 + q}{E_0 + qe^{1\beta\bar{x}}}\right] - \gamma\bar{x}, \tag{6.31}$$

where c is the biomass surface concentration in an algal cell monolayer.

Figure 6.7 illustrates the productivity and photosynthetic efficiency curves for a thin-layered culture as opposed to the usual commingling cultures.

Multilayered cultures of unicellular algae have not undergone sufficient experimental investigation. Therefore, the dependencies introduced earlier in this section will not be further employed. However, experimental findings have been published that confirm the possibility of significantly heightening algal productivity by dividing an optically dense culture into layers that do not commingle (Sidko *et al.*, 1967).

Algal Growth Dependence on Carbon Dioxide Concentration

We have earlier mentioned that microalgal culture growth is more often than not limited by light since it is precisely when light intensities are highest that the photosynthetic efficiency coefficient is maximized. But in some instances, carbon dioxide can become the limiting factor. For instance, this is true when a microalgal culture is employed in closed systems to regenerate the atmosphere and, for some reason, the carbon dioxide supply to the algae from the crew is cut off (Kirensky *et al.*, 1967a, b). In the current section, a theoretical generalization is attempted based on materials accumulated by researchers who had investigated this dependency (Kovrov *et al.*, 1969).

An algal biomass is more than 50% carbon. Carbon dioxide dissolved in the medium is practically the only source of this carbon. The carbon dioxide required to synthesize 1 g of algal biomass comes to 1 liter (1.96 g), that is, this requirement is an order of magnitude larger than requirements for all other nutritive components. For this reason, using carbon dioxide to support an intensely growing algal culture requires considerable monetary expenditures along with the knowledge of how culture growth depends on this parameter.

All researchers who have studied algal cell growth rate dependency μ on carbon dioxide concentration in the gaseous phase S_0, where this concentration is contiguous with the culture, have noted that for coordinates in μ and S_0 this dependency appears in the form of ascending curves as saturation is approached. However, initial values for the saturation region sharply vary from author to author—from 0.1 to 10% CO_2 (Myers, 1954, 1958; Meleshko and Krasotchenko, 1965; Meleshko *et al.*, 1993).

In designing the current model, the authors used their own experimental data (Kovrov *et al.*, 1969; Shtol *et al.*, 1976), as well as the data procured by Bartosh (1965). A qualitative analysis of the variation in the μ–S_0 dependency has allowed postulating a basic assumption of the model, that is, algal cell growth rate depends on dissolved carbon dioxide concentration according to the Michaelis–Menten equation:

$$\mu = \mu_{\mathrm{m}} \frac{S}{K_s + S}, \tag{6.32}$$

where μ_{m} is cell growth rate in the presence of surplus CO_2 in solution when CO_2 does not limit growth; S is the CO_2 concentration in the solution surrounding the cells; K_s is the Michaelis constant which is numerically equivalent to the concentration of dissolved CO_2 at which cell growth rate is half of μ_{m}.

The concentration of dissolved carbon dioxide depends on its concentration in the gaseous phase and on the rate at which the algal culture consumes it, and this concentration can be expressed in these terms.

From Henry's law it follows that

$$\frac{dS}{dt} = Ka(S_s - S), \tag{6.33}$$

where K is the mass transfer coefficient on the gas–liquid threshold, $m^3 \cdot s^{-1}$; a is the mass transfer area on the gas–liquid interface expressed in terms of culture volume units, m^2/m^3. When physical and chemical conditions in the culture are constant, and this is characteristic in continuous cultivation, quantities K and a are constant and furthermore can be replaced by A, where $Ka = A$. In the physical sense, coefficient A (s^{-1}) is the number of CO_2 volumes dissolved per second in a culture volume under conditions where S is subtracted from S_s ($S_s - S$). These conditions occur when S_s is the saturation concentration of carbon dioxide in the culture medium and a given concentration S_0 of CO_2 is present in the gas mixture contiguous with the culture; that is, $S_s = S_{0\varphi}$, where φ is the carbon dioxide solubility coefficient in the medium under given physical conditions. For these calculations, gas pressure in the culture is taken as constant and equal to atmospheric pressure, which allows us to operate based on the CO_2 concentration in the gas mixture instead of on its partial pressure.

When gas exchange in the culture reaches a dynamic equilibrium, as it inevitably does under constant growth conditions in a continuous culture, the gas dissolution rate into the culture medium is equal to the rate of its consumption by the cells: $dS/dt = \mu xy$, where x is biomass cell volume concentration in the culture, kg/m^3; and y is the carbon dioxide gas expended to synthesize a unit of biomass cells, m^3/kg.

Making use of the introduced equalities, Henry's law can be reformulated as

$$\mu xy = A(S_0 \quad - S), \tag{6.34}$$

from which it can be seen that the carbon dioxide concentration in the solution will be

$$S = S_0 \quad - \frac{\mu xy}{A}. \tag{6.35}$$

Substituting this carbon dioxide gas concentration into the Michaelis–Menten equation (6.32) and solving it for μ, we get the specific growth rate dependency on CO_2 concentration in the gas phase for algae:

$$\mu = \frac{1}{2xy}(\mu_m xy + K_s A + S_0 \varphi A -)$$

$$- \sqrt{(\mu_m xy + K_s A + S_0 \phi A)^2 - \quad 4\mu_m \, xyS_0 \quad A}. \tag{6.36}$$

And for culture productivity with a unit volume of $kg/m^3 \cdot$ day:

$$P_t = \frac{P_m y + K_s A + S_0 \varphi A}{2y} -$$

$$- \sqrt{\frac{(P_m y + K_s A + S_0 \varphi A)^2}{4y^2} - \quad \frac{P_m S_0 \varphi A}{y}}, \tag{6.37}$$

where $P_m = m_m x$ represents the productivity of a unit volume of culture when carbon dioxide does not limit growth.

The constants A and K_s in the above equations cannot be successfully defined by direct measurement. This is because such a measurement can be accomplished only by measuring the gas bubble surface area in an intensely bubbling culture (for A), or by measuring very low concentrations of dissolved CO_2 in the cultivator's culture medium (for K_s), and these measurements cannot be made with sufficient accuracy. Still, these constants can be defined indirectly from experimental data.

If carbon dioxide severely limits culture growth in the experiment (that is, if culture productivity is several times lower than P_m for a given illumination), then CO_2 concentration in the solution can be expected to be small compared with the saturating concentration. That is, $S \approx 0$. Then from Equation (6.34), making the substitutions $xy = P$ and $S = 0$, it is clear that

$$A = \frac{y}{\phi}\frac{P}{S_0} = \frac{y}{\phi}\tan\alpha. \tag{6.38}$$

Note that y and φ are constants. Then A is defined by the angle tangent for the slope of the initial portion of the plot of $P = f(S_0)$ (see Figure 6.8) and can be defined by one experimental point situated in the region of the ascending portion of the investigated dependency.

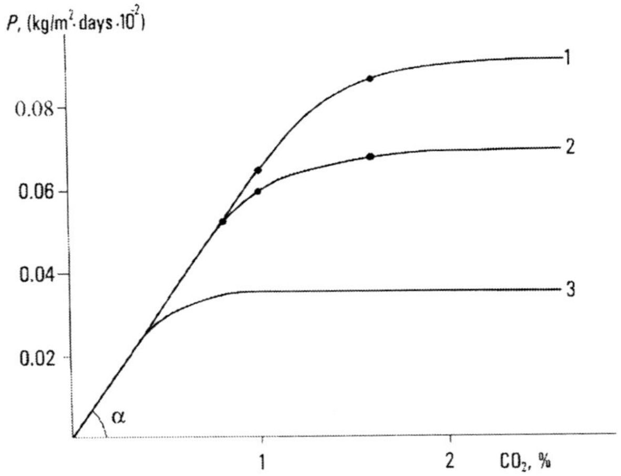

Figure 6.8 Productivity dependency of the algal culture layer that is optimum in terms of concentration on the CO_2 concentration in the mixture obtained in the cultivator under various illumination values E_0: 1—270 W/m^2, 2—170 W/m^2, 3—64 W/m^2. Culture layers of 5 mm thickness were illuminated from both sides. The productivity of each side of the layer is plotted along the y-axis. (From Shtol *et al.*, 1976.)

Quantity A is a function of both gas mixture flow rate for the gas supply to the culture and of photoreactor geometric parameters. Experimental research has shown that a photoreactor in the form of flat parallel vertically disposed cuvettes can be described by the simple relation:

$$A = \delta \frac{m}{c}, \tag{6.39}$$

where m is gas mixture flow rate through the photoreactor during bubbling, m^3/s; c is the area of the photoreactor's horizontal cross section, m^2; and the experimental coefficient is $d = 6.9 \cdot 10^4\ m^{-1}$.

In experiments, this relation turned out to be justified within the limits of specific gas flow rates from 0 to $5 \cdot 10^{-4}\ m^3/m^2 \cdot s$. When gas flow rates were large, the quantity of gas found in a unit volume of the photoreactor did not increase. This increase was inhibited by the structural peculiarities of the experimental setting—that is, by the small physical thickness of the culture layer in the photoreactor.

Relation (6.32) has limited application. It is justified only for a particular photoreactor type; but analogous dependencies can easily be procured for other devices.

In distinction to A, constant K_s describes either a given algal strain or a taxon. K_s is not connected with implementing the cultivation process. K_s can be calculated from experimental data when culture productivity in the experiment is $P_m/2$, which means that CO_2 concentration in the gaseous phase is equal to K_s. Now if CO_2 concentration in the gaseous phase is S_0^*, then Equation (6.35) takes on the form

$$K_s = S_0^* \varphi - \frac{P_m y}{2A}. \tag{6.40}$$

Substituting P_m, S_0^*, A, and constants φ and y as defined in the experiment into Equation (6.40), we have defined the concentration for the researched strain to be $K_s = 10^{-4}\ m^3/m^3$. This is an exceedingly low concentration for direct analysis in the photoreactor. Interestingly, this concentration is close to that of the saturating concentration that obtains when water is in contact with atmospheric air ($S_0 = 0.0003$).

The growth model presented here is but the first approximation. It has been assumed that the gaseous CO_2 concentration at the photoreactor output differs in a negligible way from its input concentration. In the opposing case, the average concentration ($S_{in} + S_{out}/2$) must be used in the capacity of S_0. Although this average concentration is not absolute, it is more precise, and it will reflect the true picture. The model also does not account for the decrease in culture quantity per photoreactor unit volume that is due to the presence of gas pockets. Therefore the productivity calculation should be executed with the volume of the photoreactor's liquid phase in mind, because the liquid phase volume is always less than the photoreactor volume.

As already noted, quantities μ_m and P_m used in the current model characterize culture growth when carbon dioxide is present in surplus, that is, when growth is not limited by this substrate. In this instance, culture growth can be defined by either mineral components in the medium or by physiological (genetic) mechanisms in the cells, or by, which is more interesting from a practical point of view, the limiting factor of illumination E_0. So we see that it is much more characteristic that $P_m = f(E_0)$. When describing $P_m = f(E_0)$, biomass surface concentration x is used, as is surface productivity P, kg/m^2 · day, whereas in the model $P_{vol} = f(S_0)$ and P and P_m are culture volume productivities, kg/m^3 · day. The relation between volume productivity P_{vol} and surface productivity P is: $P = P_{vol}d$, where d is the thickness of the culture's illuminated layer, m. Then the general productivity equation for a unit volume of culture illuminated surface can be formulated as:

$$P = \frac{P_E y - d(K_s A + S_0 \varphi A)}{2y} - d\sqrt{\frac{\left(\frac{P_E}{d} y + K_s A + S_0 \varphi A\right)^2}{4y^2} - \frac{\frac{P_E}{d} S_0 \varphi A}{y}}, \tag{6.41}$$

where P_E is the illumination-dependent surface productivity that is described in Equation (6.9) for a layer with one-sided illumination, and which is also described in Equations (6.16) and (6.17) for a layer with two-sided illumination.

Equation (6.41) is plotted in Figure 6.8 for several illumination intensities. These plots coincide with experimental curves, which speaks to the veracity of the initial assumption made with respect to the applicability of the Michaelis–Menten equation when describing the dependency $\mu = f(S_{CO2})$.

The current model for cell growth dependency on illumination and carbon dioxide concentration demonstrates that when CO_2 concentrations are low, then cell growth depends only on this parameter. And it demonstrates the converse: when CO_2 concentrations are high, that is, in the saturated zone, growth depends only on illumination. However, there is a certain average field where both these factors influence growth. The current mathematical model also allows us to compute parameters for cultivators with a definite productivity value, photosynthetic efficiency coefficient, and carbon dioxide saturation level.

6.3 OPTICAL PROPERTIES OF A SUSPENSION OF MICROALGAE AND THE LIGHT REGIME USED IN THEIR CULTIVATION

Light is the source of energy for microalgae, as it is for all green plants. Their growth is more often than not limited by the lack of light energy.

Light is the most expensive resource in cultivation. To ensure that light is fully absorbed and not wasted, cell suspension density is maintained at a high level in the photoreactors. In dense cultures, each cell not only consumes light energy, but it functions as a screen that scatters and selectively absorbs light. In the process of light absorption, some cells shade others from the light. To ensure that all cells have an equal chance at being optimally exposed to light over the cultivation period, the cells are mixed in the photoreactor. This ensures that the cells continually change places, sometimes chancing to fall under the bright light of the surface, and sometimes chancing to be hidden from the light in the depths of the culture. The interaction of all these factors creates a highly complicated three-dimensional scheme of light distribution in the photoreactor volume. In the end, it is precisely this scheme that determines light usage efficiency and algal culture productivity.

Special note should be given to the fact that optimizing light distribution in the photoreactor is not equivalent to optimizing light conditions for each cell in the photoreactor. On the contrary, there are conditions where some cells can be strongly limited by a lack of light, although the photoreactor itself utilizes light in a most effective manner, that is, the cell mass as a whole is assimilating the light in a maximal way. So the problem of optimal light consumption is resolved not for individual cells, but for their entire population.

From this it follows that the most important task in optimizing algal productivity is first and foremost formulating the way a three-dimensional light field forms in a suspension-packed photoreactor and how this light field depends on cell optical characteristics and concentrations in the population.

These principles are applied concretely to the problem of determining how the light field is developed and utilized in the cells individually and the suspension as a whole in order to maximize photosynthetic efficiency, including an examination of the following topics: optical characteristics with respect to the cell itself; the relationship between absorption and light scattering in suspensions as it depends on cell concentration and their pigment content; how light spectral composition determines its penetration depth into the algal suspension; and algal growth rate dependence on illumination. Finally, all of these factors are taken as the basis for proposing engineering methods for designing microalgal cultivators.

When photosynthesis is the main means for sustaining the microalgal culture, the most fundamental difficulty met while trying to achieve the most highly effective cultivation results lies in supplying the culture with light energy. Fully-absorptive, that is, very dense, cultures utilize light in a maximal way. But very dense cultures come with their own set of problems. The cells lying on the surface can experience surplus illumination, and when they do, their light usage coefficient (the photosynthetic efficiency coefficient) falls off. Cells deep in the cultivator can be so shaded by the layers lying atop them that less light reaches them than is needed to compensate for the amount of energy they expend on respiration. That is, cells in the lower layers are living in conditions below the compensation point of photosynthesis. Below this point, cells only expend energy, and do not create new cells.

Therefore, a knowledge of algal optical characteristics is very important in the creation of an optimal light regime for a dense algal culture. A knowledge of a suspension's optical characteristics is required to compute the light field present in a microalgal suspension, and to make a quantitative estimate about the efficiency of radiant energy usage during photosynthesis.

Optical characteristics are examined taking *Chlorella* as our sample culture. Over the course of many years in our Institute, F.Ya. Sidko and his colleagues conducted a systematic investigation of the optical characteristics of *Chlorella* microalgae. Materials were published by them in the book *Experimental Ecological Systems Designed for Human Habitation* (Gitelson *et al.*, 1975).

Utilizing Algal Optical Characteristics to Maximize Photosynthetic Efficiency

The principal optical properties of microalgal suspension are summed up in graphs presented in Figure 6.9—absorption spectra of *Chlorella* suspension versus spectra of emitted chloroplasts and chlorophyll solution; 6.10—transmission spectra of *Chlorella* suspension relative to concentration; 6.11—spectral dependence of light attenuation in *Chlorella* suspension on suspension concentration; 6.12—optical density of *Chlorella* suspension for the spectral regions of maximum *Chlorella* absorption—680 nm—and minimum—555 nm; 6.13—a set of graphs depicting distribution of light energy in a *Chlorella* suspension depending on spectral composition of light sources and suspension cell concentration.

These data provide a reliable foundation for calculating the algal photoreactor design, light conditions, and expected productivity.

Optical characteristics in the microalgal cell suspension as they were described above can be used to determine photosynthetic efficiency. The rest of this section is devoted to applying these optical characteristics to do just that.

The Effects of Several Photoperiodic Regimes on *Chlorella* Productivity

If one or more of the parameters that support microalgal photosynthesis are periodically altered, then cell photosynthetic efficiency will vary in relation to the frequency and amplitude of these fluctuations. Culture productivity depends to a significant extent on the duration and amplitude of transitional processes.

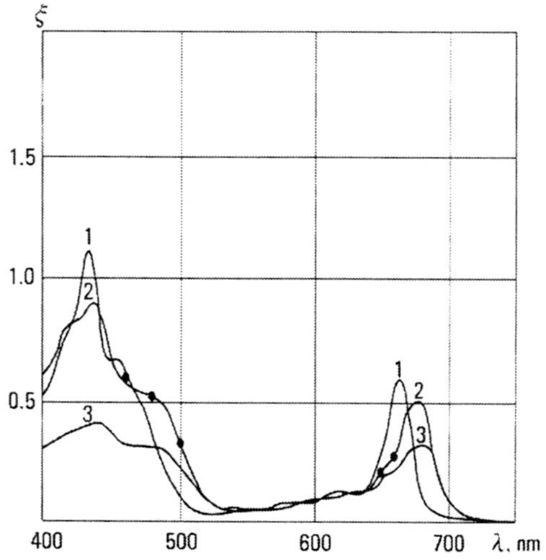

Figure 6.9 Absorption spectra for a *Chlorella* suspension (3), microparticles of its destroyed chlorolasts in the intercell medium (2), and pigments in alcohol solution (1). ξ—extinction.

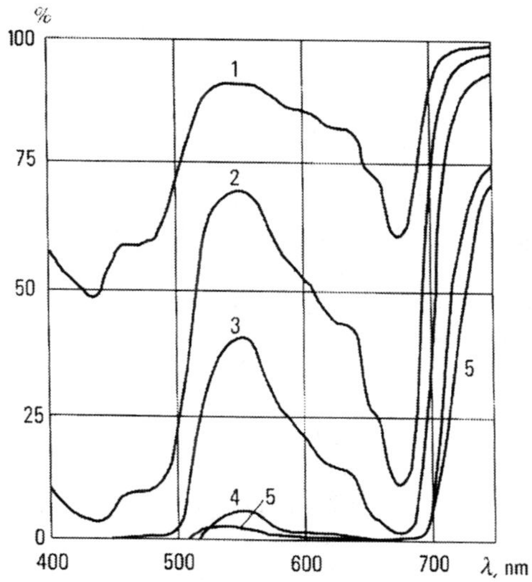

Figure 6.10 Transmission spectra (1–4) and reflection spectrum (5) for the *Chlorella* algal suspension (Belyanin *et al.*, 1964).

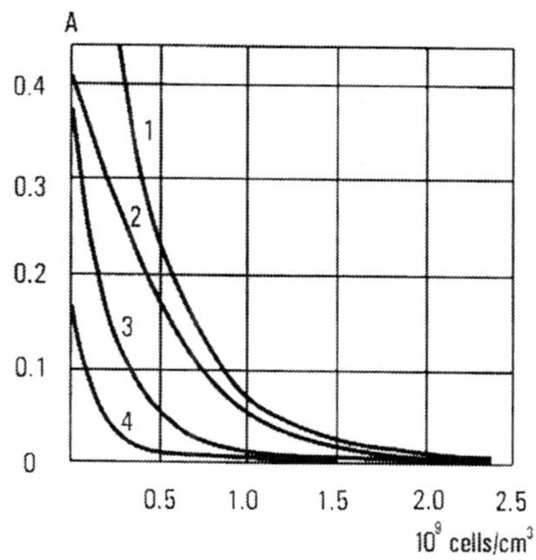

Figure 6.11 Attenuation of blue (4), red (3), and green (2) light and the total illumination intensity for a mercury luminescent bulb (1). All attenuation values are plotted as they depend on culture density. The total illumination intensity is arbitrarily taken to be 1 at the surface.

For instance, when growing microalgae in orbital station cultivators, culture illumination can periodically change with a change in orbital parameters. Therefore it is only natural to research the photoperiodic effects on microalgal productivity that can arise as a satellite rotates around the Earth.

Belyanin and his coauthors (1964) researched the photoperiodic effect on *Chlorella* productivity and photosynthetic efficiency for various irradiation levels. In their study, a photoperiodic condition was chosen that simulated possible conditions on an orbiter that rotated around the Earth every 90 min. In this rotational period, 54 min are light, and 36 min are dark.

Various radiation levels (from 85 to 310 W/m^2 PAR) were used to examine the effects of periodically altering light and dark exposure times for the culture. Culture temperature for each selected radiation level was maintained to within ±0.2°C of the optimal level. We observed practically no temperature gradients during the dark periods. This excluded temperature effects from being a possible influence on microalgal productivity. During light periods, the culture was intensely sparged with air containing 3% carbon dioxide. During periods of darkness, culture aeration continued but carbon dioxide was withheld. Culture pH was maintained constant at 7 ±0.5. After each dark phase, restoring the light was simultaneously accompanied by starting up the carbon dioxide supply again.

We can trace the dynamics of algal productivity changes effected during light and dark cycle alterations from readings produced by the continuous optical density sensor in each of those periods. Figure 6.14 shows that algal culture growth begins almost immediately after the cells are exposed to light again. However, during the dark period, algal suspension optical density decreases as cells lose biomass due to the fact that respiration is not compensated by photosynthesis in the dark periods. In the graph this process of cell biomass loss in the dark is shown as a decrease in the optical density during the dark period as a result of substrate being spent on cell respiration.

Therefore some compensation time is required during the transition from dark to light periods in order to reach the initial optical density level. This compensation time leads to a productivity decrease when compared with culture productivities for cultures exposed to continuous illumination. The compensation time value depends on the radiation intensity and increases with a decrease in the latter.

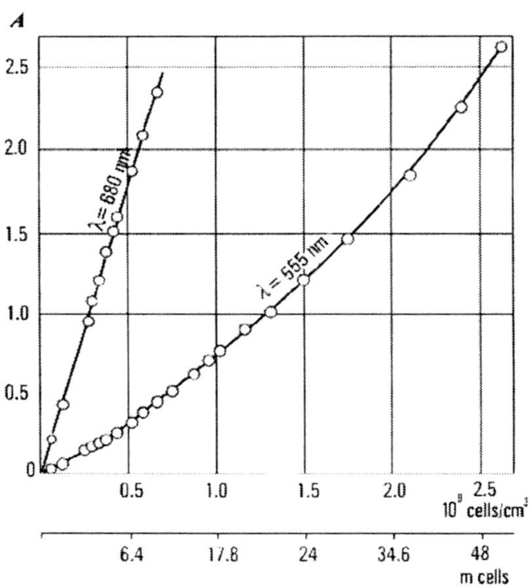

Figure 6.12 Dependence of *Chlorella* optical density *A* on cell concentration. Suspension layer thickness is 3 mm.

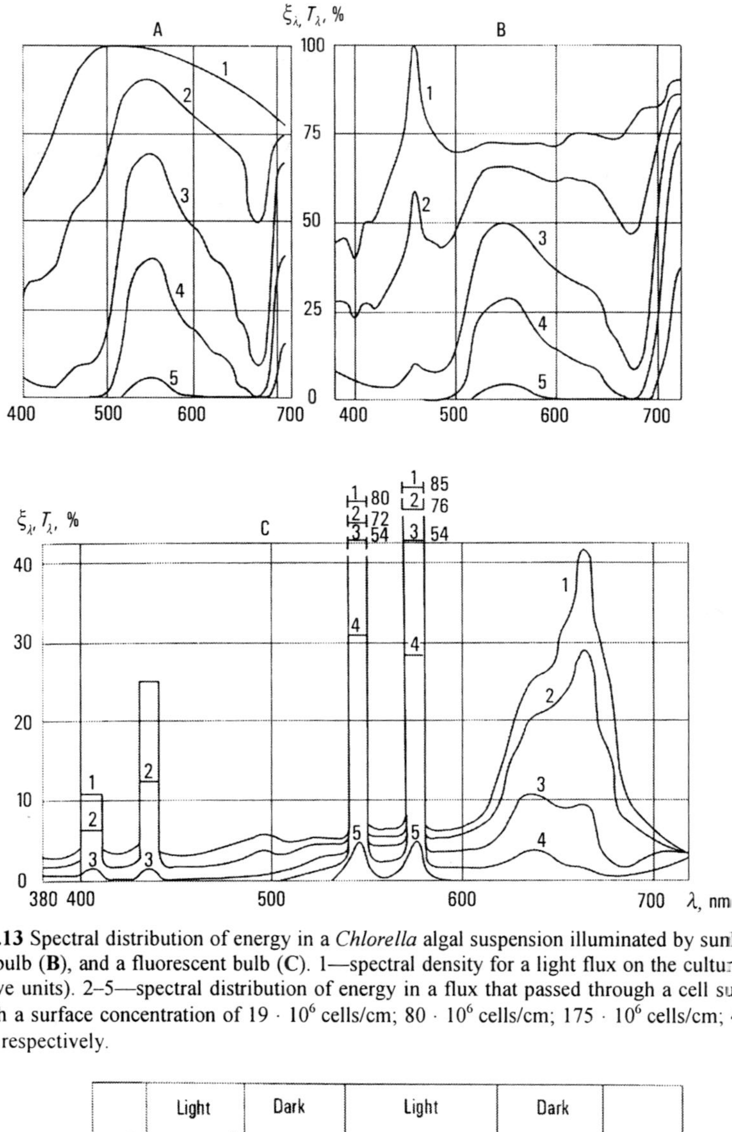

Figure 6.13 Spectral distribution of energy in a *Chlorella* algal suspension illuminated by sunlight (**A**), a xenon bulb (**B**), and a fluorescent bulb (**C**). 1—spectral density for a light flux on the culture surface (in relative units). 2–5—spectral distribution of energy in a flux that passed through a cell suspension layer with a surface concentration of $19 \cdot 10^6$ cells/cm; $80 \cdot 10^6$ cells/cm; $175 \cdot 10^6$ cells/cm; $450 \cdot 10^6$ cells/cm, respectively.

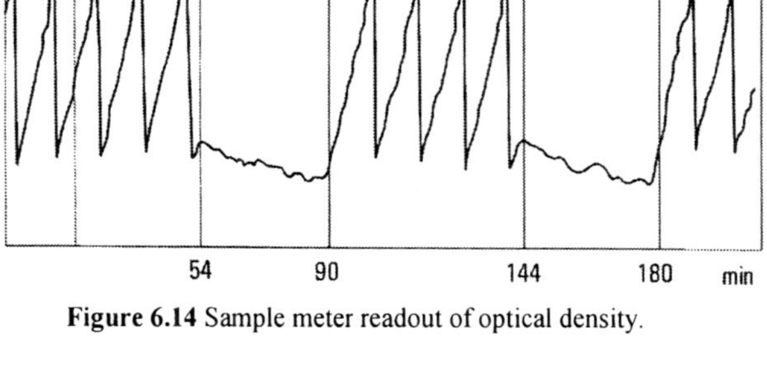

Figure 6.14 Sample meter readout of optical density.

Comparative results of microalgal culture productivity are illustrated in Figure 6.15 for microalgae cultivated in a light–dark cycle and for those receiving varying radiation levels of continuous illumination. In this figure, illumination intensities lie along the *x*-axis, and along the *y*-axis we have 1) the relative difference in the productivities which were procured under continuous illumination (P_{light}) (curve 1) and 2) those productivities reached during light–dark alterations ($P_{light-dark}$) and which were calculated only on the basis of light periods. When plotting curve 2, biomass losses due to respiration during dark periods are considered to be 14% of the biomass per day, and these losses are added to the culture productivity achieved by the dark–light regime.

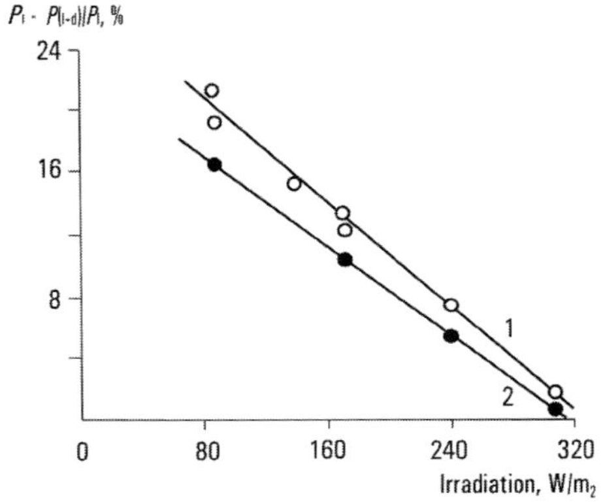

Figure 6.15 Explanation found in the text.

Experiments show that after the dark period subsides, there is an interval in which algal productivity level is restored to the level that was obtained during the previous light period. The duration of the restoration period depends on how long the cells were subjected to darkness. For instance, after an 8-h exposure to darkness, the culture restored its growth rate in 2 to 3 h, but after a 20-h exposure to darkness, it restored its growth rate only after 9 h.

In general, when dark periods last from 3 to 8 h, the restoration period does not exceed 4 h.

6.4 DESIGN OF AN ALGAL CULTIVATOR

The term cultivator implies an equipment complex designed to support prolonged cultivation of unicellular algae when the basic parameter values of the growing process are maintained within the necessary limits dictated by the goals of an experiment. The main part of the cultivator is the photoreactor (photobioreactor). It is in the photoreactor that microalgal cells are subjected to light effects and that light energy is used as photosynthesis takes place.

The current section is based on practical knowledge gained in the building of many cultivators. These cultivators and the method of their construction were developed under the leadership of B.G. Kovrov. Their design specifications and implementation details were published by him and his coworkers (Shtol *et al.*, 1976).

Some Preparatory Remarks on Cultivator Design

The maximum photosynthetic efficiency coefficient for algae is known to correspond to an illumination intensity of about 50 W/m^2, and at this intensity it approaches 14% (Belyanin *et al.*, 1964; Kovrov *et al.*, 1969). When illumination intensities climb even higher, the photosynthetic efficiency coefficient quickly decreases, and when illumination intensity reaches 400 W/m^2, the photosynthetic efficiency coefficient is only 6%. Even so, when illumination intensity reaches 400 W/m^2, that is, an intensity significantly higher than that of sunlight on the Earth's surface, then cultivation productivity for a unit area of irradiated algal culture continues to grow to very high values. When illumination intensities are high and the photosynthetic efficiency coefficient is low, the electric energy expended on synthesizing a unit of biomass is very great. Therefore, if the experiment does not require high illumination intensities it makes more sense to conduct the process with an illumination intensity not higher than 300 W/m^2.

Materials

Selection of materials for parts and unit assembly manufacture is very important and is often a deciding factor in whether cultivator design proves successful or not. Some materials are toxic to the algal culture and therefore cannot be used in parts manufacture for those parts that come into direct contact with the algae. Such materials include copper and its alloys, lead and its alloys, many brands of black and red rubber in sheets and rolls, and alloys of certain heavy metals. The second group of undesirable materials are those that corrode or otherwise undergo destruction as a result of prolonged contact with algae. These materials do not decay from contact with the algal culture itself, but from various concomitant microorganisms. Several brands of plain carbon structural and tool steel and aluminum alloys do not prove durable in a nonsterile culture. Experience has shown that the most suitable materials are alloyed, stainless steel, titanium alloys, fluoroplastic, and white vacuum rubber. Aluminum alloys precoated with an anodic treatment can be used in the manufacture of those thick-walled parts that come into contact with culture surfaces such as casings and flanges. The best material for the transparent light-receptive walls is plexiglass. Simple or quartz glass can sometimes be used if the design and load allow for it. Copper alloys, such as brass or bronze, magnesium alloys, and other alloys can be used to manufacture parts that do not come into direct contact with the algal culture or its nutritive medium.

Mixing

Algal cultivation requires constant mixing of the culture to provide the cells with nourishment and to wash away their excretions. One peculiarity of algal cultivation lies in the fact that large light-receptive areas are necessary when dealing with a thin culture layer, and this complicates the installation of any mechanical mixers that are meant to directly mix the algal culture. In practice, there are two basic methods for mixing the algal culture.

If the cultivator is designed for operation under normal terrestrial conditions, that is, when gravity is an active force, the simplest method of mixing is sparging by means of a gas mixture of air and carbon dioxide. This method serves two purposes: the gas mixture provides for culture gas exchange and mixes the culture. The intensity with which the culture is mixed can be changed by regulating the amount of gas introduced into the culture. Under normal conditions, 0.1 liter of gas per min is sufficient to mix 1 liter of culture. If more gas than this is introduced, the whole photoreactor fills with foam, and this in turn increases light loss due to its transmission out of the photoreactor through the numerous gas pockets that form between the two cuvette walls and that travel up their whole length.

The second mixing method entails the creation of a closed loop through which the culture is turbulently pumped at a specified high rate. For this method, the light-receptive areas are washed by the culture stream. This method can be used in conditions where gravity is present, as well as in conditions of weightlessness. The two mixing methods can also be combined as necessary.

Algal Cell Surface Adhesion

Chlorella cells, like other algal cells, are capable of adhering to contact surfaces. They especially like adhering to illuminated surfaces. This adhesion is all the more aggravated when unfavorable factors are at work on the algae—overheating, abnormally high or low pH levels, the limitation of one or another nutritive component, and so on. Surface "fouling" of light receptive areas by algal cells decreases culture productivity, adversely affects heat and gas exchange, and without doubt is a generally deleterious phenomenon. Practically all materials are subject to fouling: metals, plexiglass, rubber, glass, and plastics, including fluoroplastic. The most effective methods to combat fouling are the same as those used to mix the culture, that is, sparging or the creation of some closed loop in which the culture is constantly moving. In the first case, as the gas bubbles rise upwards through the culture volume, they remove algal cells that have settled on the walls. In the same way, the culture stream inhibits cell accumulation on the walls as it moves along the loop at a rate of not less than 1 to 2 m/s. When designing the cultivator, special attention should be paid to keeping the number of hydrodynamic shadows to a minimum in the light-receptive and closed loop assemblies. This will ensure that culture stream rates will not be slowed and that gas bubbles will not be trapped by the surfaces of these assemblies.

Foam and Foam Breaking

Microalgal cells release a large quantity of organic substances into the ambient medium—in some conditions a volume equaling up to 20% of the synthesized biomass. These substances act like emulsifying agents, creating a stable foam. When the culture is intensely sparged and/or when the culture undergoes turbulent streaming motion, a great deal of foam forms in the cultivator. This foam is very durable and unless special measures are undertaken to break it up, it fills the gas pipes, storage vessels, chambers containing the sensors, gas coolers, and so forth.

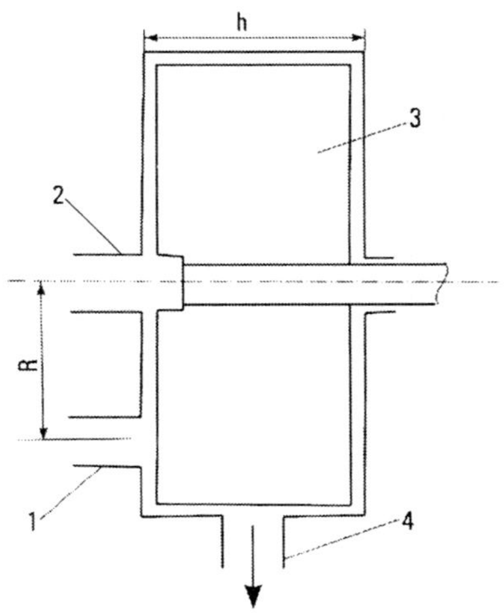

Figure 6.16 Diagram of the separator/foam-breaker. 1—input coupling; 2—gas output coupling; 3—turning rotor; 4—liquid output coupling.

Various methods exist to effect foam destruction, and these include chemical foam breakers, built-in blades, other mechanical devices, liquid or gas streams, and so on, as experimental evidence has shown. The most dependable of these methods turns out to be the inclusion of a centrifugal separator/foam-breaker in cultivator design. This separator/foam-breaker is installed outside the light-receptive area. The separator destroys the foam that accompanies the gas stream and separates it into its gas and liquid parts.

The separator/foam-breaker is illustrated in Figure 6.16. The foam, which consists of a liquid and a gas phase, enters the separator through input coupling (1) and falls on a spinning rotor (3) where it is broken up by the blades into individual droplets.

All algal cultivator designs must include the following parts and systems:

- nutritive system
- biomass harvesting and selection system

- gas exchange and mixing system
- thermostatic system
- medium control system
- photoreactor
- illumination system
- auxiliary equipment.

A block diagram of the cultivator is shown in Figure 6.17.

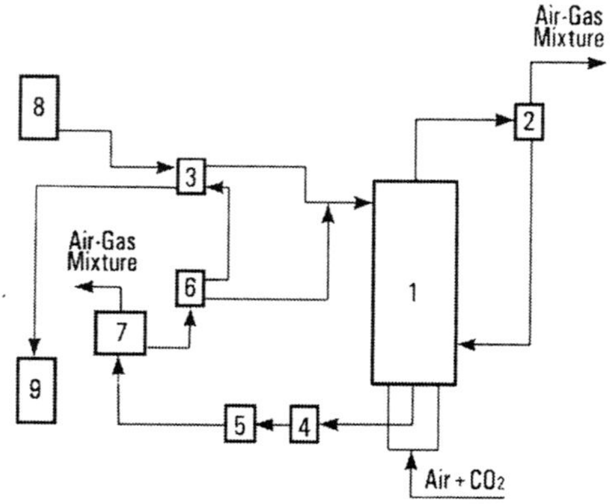

Figure 6.17 Cultivator block diagram. 1—photoreactor; 2—foam-breaker; 3—pump-dosage assembly; 4—temperature sensor; 5—pH sensor; 6—optical density sensor; 7—main pump; 8—nutritive medium vessel; 9—harvest vessel.

Photoreactor

The main unit in the cultivator is the photoreactor. It is the vessel in which algal culture growth takes place since it is here that light energy is absorbed. In the first cultivators ever built, glass flasks, open vats, etc., served as photoreactors (Shtol *et al.*, 1976). Later, several structurally different photoreactor types were developed that enabled a more prolonged process of algal cultivation.

Photoreactors with flat and parallel cuvettes proved simple and easy to use. As apparent from Figure 6.18, the main photoreactor element, the cuvette, consists of back wall (3) and front wall (5) which are separated by a spacer (4). The culture occupies volume (6) which lies within the spacer and walls. When the illumination is two-sided, both walls are made of transparent material, as a rule, from plexiglass. Metal shells (8) are laid over the outside of the assembly to reinforce it and to provide even distribution of the two-sided stress that arises during assembly and use of the cuvettes. The whole package—the walls, the shells, and the spacer—is secured by bolts (1) and nuts (2) at various intervals. This cuvette design makes for easy assembly and disassembly, thereby simplifying the replacement of those cuvette parts removed from operation and their cleaning. Sparger (11) is located in the lower half of the cuvette. The sparger introduces the gas mixture into the cuvette in the form of fine bubbles. Gas and foam outlet (7) is located in the cuvette's upper half. Besides this, the cuvette has input (9) for nutritive medium supply and output (10) for removing the culture from the cuvette. Input (9) and output (10) also serve as the place where the cuvette is joined to the closed housing in which the culture is circulated through a system of temperature sensors, pH sensors, and other sensing devices.

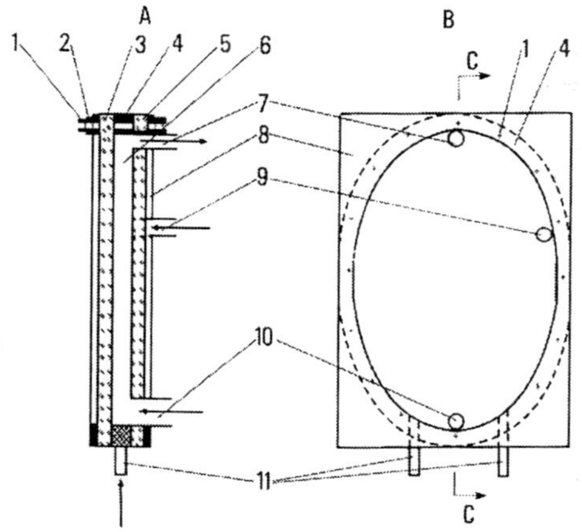

Figure 6.18 Diagram of cuvette assembly. **A** is cross section view of **B**, from **C** to **C**. 1—bolts; 2—nuts; 3—back wall; 4—spacer; 5—front wall; 6—inner space occupied by the algal culture; 7—gas and foam outlet; 8—metal shells; 9—nutritive medium supply input; 10—output for removing the culture from the cuvette; 11—sparger.

Culture thermostatic control is accomplished either with the aid of heat exchangers that can be installed anywhere in the system or by means of water jackets that are situated right on the outside of the cuvette walls.

Calculating Photoreactor Design Parameters with Flat, Parallel Cuvettes

There are two ways to calculate the illumination intensity of the cultivator's light-receptive surface. Which method is chosen depends on the purposes and conditions of the experiment at hand. The first method focuses on procuring a high photosynthetic efficiency coefficient, and the second focuses on obtaining high productivity levels.

If it is necessary to get as high a photosynthetic efficiency coefficient as possible, algal culture surface illumination is calculated using Equation (6.14). In this case, light is used by the culture in the best of ways, but due to the relatively low illumination intensity, the productivity yielded by a unit of light-receptive surface will be small.

To receive the maximum productivity from a unit of light-receptive surface, the illumination intensity, as was shown above, should reach 250–300 W/m^2 PAR and higher. Its maximum value depends on the construction materials used and the work schedule planned for the light-receptive surfaces so that they can handle intense streams of light and heat. The photosynthetic efficiency coefficient decreases under such conditions. To enable the same productivity levels using the second calculation method, a significantly smaller light-receptive surface area is required.

In other words, in procuring equal productivity levels, the first calculation method is more economic in terms of energy consumption, and the second method entails smaller size and mass. In applied design, the illumination intensity is usually between the aforementioned extreme values, and the choice of illumination intensity depends in each individual case on the concrete conditions and cultivation purposes.

When cultivator design calls for two-sided illumination, optimal productivity is calculated using a formula derived from Equation (6.18):

$$P_{opt} = \frac{kE_0}{E_0 + q} - \frac{\gamma}{21\beta}\left[1 + \ln\frac{21k\beta E_0}{\gamma(E_0 + q)}\right]. \tag{6.42}$$

Now, dividing the general, targeted productivity P_{gen} by P_{opt}, we find the general area of the photoreactor's light-receptive surface. The area thus arrived at must be increased by 10–15%, since during work 10–15% of photoreactor light-receptive surface is occupied by the largest of the gas bubbles that arise from sparging and that grow into gas pockets contacting both the front and back walls simultaneously as they travel through the photoreactor.

If the carbon dioxide concentration of the input gas is limited, then photoreactor design should take this parameter into consideration as well. This calculation depends on a minimal culture layer thickness in the photoreactor.

Photoreactor Design when Carbon Dioxide is at Saturation Level

Approximate Equation (6.39) that defines photoreactor gas exchange parameter A was introduced in Section 6.1. The bigger this parameter value, the lower the carbon dioxide concentration in the medium needed to produce a given photosynthetic level. Parameter A is about 5000 s^{-1} for a photoreactor built in the form of a vertically-disposed flat, parallel cuvette and which consumes a gas mixture amount for sparging approximately equal to $7 \cdot 10^2$ m^3/s per 1 m^2 of photoreactor horizontal cross section area.

It theoretically follows from the Michaelis–Menten equation that cell growth rate reaches its limit only when there is infinite substrate concentration—which in this case would be a medium infinitely saturated with carbon dioxide. For this reason, the calculation is performed for a carbon dioxide concentration that yields a 90% saturated process, that is, for a carbon dioxide concentration that enables a growth rate (or productivity) equal to 0.9 of the maximum ($P = 0.9\ P_m$). Substituting the given value into Equation (6.36), and having designated the carbon dioxide concentration in gas mixture S_{OH}, we get

$$S_{OH} = 9\frac{K_s}{\varphi} + 0.9\frac{P_m}{\varphi A}. \qquad (6.43)$$

Substituting the numerical values of the constant, and taking 37°C as the cultivation temperature, we get

$$S_{OH} = 1.55 \cdot 10^{-3} + 1.55\frac{P_m}{A}. \qquad (6.44)$$

Since the second term is substantially greater than the first, from Equation (6.44) it follows that the lower the saturating carbon dioxide concentration in the gas phase, the lower the culture's volume productivity.

In Section 6.2 we showed that when an algal culture is limited by light, culture layer physical thickness does not bear upon productivity per unit of photoreactor illuminated surface. Consequently, physical layer thickness can randomly change without a loss in cultivator productivity or the photosynthetic efficiency coefficient. Physical layer thickness is inversely proportional to both the productivity of a unit volume of the culture and to the amount of carbon dioxide needed to produce maximal photosynthesis. This means that if gas exchange parameter A of the planned cultivator is assigned, and if photoreactor illumination is assigned (or the productivity of a unit area of light-receptive surface), then we can calculate the culture layer thickness at which photosynthesis will be maximal for a given carbon dioxide content in the gas phase.

For instance, a culture layer illumination intensity for a culture illuminated from two sides is targeted to be 300 W/m^2. Under conditions when carbon dioxide concentration is 1% in the gaseous phase, the goal is to reach a photosynthetic productivity that is 90% of the potential achievable productivity. The following steps must be taken:

1. Culture surface productivity is defined for an illumination intensity of 300 W/m^2 according to Equation (6.17), assuming that the layer is fully-absorptive $x = 3 \cdot 10^{-2}$ kg/m^2, and that illumination is two-sided. Culture surface productivity comes out to 0.202 kg/m^2 per day.
2. Knowing that the gas exchange parameter for a given type of A sets $A = 5000$, using Equation (6.44) the volume productivity is defined at which maximal photosynthesis is reached given a 1% carbon dioxide concentration and at which cell growth rate is not limited. This volume productivity comes out to 27.3 kg/m^3 per day.
3. Dividing surface productivity by volume productivity, culture physical layer thickness in the photoreactor turns out to be: $d = 0.0074$ m, or 7.4 mm.

Endeavoring to work at low carbon dioxide concentrations leads to a thickening of the culture layer in the photoreactor, and this means an increase in culture volume.

System for Nourishing and Harvesting the Culture

In the photosynthetic process, algal cell biomass constantly increases, and there is a constant demand for mineral nutrients dissolved in the culture medium. There is an optimal algal biomass concentration that must be maintained for each irradiation intensity. It is also important to maintain the dissolved nutrients in the medium within limits that do not limit or inhibit algal growth. The principles necessary to calculate proper nutritive medium levels are described below in Section 6.5.

To solve the cultivator nutrient supply problem, a dosage assembly is needed. Its function is to periodically add a definite nutritive medium volume to the photoreactor and to simultaneously remove from the photoreactor the same volume of algal culture. In a design sense, this assembly (Figure 6.19) consists of dosage meter (7), valve mechanism (5) and vessel (3) containing the nutritive medium and vessel (10) containing the algae that were skimmed off and removed from the culture mass. The dosage meter is a closed vessel usually made from plexiglass, which is divided by a rubber membrane (8) into two chambers. One of these chambers (9) is joined through valve (11) to vessel (10) containing skimmed off algae and through valve (12) to photoreactor (1). The other chamber (6) is joined through valve (2) to photoreactor (1) and through valve (4) to vessel (3) containing the nutritive medium. The valve mechanism consists of a drive. The drive can be one of two things. It can be mechanical and composed of a low-power slow-speed electric motor which sets the cam in motion. Or the drive can have electromagnetic or pneumatic valves. The drive enables shutting and opening the valves in pairs.

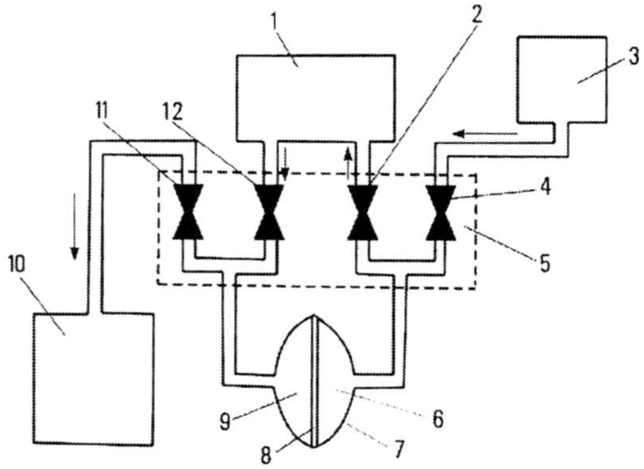

Figure 6.19 Diagram of dosage meter assembly. 1—photoreactor; 2—nutritive medium supply valve from the dosage meter to the photoreactor; 3—vessel for nutritive medium 4—nutritive medium supply valve from the nutritive medium vessel to the dosage meter; 5—valve mechanism; 6—chamber for algal culture in the dosage meter; 7—the dosage meter proper; 8—membrane; 9—chamber for algal culture in the dosage meter; 10—vessel for culture collection; 11—culture supply valve from the dosage meter to the collection vessel; 12—culture supply valve from the photoreactor to the dosage meter.

The work period frequency is directly proportional to algal culture growth rate and inversely proportional to dosage meter capacity. Dosage control is automatically regulated by an optical density flow sensor.

Densitostat System: Automatic Regulation of Optical Density to Stabilize Biomass and Nutritive Medium

Optical density (turbidity) regulation is the key to the continuous cultivation process because if optical density is held constant, then all other parameters stabilize due to their correlation with this density. In a densitostat system, the portion of nutritive solution supplied to the culture exerts a regulating effect on optical density: the optical density decreases as nutrient substance concentration increases. It is clear from the structural design drawings for a densitostat system that suspension optical density is the single regulating parameter that acts by means of negative feedback. Nutrient element concentration in the culture medium is regulated by an open cycle with disturbance compensation—that is, cell growth.

It is a good idea to apply the simple two-position law of regulation that together with the actuating mechanism—the dosage meter—enables supply of the nutritive medium to the photoreactor.

One of the most complicated and crucial parts of the densitostat system is the optical density sensor. Both continuous and discrete optical sensors are used. The flowthrough cuvette of the continuous sensor is mounted to the photoreactor. A pump suctions the culture out of the photoreactor and it flows through the sensor cuvette where various measurements are performed on the culture before it flows back into the photoreactor. The discrete sensor is usually equipped with a sampling device and therefore it can be mounted anywhere on the photoreactor or its tubes. Sensor signals travel to a secondary device where the signals can be read and regulated.

In other words, the process in the cultivator remains functionally continuous, although it is technically realized by a discrete, impulse-driven solution. Such a process is called *quasi-continuous*. The advantages of a quasi-continuous cultivation process in comparison to a continuous process happen to be the same as those of a discrete, digitally-controlled computer over an analog computer.

Just this simple solution allowed the self-sustaining operation of the cultivators over a period of many months, and regenerating the atmosphere in closed ecosystems to provide for human respiration as described in the subsequent chapters of this book. The quasi-continuous process can be confidently recommended for widespread use in continuous cultivation systems within the microbiology industry.

Culture Thermostatic System. *Chlorella* Photosynthesis Efficiency and its Dependence on Temperature

Maintaining algal culture temperatures within optimal limits can have a decisive effect on cultivation, as is seen from Figure 6.20 (Batov, 1967), and it is therefore imperative to include a heat exchanger in an algal cultivator design. This algal cultivator removes excess heat originating from the light sources.

Although photosynthesis takes place within a wide range of temperatures, research on which this book is based examined only that temperature range for which the photosynthetic rate was relatively high.

It should be expected that the nonuniform way in which the cells were irradiated in dense suspensions reflects on photosynthetic temperature dependency.

Five curves depicting the dependence of culture productivity on temperature are illustrated in Figure 6.20. These curves were obtained when various radiation levels were projected onto the photoreactor surface that housed a dense microalgal suspension. The cell volume cells made up 2–5% of the algal suspension volume.

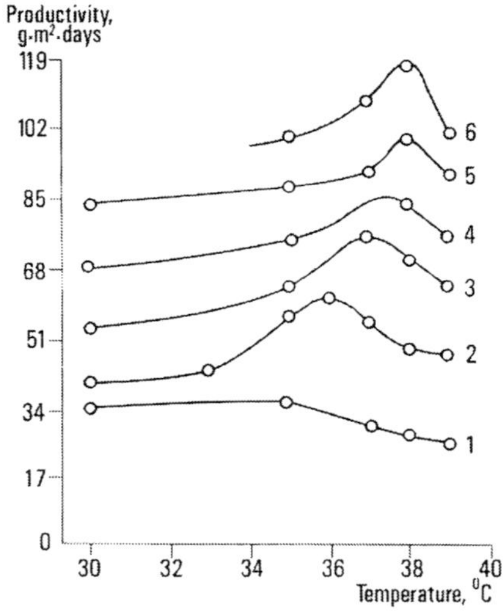

Figure 6.20 *Chlorella* productivity dependence on temperature for various surface irradiation levels in dense culture (in W/m^2). 1—120; 2—174; 3—318; 4—270; 5—435; 6—755.

The figure clearly shows that when the temperature increases, culture productivity also increases and the curves pass through their maximum. When culture irradiation increases, optimal temperatures shift to the higher side without exceeding 38.0–38.5°C. When the temperature rises above the optimal value, culture productivity sharply falls. When the temperature reaches a critical level, microalgal photosynthesis stops altogether. The temperature curves can be broken down into two parts. Of these two parts, the left side characterizes temperatures that stimulate photosynthesis, and the right side characterizes the process of photosynthetic inhibition.

In a highly productive cultivation regime, the photosynthetic efficiency coefficient only reaches 5–8%. For this reason, unused light energy and all infrared radiation energy must be removed from the cultivator in the form of heat. (For details see Shtol, 1976.)

It is worth mentioning that the required accuracy in suspension temperature stabilization depends on cultivator productivity (see Figure 6.21). Supposing the magnitude of cultivator productivity decreases due to temperature instability to be 2%, then an accuracy of ±2.5% is found to be adequate for an illumination intensity of 118 W/m^2. When illumination intensity is further increased, the required accuracy also grows. The choice of the required accuracy in suspension temperature stabilization depends on the magnitude of allowable cultivator productivity losses and the real characteristics of the applied regulation system.

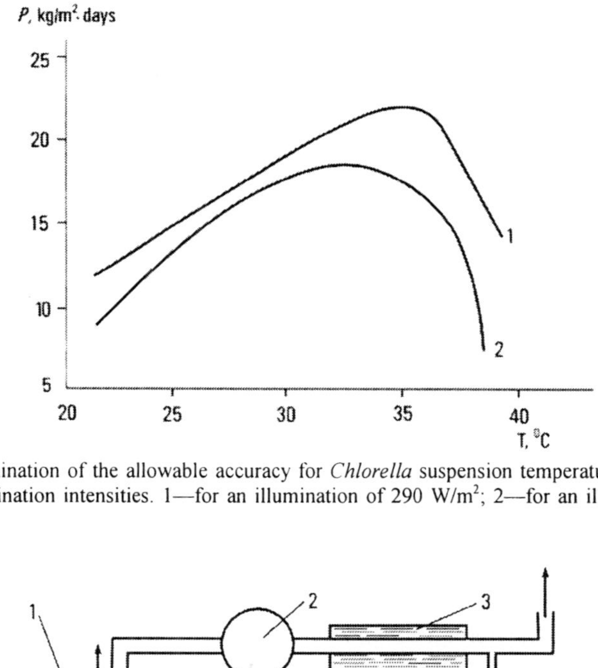

Figure 6.21 Determination of the allowable accuracy for *Chlorella* suspension temperature regulation under various illumination intensities. 1—for an illumination of 290 W/m^2; 2—for an illumination of 118 W/m^2.

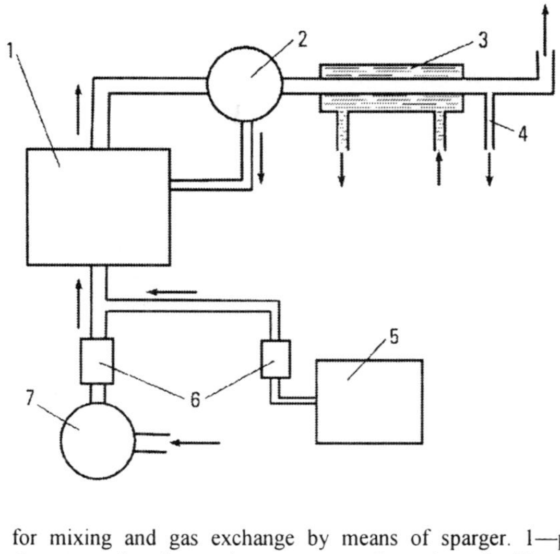

Figure 6.22 Diagram for mixing and gas exchange by means of sparger. 1—photoreactor; 2—separator/foam-breaker; 3—gas cooler; 4—condensate output; 5—carbon dioxide source; 6—air and carbon dioxide gas consumption flowmeters; 7—air compressor.

Mixing and Gas Exchange System

If culture mixing and gas exchange are performed using a sparger, then the following components are installed in the part of the system that precedes the photoreactor (see Figure 6.22): carbon dioxide source (**5**) (these sources can be biological objects such as animals or humans); compressor (7) and flowmeter (6). Components that come after the photoreactor in system flow are separator/foam-breaker (**2**) and gas cooler (**3**).

Flowing through the photoreactor volume, the gas mixture grows rich with oxygen and loses carbon dioxide. Figure 6.16 illustrates the separator/foam-breaker device. The mixture of foam and gas separates in the separator/foam-breaker into gas and liquid. The gas next enters the cooler, and the liquid (the culture) is directed back into the photoreactor. The gas gives off moisture (condensate) in the cooler, and then leaves the cultivator altogether. The coolers can be of various design. Their sizes and heat absorption rating in each case are defined by calculations that depend on the gas quantity that passes through the cooler, and on gas and coolant temperatures.

6.5 DEVELOPING A NUTRITIVE MEDIUM FOR THE CONTINUOUS ALGAL CULTURE

The most widespread nutritive media for microalgal cultivation were developed by Tamiya, Myers, Pratt, and Knopp. These media were developed for periodic microalgal cultures. For this reason, it was necessary to critically examine the applicability of these mediums when transitioning to continuous intensive cultures.

The idea of using two media in tandem that differ in composition was introduced for purposes of continuous cultivation. During the process of cultivation, the first of these media is supplied to the photoreactor from without either continuously or in very small doses. This medium is the so-called correcting or compensating nutritive medium. The second medium surrounds the cells in the culture and is likewise removed from the photoreactor continuously or in pulses together with the cells. The second medium, the so-called background or culture medium, differs from the nutritive medium by its reduced concentrations of nutrient elements.

To find the connection between the element concentrations in the nutritive and background media, the following equation describing a continuous culture is presented:

$$\frac{dS}{dt} = D(S_0 - S) - g\mu X, \tag{6.45}$$

where S_0 is the concentration of any element in the nutritive medium; S is its concentration in the background medium; g is element consumption during synthesis of a biomass unit (g/g); X is biomass concentration in the culture; D is dilution; and μ is cell growth rate.

When the continuous process is at steady state (in equilibrium), then $dS/dt = 0$, $D = \mathrm{m}$.

Substituting these into Equation (6.45), it can be solved for S_0 as follows:

$$S_0 = S + gX. \tag{6.46}$$

This equation shows that the higher cell biomass concentration in the culture and the higher the amount of elements consumed on the synthesis of a biomass unit, the more the nutritive and background media differ in their element concentrations.

From Equation (6.46) it is easy to procure the maximum cell biomass concentration for each known medium that is used as a nutritive medium as follows: $X = (S_0 - S)/g$.

If a medium element is completely used up by the cells, that is, $S = 0$, then $X_{max} = S_0/g$.

So, for example, if in Tamiya's standard medium, the concentration of KNO_3 is 5 g/l, then the nitrogen concentration itself, that is the S_0 concentration, is 0.7 g/l. At this nitrogen concentration, maximum algal biomass concentration turns out to be $X_{max} = 0.7/0.08 = 8.75$ g/l. That is, the Tamiya medium does not allow continuous cultivation to be conducted with a biomass concentration higher than 8.75 g/l.

Medium characteristics that produce a maximum biomass concentration level are listed in Table 6.1. Use of these characteristics to obtain maximum biomass concentration assumes that nitrogen supplies do not limit cell growth.

As apparent from the table, even the medium richest in nitrogen—the Craig–Trellis undiluted medium—allows microalgal cultivation with a biomass concentration not higher than 13.2 g/l of dry biomass.

Table 6.1 Maximum biomass concentration values in a continuous culture for various media. (Calculations are based on nitrogen consumption.)

Medium	Nitrogen source	Nitrogen source compound concentr., g/l	Nitrogen concentr.	Nitrogen consumption per g of biomass	Maximum biomass concentr., g/g
Tamiya	KNO_3	5.000	0.6940	0.08	8 700
Myers	KNO_3	1.213	0.1680	0.08	2.100
Pratt	KNO_3	0.100	0.0140	0.08	0.174
Benecke	$Ca(NO_3)_2$	0.500	0.0850	0.08	1.070
ChO-10	$Ca(NO_3)_2$	0.040	0.0068	0.08	0.085
Knop	KNO_3	1.000	0.1570	0.08	1.970
Knop (undiluted)	$Ca(NO_3)_2$	0.100			
Yaguzhinsky	KNO_3	0.500	0.7000	0.08	0.870
Craig–Trellis (undiluted)	KNO_3	7.600	1.0500	0.08	13.200

An analogous calculation can be made to determine maximum biomass concentrations with respect to other medium elements. This calculation demonstrates that supply levels for each of the elements is different, and that when the medium is put to use, some elements are completely consumed and stop culture growth, whereas other elements are not at all used up and are simply eliminated from the photoreactor with the background medium. A sample calculated for various elements was performed on Tamiya's medium and the results are presented in Table 6.2.

Table 6.2 Basic element usage for a maximum biomass quantity (8.7 g/l) cultivated in a Tamiya medium.

Element	Element concentration in the medium, g/l	Amount of element consumed by the cells, g/g	Amount of element consumed by a maximal biomass, g/l	Utilization factor, %
N	0.694	0.0800	0.6900	100.0
S	0.325	0.0032	0.0278	8.5
Mg	0.244	0.0050	0.0435	18.0
K	2.290	0.0120	0.1040	4.5
P	0.285	0.1180	0.1570	55.0
Total	3.838	0.118	1.0270	27.0

Table 6.2 shows that when microalgae are cultivated using Tamiya's medium, the nitrogen supply is exhausted first, after which cell growth should stop. However, the other medium elements are by no means used up, and only 27% of their collective quantity is consumed. The remaining 73% is discarded with the background medium.

For the cells to use the medium completely up, all standard mediums must be replaced by a balanced media where all biogenic elements are consumed in equal degrees of completeness.

Nitrogen Sources

Nitrogen is a very important biogenic element in the medium. Its consumption reaches 80–100 mg per gram of synthesized biomass, which exceeds the consumption of all other mineral elements in the medium by several times. A comparative study of all nitrogen sources was conducted to determine their suitability for continuous densitostat cultivation for a given species of microalgae (Gitelson *et al.*, 1964).

KNO_3, NH_4NO_3, NH_4HCO_3, and $(NH_2)CO_2$ were studied as nitrogen sources. In experiments with potassium nitrate, medium alkalinization was immediately discovered due to an accumulation of magnesium in the medium. This was to be expected since cells consume much more nitrogen than magnesium, and the unabsorbed magnesium accumulates. This accumulation leads to a pH level of 11, which in turn lowers culture productivity at first, and then leads to growth stoppage altogether. Unfortunately, the most important component—nitrogen—only constitutes 13.6% of the compound's weight.

An attempt was made to replace the magnesium cation with another cation, which the culture would much more intensively absorb—ammonia nitrate, containing a much better 35% nitrogen. However, it turned out that the cells had a marked preference for ammonium over nitrate. This problem was caused by the fact that ammonia nitrate is physiologically an acid salt that elicits a significant reduction in the medium pH level. In experiments, the pH level decreased to 3.5 and this led to the cessation of culture growth.

In the next variant, the anion NO_3^-—which acidifies the medium—was replaced by the more adequate anion HCO_3^-. The resulting compound NH_4HCO_3 was then tested.

For NH_4HCO_3, a significant amount of ammonia was eliminated in exhaust gas leaving the photoreactor. This fact made this compound completely unacceptable for use in systems involving humans.

The next compound studied was the carbamide form of nitrogen, although during batch cultivation it did not yield satisfying results (Pinevich *et al.*, 1961). The use of urea as a nitrogen source in the nutritive medium met with no obstacles. Since nitrogen in this compound was not in an ionic form, its consumption did not significantly affect medium pH. This circumstance, in the first place, allowed cultivating with biomass concentrations up to 30 to 40 g/l. In the second place, it enabled foregoing continuous monitoring of culture pH, and this in turn simplified the apparatus needed for the process.

These experiments found that there is a wide range of nitrogen concentrations that can obtain in the background medium at which nitrogen exerts little influence on the growth rate. Maintaining nitrogen concentrations in the neighborhood of 70–400 mg/l provides for a prolonged and continuous cultivation process.

Moreover, urea is especially suitable in a closed system since human excretions of nitrogen are mainly in the form of that compound.

The Concentration of Basic Biogenic Elements in the Background Medium

The minimum requirements for the remaining biogenic elements needed by the microalgal culture were determined to be the following: potassium, phosphorous, magnesium, and sulfur. The amount of each element needed for the growth of a unit of cell biomass was determined, and the most advantageous background concentrations were found for the specific conditions that obtain during intense continuous algal cultivation.

In order to conduct this experiment, the concentration of one element had to be manipulated, leaving the others in the range of a previously known optimum. In standard nutritive media, this was impossible to accomplish since element anions and cations are paired when standard salts are used. In order to circumvent this difficulty, other chemical compounds—NaH_2PO_4, $MgCl_2$, Na_2SO_4—were used that would allow "unbinding" of their elements. Consequently, media were employed that were constituted from the corresponding acids and bases (Gitelson *et al.*, 1967). Experimental results are listed in Table 6.3.

Table 6.3 Minimal requirement and minimal background concentration of biogenic elements for continuous densitostat microalgal cultivation.

Element	Minimal requirement, g/g	Minimal background concentration not limiting growth, g/l	Utilization factor when X=15 g/l, %
P	0.0119	0.000	100
Mg	0.0033	0.015	92
S	0.0043	0.000	100
K	0.0107	0.015	92

As Table 6.3 indicates, background concentrations that support cell growth are very low. For phosphorous and sulfur, these concentrations are at the sensitivity limit of the methods used to determine them. This fact enables us to conduct cultivation with very high biogenic element utilization factors, since the cells absorb practically all of these substances. However, some well-known methodological difficulties are encountered in such a cultivation process. These difficulties are connected with the need for a very accurate biomass cell-concentration stabilization. So in many cases full usage of medium biogenic elements must be foregone in favor of allowing an expected increase in background element concentrations in order to increase the sustainability and reliability of the process as a whole.

The background concentrations of biogenic elements produced by these experiments confirm that in the given case cell growth is limited by another factor—namely, light energy—since all experiments were conducted with optically dense culture layers that are fully light-absorptive.

Special experiments were not conducted to determine the bandwidth of optimal background concentrations. This is because when the closed system meant for "human–microalgae" cohabitation was being created, the lower boundary of background concentrations was the main interest. However, on the basis of longtime work experience with continuous *Chlorella* cultures it can be concluded that an increase in background concentrations up to the hundredths of milligrams per liter for all basic elements does not adversely affect culture growth, but instead their consumption in the synthesis of a unit biomass increases.

Utilization of Human Excrement by the Microalgae

For a closed "human–microalgae" system, an extremely important question is whether or not the microalgae can accept unprocessed human excrement into its nutritive medium. Or do the microalgae require that this excrement be more or less preliminarily processed? For instance, must this excrement undergo some kind of biological or physicochemical mineralization?

Towards the beginning of work on microalgal cultivation, media that include human excrement were not employed under laboratory conditions, however microalgae had already been used for many years in the practice of purifying sewer water (Vinberg, 1960, Vinberg *et al.*, 1966)

Experience in purifying sewer water in biologic ponds seeded with microalgae indicated that microalgae can assimilate human wastes. This could not, however, be taken as evidence that this process could be transferred "as is" to a closed system.

Water purification intensity in biologic ponds is exceedingly low. A whole assortment of microalgal, bacterial, fungal, and protozoan species live in the culture within the pond. Not only this, but the purpose of the process is not high culture microalgal productivity, but water purification. In a closed system, both of these problems must be solved simultaneously.

Work that sought to investigate employing human excrement for *Chlorella* cultivation under laboratory conditions began in 1961. In our experiments, the only source of biogenic elements were taken to be human excrement and gaseous carbon dioxide (Rerberg and Kuzmina, 1964; Rerberg *et al.*, 1968). These experiments demonstrated that water compliant with state standards for drinking water can be procured from human excrement.

Experiments using urine as a nutritive medium component yielded positive results. The *Chlorella* grew partaking of this medium just as well as they did on the purely synthetic medium. The maximal urine dosage applied in the nutritive medium during experiments was calculated to completely satisfy the culture's requirement for nitrogen.

The experiments conducted have also demonstrated that human urine can be accepted by the microalgae in a closed system without any kind of preliminary processing.

Microalgal Metabolites in the Background Medium

When using microalgae as a link in a "human–microalgae" system, the problem of sustaining this link with water arises. This is the so-called technological water. The demand for the water used to prepare the medium substantially exceeds all other requirements for water in the closed system. In addition to this, almost the same amount of water is given off during the microalgal cultivation waste withdrawal process when cells are separated from the medium in the separator centrifuge. This is the only source of water that can be reused to satisfy the culture's requirement for technological water.

Reusing the medium many times can result in an accumulation of metabolites. If the presence of metabolites in the medium does not retard cell growth the first time around, it does not mean that it will not inhibit cell growth during subsequent reusages, since the metabolites accumulate with each usage. Since accumulated metabolites would retard cell growth with repeated medium usage, an auxiliary device for water purification would have had to be introduced into the system if this problem were not resolved.

It therefore became imperative to undertake research into the possibility of repeated use of the medium. The experiments were complicated by the addition of human urine into the nutritive medium. This complication was dictated by the fact that it was optimally efficient to utilize human urine in just this capacity in the closed system, which meant that it was therefore reasonable to combine researching the effects of both *Chlorella* metabolites and human metabolites that enter the system from urine.

Experimental results demonstrated that the retardation effects of metabolites on cell growth were absent or were lower than the limits of measurement accuracy. By the end of the experiment, culture productivity had fallen off 15–17% from its initial value, and this could be fully explained by a decrease in medium transparency. During the course of the experiment, the medium took on a cloudiness and became dirty yellow in color. It was found that overall 15–20% of incident light falling on the culture was absorbed by the impurities in the medium.

The experiment did not demonstrate the expected growth in medium oxidizability.

The concentration of NaCl introduced into the medium by human urine should grow together with the oxidizability.

Experimental and calculation results are illustrated in Figure 6.23. They confirm the correctness of the performed computations.

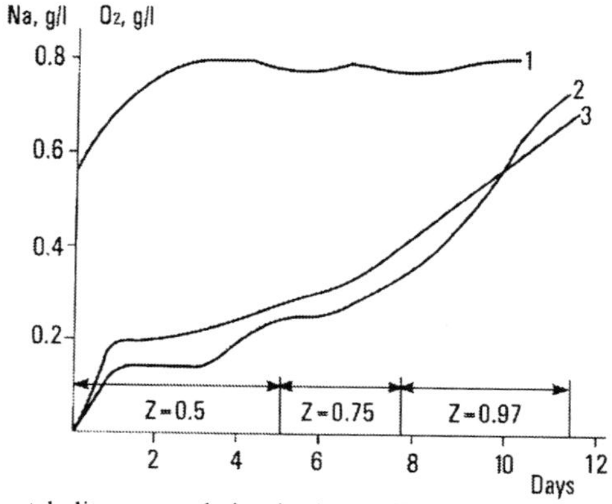

Figure 6.23 Sodium and metabolite accumulation in the medium under repetitive usage. 1—oxidizability; 2—observed sodium concentration, g/l; 3—calculated sodium concentration, g/l. Z is medium reuse factor.

An absence of metabolite accumulation, which is fixed by the absence of oxidizability growth, forces the assumption that not only a metabolite source exists in the culture, but a metabolite consumer exists there as well. Such a consumer might be the microorganisms that cohabit with the *Chlorella* in a nonsterile culture. This assumption is confirmed by the active reaction of bacterial microflora to changes in the medium during the course of the experiment: the genus *Pseudomonas* began to predominate over *Bacterium*.

So the conducted experiment showed that reuse of the medium in principle is possible even when human urine has been introduced into it. Human metabolites and those of the culture itself do not retard cell growth, although they do decrease productivity for a purely physical reason—the medium becomes non-transparent and light-absorptive. When the reuse factor is high, there is also the danger of sodium chloride accumulation in the medium.

To determine the value at which the medium reuse factor is limited by the accumulation of sodium chloride, the effect of various NaCl concentrations on microalgal culture growth was researched (Rerberg and Vorobyeva, 1967).

Research has shown that steady-state culture photosynthesis of the strain *Chlorella vulgaris* is possible up to concentrations of 10 g/l NaCl, although there is a slight decrease in productivity (Lisovsky and Sypnevskaya, 1969).

Taking the limiting concentration of NaCl in the medium to be 10 g/l, it was easy to compute the reuse factor for a given medium

It allows calculating the daily load on the vacuum still or other device used in the system to remove NaCl from the cultivator.

6.6 CONTROLLING BIOCHEMICAL COMPOSITION OF MICROALGAE

Conditions in a closed ecological system designed for human life support lay strict demands on the material cycle between the organism populations that comprise the system. This means that all biosynthesis products from one link should be consumed by the other link(s) of the closed system. It is not likely that such an harmonic, complementary combination of biosynthetic genetic programs for various organisms would be absolute, but the extent to which this can be achieved depends on the extent to which the ecosystem is closed.

From the comparison of *Chlorella* biomass biochemical composition with human food rations, it is easy to see that they are very far from complementary. For instance, *Chlorella* cells are more than 60% protein, and the ratio of protein in the human diet should not exceed 23–25%. On the other hand, carbohydrates, which make up more than half of the human diet, all make up not more than 20% of the *Chlorella* biomass.

Even taking into account the doubtful or, at the very least, the remote possibility of actually using algae for food, then there is still the fact that algal culture products and human food requirements do not coincide. And there are other problems besides this incongruity that create problems for their cohabitation in a closed system. Suppose that the algae only regenerate the atmosphere and water for humans, and are not used for food. Because of the difference in biochemical composition, the human respiration quotient does not coincide with the algal assimilation factor, and an overproduction of oxygen arises when there is a carbon dioxide deficit. On the other hand, the elemental composition of human excrement, which in a closed system should serve as a source of biogenic elements for microalgae, does not correspond to the latter's requirements. For instance, there will be a significant nitrogen deficit.

Proceeding from these considerations, the possibilities to control the biochemical directionality of biosynthesis in *Chlorella* cells will be examined using parametric methods, that is, not by directly transforming the genome, but by realizing specific phenotypes through the changing of various external biosynthesis parameters. The task is to make manifest the plasticity limits of *Chlorella* biosynthesis and to determine how to render them greater complements to human needs. The experiments investigate how *Chlorella* biosynthesis directivity changes as the culture become limited by each of its biogenic elements.

Experiments that concentrate on changes in the biochemical composition of the *Chlorella vulgaris* cell were conducted during continuous cultivation using stabile physical parameters.

A background stage preceded the investigation of the effect of each biogenic element on the culture.

The composition of the background and nutritive media were altered in such a way as to block biosynthesis and thereby enable studying the effects of one of the biogenic elements on the culture. To induce phosphorous starvation, the salt KH_2PO_4 was replaced by an equimolar quantity of the salt $KHCO_3$. To induce potassium starvation, this latter salt was replaced by NaH_2PO_4. To induce sulfur starvation, $MgSO_4 \cdot 7H_2O$ was replaced by an equimolar quantity of $MgCl_2$, and then to induce magnesium starvation, this salt was replaced by $NaSO_4$.

After the background stage, the cell suspension was thrice washed with water and resuspended in a medium lacking one of the biogenic elements. Then it was placed back in the cultivator, which had been flushed thoroughly in preparation for this. Conditions were those meant for intensive biosynthesis.

Every 6 h readings were made to analyze the medium and the biomass.

- *Biomass biochemical transformation when biosynthesis is nitrogen-blocked.* When nitrogen is removed from the medium, the biosynthesis rate starts to decrease in the very first hours. After 6 h, it decreases by 29% from its initial level, and it falls to zero during the course of the next 30 h. At the same time, deep changes in biomass biochemical composition are effected.
- *Biomass biochemical transformation in the absence of sulfur.* Removal of sulfur from the medium in the first 11 h elicits a 17% reduction in the biosynthesis rate. After 30 h, the biosynthesis rate falls to zero in the absence of sulfur.
- *Biomass biochemical transformation in the absence of phosphorous.* During the first 18 h, the absence of phosphorous in the medium does not reflect on the biosynthesis rate, but after this period, the biosynthesis rate gradually decreases. In the course of the next 28 h, the biosynthesis rate falls off to zero. Cell biochemical composition in the biomass undergoes deep changes with respect to the ratio of its proteins, lipids, and carbohydrates.

- *Biomass biochemical transformation in the absence of potassium.* When potassium is removed from the medium, the biosynthesis rate decreases by 9%. For the next 7 h, a sharp fall to 67% of the original biosynthesis rate is observed, which is replaced by a rise in this rate up to 78% of the initial growth rate. For the next 29 h, the biosynthesis rate falls to zero.
- *Biomass biochemical transformation when biosynthesis is limited by magnesium.* When magnesium is removed from the medium, the biosynthesis rate falls by 26% for the first 10 h. After 22 h, it falls to 67% of its initial value. In the next 6 h, a short-lived increase in biosynthesis up to 70% of the initial growth rate is observed. This is replaced by a gradual decrease in the growth rate until it reaches zero after another 38 h.

These results testify to the fact that excluding one of the biogenic elements from the nutritive medium does not elicit an immediate cessation in the biosynthesis process, but it leads to its gradual retardation.

Biosynthesis stops only after a significant increase in the biomass. When nitrogen is removed from the nutritive medium, the supply of nitrogen within the cells is sufficient to sustain a 1.8 time increase in the biomass.

Specific changes in cell biomass were uncovered that depended on the limiting element. The fastest and deepest shifts in biosynthesis direction occur when nitrogen is the limiting element.

The restructuring of the internal cell apparatus elicited by the absence of nitrogen, an element that is key to protein structure, reduces the synthesis of proteins by 2.3 times. During nitrogen starvation, the cycle in which carbon is restored as carbohydrates probably obtains a special significance. The content of carbohydrates during this process grows by 5 times. Moreover, a significant portion of the carbon is spent on lipid formation. In connection with this, the cell fat store grows almost 2 times.

The remaining biogenic elements can be listed in descending order with respect to the effect their absence has on the biosynthesis rate: sulfur, phosphorous, magnesium, potassium.

When sulfur is lacking, preference is given to the synthesis of carbohydrates. Their general content increases almost 4 times while the quantity of protein substances and lipids falls 1.5 and 2 times, respectively.

During phosphorous and magnesium blocking, *Chlorella* metabolism shifts in the direction of heightened carbohydrate synthesis. The absence of phosphorous in the medium increases their general carbohydrate content by 2 times. At the same time, their protein content decreases by 1.5 times and their lipid content decreases by 2 times. Magnesium blocking increases cell carbohydrate content by 2.5 times although there is almost no change in the lipid content, and the quantity of protein substances decreases by 20%.

From the described experiments, it follows that when N, P, S, or Mg are missing from the nutritive medium, and when all other biosynthesis parameters are held constant, then biosynthesis shifts in the direction of heightened carbohydrate synthesis. The carbohydrate content basically increases as a result of hemicellulose accumulation. The quantity of hemicellulose towards the end of the experiment increases by 4 times when nitrogen is blocked, by 5 times when sulfur is blocked, by 6 times when phosphorous is blocked, and by 3 times when magnesium is blocked.

Cultivation using a medium that is deprived of potassium demonstrates that the absence of this biogenic element leads to biomass synthesis with a twofold increase in lipid content and with a small 12% decrease in protein substances and carbohydrates. Within the carbohydrate group, a quantitative increase occurs only with respect to water-soluble sugars, while starch and hemicellulose synthesis decreases.

Removal of one of the biogenic elements from the medium elicits a significant reduction in cell chlorophyll content. During nitrogen blocking, cell chlorophyll content is reduced by 57%; during sulfur blocking by 62%; during phosphorous blocking by 75%; during potassium blocking by 44%; and during magnesium blocking by 31%. The lack of phosphorous exerts the most detrimental effect on chlorophyll synthesis. The lack of magnesium has the least detrimental effect on chlorophyll synthesis.[1]

The absence of any one of these biogenic elements reduces the incorporation of other elements in the biomass. The way in which elements are excluded from incorporation in the biomass depends upon which biogenic element is lacking.

As these results show, not only is the microalgal growth rate subject to control, but a shift in the targeted direction of the course of biosynthesis can be evinced within wide limits when the cell genetic code is taken into account. Control of biosynthesis growth is not targeted by means of direct interference with its genes, but by a studied change in external parameters. This method for controlling biosynthesis growth is called *parametric*. As can be seen by the results of this section, parametric control is quite effective. However, it must be kept in mind that a deep change in cell biochemical composition must be paid for by a decrease in the biosynthesis rate.

In addition to its application in the task of creating a bioregenerative human life support system based on microalgal cultivation, parametric control of directed biosynthesis allows solving two other problems: 1) resolving the microalgal assimilatory quotient with the human respiration quotient, and 2) bringing algal biomass biochemical composition and its protein, carbohydrate, and lipid ratio closer to the ratio of these same substances as they are required in the human diet.

6.7 THE SUSTAINABILITY AND RELIABILITY OF CONTINUOUS CULTIVATION OF ALGAE FOR BIOLOGICAL LIFE SUPPORT SYSTEMS[2]

When discussing life support system design alternatives, it is important to resolve which systems are more reliable: those based on physicochemical processes, or those based on biological processes. Space technology designers know that physi-

---

[1] Note: though magnesium is essential for chlorophyll synthesis, being its component, the lack of magnesium affected the general rate of *Chlorella* biosynthesis less than the lack of nitrogen or phosphorus. The reason seems to be a lower magnesium requirement per unit synthesized biomass.
[2] In collaboration with E.K. Rodicheva.

cal and chemical processes are simpler than biological ones. From this, they intuitively conclude that they are more reliable—but this is untrue, or at least open to debate.

The principle of self-restoration is one of the general principles that determines how life is organized. This principle is the one that potentially enables immortality. The self-restoration principle imparts a reliability to the biological system that cannot be attained by any modern technical means unless and only insofar as the technical means copies one of the principles by which living matter is organized.

For that reason, this section describes an experiment that examines self-restoration in the context of a microalgal cell population.

Each living cell, with the exception of highly specialized cells in a multicellular organism, possesses the ability to reproduce itself. This ability is enabled by the completeness of the information contained in each cell genome necessary and sufficient for self-reproduction. The cell also contains the entire apparatus needed for reading and realizing this information. From outside, the cells need only the signal that stimulates or prohibits multiplication.

When speaking of reproduction, the cell is autonomous. This means that if a cell population experiences some kind of catastrophe and only one cell survives, that this one cell is enough to restore the entire population: it will repopulate the culture with its descendents by means of a series of cell divisions. This is not only an emergency mechanism. Life uses this mechanism in its everyday operation.

A simple example illustrates this point. Each milliliter of our blood contains 5 billion red blood cells, which means there are about $25 \cdot 10^{12}$ cells in a healthy adult human body. The lifespan of each erythrocyte is 100 days in all. This means that every day about 1% of their population—or $25 \cdot 10^{10}$ cells perish and are replaced by new ones, which appear as bone marrow cells and constantly divide. We do not even notice this process. The blood system works in a well-tuned regime so that it is impossible to notice a fluctuation in the blood's erythrocyte concentration. But if the destruction of cells exceeds the norm, for instance, if there is loss of blood, then cell reproduction accelerates to compensate for the loss of blood and restore the normal composition of the blood. This mechanism of regulated self-restoration lies at the basis of the sustainability of living systems.

Technical means are never constructed in this manner. Each technical instrument is intended for a special function, and if it breaks, then a completely different instrument is used to restore it.

There is only one mechanism of autonomous cell self-reproduction and self-restoration, but it is used in two situations, thereby providing for the unsurpassed reliability of biological systems: it maintains normal cell concentration and it restores normal cell concentration when cells have been lost. It is not at all surprising that this mechanism is more reliable than technical ones. As a solace to designers, evolution simply had a lot more time than inventors to perfect its creations.

Stability is one of the most important criteria in the creation and evaluation of human life support systems intended for use during space flights or in other hostile environments.

A continuous culture of microalgae affords us the opportunity to verify these qualitative judgments in statistically-supported quantitative experiments allowing quantitative evaluation of sustainability and reliability of biotechnological processes used in the biological life support system. The parameters of a continuous culture can be quantitatively estimated by inflicting damaging substances on it in doses and then observing its restoration process. This yields a measure for the sustainability and reliability of the biological system, and more precisely, of the microalgal population intended to function as the means by which the human habitat can be regenerated.

The authors performed these experiments on a thermophilic strain of *Chlorella vulgaris*. All experiments were conducted using the experimental complex described in Section 6.2 (see Figure 6.24).

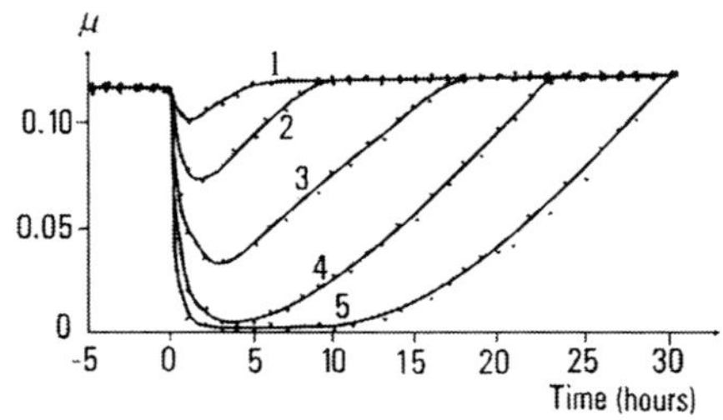

Figure 6.24 Dynamics of specific growth rate changes (h^{-1}) for a culture exposed to different UV doses. 1—2.8 · 10^2 J/g; 2—5.6 · 10^2 J/g; 3—11.2 · 10^2 J/g; 4—16.8 · 10^2 J/g; 5—22.4 · 10^2 J/g.

A Description of the Experiments Performed

The process of self-restoration can be experimentally investigated and a mathematical model of the process can be devised on the basis of steady-state continuous microalgal cultivation. For this purpose, *Chlorella* was continuously cultivated using a densitostat regime. The control device automatically applied the appropriate nutrient dilution rate, which was calculated based on culture growth rate.

To study how various physical and chemical factors affect not only intracellular processes, but changes in population as well, all biosynthesis parameters had to be maintained constant. The experimental complex enabled controlling cell habitat conditions and maintaining the following cultivation parameters constant: cell suspension concentration, nutritive medium composition and concentration, and the percentage of carbon dioxide supplied in the gas mixture. For a given cultivation condition and cell state in these experiments, cell concentration was automatically maintained constant, irrespective of growth rate variations.

Each experiment consisted of three periods: 1) the background period, in which steady-state growth conditions were established (20–24 h), and the initial growth rate is the growth rate established during this stage; 2) the effect period, where the algae were exposed to a damaging effect for about 25 to 40 min (the dosage was different for different experiments); and 3) the after-effect period, where exposure after-effects were studied for a minimum of 24 h and for a maximum of 2–3 days.

Under these conditions, ecosystem response to the effects of the various factors were reliably and unambiguously tracked. Since the cultivation process is highly stable, the results can be reliably replicated.

The damaging effect in our experimental complex was an ultraviolet (UV) emission source and a flowthrough quartz tube housed in a special protective metal casing having a volume of 200 cm^3, which was 25% of the culture volume in the cultivator. The UV emission source, a mercury–quartz lamp, was situated 20 cm from the casing. The casing was integrated in the culture circulation loop so that the culture was exposed to UV radiation as it flowed through the cultivator assembly. The culture moved through the circulation loop and exposed photoreactor volume at a constant rate and was continually mixed. During the exposure time all cells received an equal dose of UV radiation, and the power dosage was 200 $kerg \cdot cm^{-2} \cdot s^{-1}$.

To compensate for the fact that the cells were to be exposed to UV emissions only in the effect period whereas all parameters were to be held constant in all three periods, the UV emissions source was always left turned on and a light filter 2.5 mm thick situated in the casing was switched on during the background and after-effect periods to transmit the visible spectrum, but to absorb the UV radiation.

After the continuous *Chlorella* culture was exposed to the damaging effect of ultraviolet emissions, both the resulting decrease that first appeared in the algal growth rate and the subsequent growth-rate increase that obtained during self-restoration were measured.

Results

In an asynchronous *Chlorella* population, individual cells exhibit varying degrees of sensitivity to the damaging UV emissions factor. Some of the cells are lethally damaged, some of the cells experience reversible damage, and some cells are undamaged. The ratio of these three cell groups depends on the emission dose and cultivation conditions. The effect of UV exposure experienced by individual cells in the population manifests itself as a decrease in cell growth rate. During self-restoration in the after-effect period, damaged cells restore themselves while undamaged cells reproduce. The faster the undamaged cells reproduce, the quicker the population replaces itself. Since this restorative process takes places in a continuous culture, the flow eliminates dead and damaged cells and the population spontaneously retains and accumulates the faster growing cells. In this respect, a flow densitostat culture is a direct analog to the natural Darwinian selection process.

The ratio of initial growth rate to the minimal specific growth rate after UV emission was used as an indicator to estimate the damaging effects of UV emissions on the population. The intensity of the restoration process was measured by the time it took for the culture to regain the specific growth rate typical for an unexposed culture, that is, the time it took for the culture to attain to the initial growth rate.

Different experiments in this series were run with different constant values for basic parameters that affect how UV over-emission influences the cell population—illumination, temperature, carbon dioxide concentration—and it was found that the parameter that enabled the greatest variation in damage effects was the cell growth rate. The damaging effect increases with culture growth rate, and it decreases as culture growth rate decreases. A population growing at the optimal growth rate is subject to the largest decrease in growth rate. This confirms the idea that the most active cell growth condition is the most susceptible to damaging factors of various kinds.

The ultimate effect of UV radiation substantially depends on an entire set of causes and factors, including: genetic—the ploidy and nucleotide composition of DNA in a cell; and physiological—varying growth phases, varying nutritional conditions.

Figure 6.24 illustrates the dynamics of the specific growth rates obtained after exposing the culture to different UV dosages. The damaging effect of UV radiation depends on the dosage. The dosage-effect curves derived from different experiments are all S-shaped (see Figure 6.25).

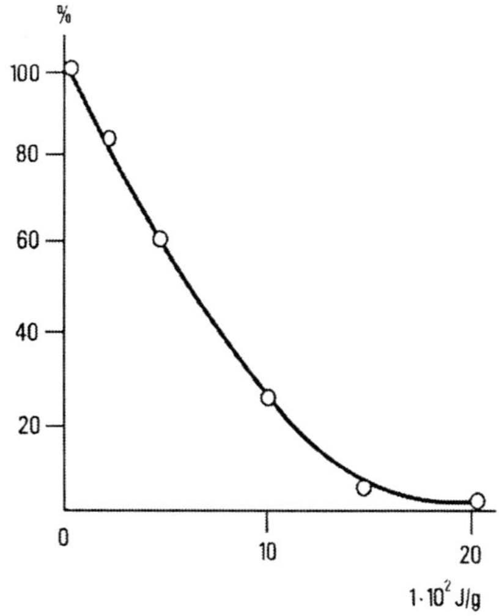

Figure 6.25 Minimal specific growth rate after exposure (in % of initial growth vs. growth rate after UV dosage).

The survival of exposed cells always depends on at least two competing processes: damage realization and the restoration rate (Korogodin, 1991). In our experiments cultivation was conducted under densitostat conditions, which means that biomass concentration was maintained constant. The medium supply rate, and consequently, the rate at which damaged cells were removed from the cultivator, depended on culture growth rate as expressed by the following equation:

$$N = N_0 e^{-\mu t}, \tag{6.47}$$

where N is the quantity of cells at specified time t, N_0 is the quantity of cells at the initial moment, and μ is the relative increment in biomass volume per unit time. The removal rate of individual cells depends on the amount of medium passing through the cultivator. The higher the culture growth rate in the after-effect period, the more intensively the population should be restored. Simple calculations show that at an after-effect growth rate μ, three volumes of nutritive solution are needed to remove 95% of the initial cells, in other words, 3 μt. So, at the optimal temperature, when the growth rate is at its highest, the population should restore faster than at other temperatures.

Restoration time was expressed in terms of the cell removal scale (the μt scale, where μt is the amount of medium equal to total cultivator volume that passes through the cultivator) in order to separate the part of the population restoration process associated with removing damaged individuals from the suspension from the part where cell restoration actually takes place, if any. Figure 6.26 illustrates specific growth rate restoration dependencies on the initial growth rate value (100%) for three temperatures in the after-effect period when the UV applied in the effect period was $11.2 \cdot 10^2$ joules/g. The x-axis represents restoration time expressed in terms of the μt scale.

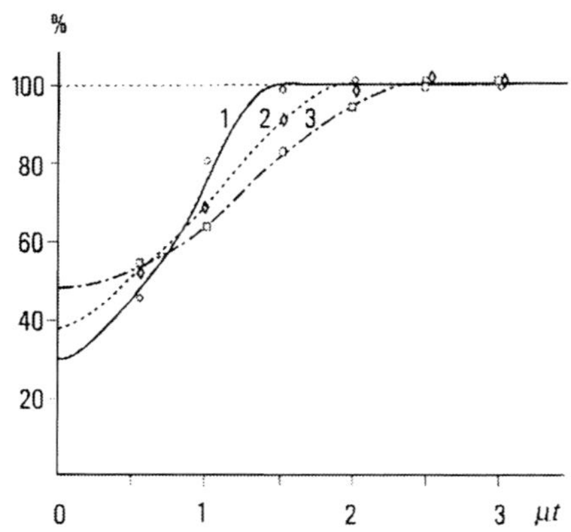

Figure 6.26 Specific growth rate restoration after exposure to a 11.2 · 102 J/g dosage: % for 3 post-exposure cultivation temperatures. 1—37°C; 2—40°C; 3—26°C.

These plotted dependencies show that in all three cases the faster damaged cells are removed from the culture, the faster the specific growth rate is reestablished. This suggests that the population removes damaged cells and engages in cell restoration simultaneously. As Figure 6.26 shows, restoration is faster at 37°C than at 40°C, and at 40°C faster than at 28°C.

Mathematical Modeling of UV Experimental Results

Taking the above observations into account, a mathematical model describing the way in which the biosynthesis process restores after UV exposure was proposed.

The condition of the population was characterized by the initial specific growth rate—μ, defined above as the relative increment of biomass per unit time. This specific growth rate can be changed by altering one or more cultivation parameters or by external factors acting on the culture. The characteristic time scale for the system is $1/\mu$. When the effect takes much less time than $1/\mu$, its impact can be considered instantaneous.

After being exposed to UV radiation, part of the population dies, which means that it yields no biomass increment and that the constant flowthrough eliminates it from the cultivator. Another part of the population remains unaltered (normal) for the first part of the effect period, that is; it continues to grow at the initial growth rate of $\mu = \mu_d$. The third part of the cells are damaged to one degree or another. Later in the after-effect period, the damaged cells begin to act like normal cells: some of them die and some of them reproduce at a certain rate, either slower or faster than initial growth rate μ.

As a first approximation, damaged cells can be considered to conform with: a specific transition rate into normal cells, α; a specific death rate, β; and a specific reproduction rate, γ. In other words, it is assumed that when no cells are removed from the entire mass of damaged cells, N_{dam}, then the number of cells that revert to normal operation per unit time are a N_{dam}, and that βN_{dam} die, and γ N_{dam} reproduce.

Coefficients α, β, and γ depend on the magnitude and the nature of the UV impact pulse, and in general, these coefficients are functions of time. In the following approximation, a series of cell groups having sustained varying degrees of damage are introduced as α_i, β_i, and γ_i.

Since the cultivation process is a densitostat continuous process, a certain amount of cells in each of these categories will be removed from the cultivator at the rate $\mu_d = \mu_d(t)$.

Given the above conditions, population behavior can be described by the following equations:

$$\frac{dN_n}{dt} = \mu N_n - \mu_d N_n + \alpha N_{dam}; \quad (6.48)$$

$$\frac{dN_{dam}}{dt} = \gamma N_{dam} - \mu_d N_{dam} - \alpha N_{dam} - \beta N_{dam}; \quad (6.49)$$

$$\frac{dN_{dead}}{dt} = -\mu_d N_{dead} + \beta N_{dam}, \quad (6.50)$$

where N_n, N_{dam}, and N_{dead} are the number of normal, damaged, and dead cells per culture unit volume, respectively.

The system is such the biomass concentration in it remains constant, that is, for the asynchronous culture:

$$N_n + N_{dam} + N_{dead} = N = \text{const.}$$

Hence,

$$\frac{d(N_n + N_{dam} + N_{dead})}{dt} = 0.$$

Summing up (1) + (2) + (3):

$$\frac{d}{dt}(N_n + N_{dam} + N_{dead}) =$$
$$\mu N_n + \gamma N_{dam} - \mu_d(N_n + N_{dam} + N_{dead}) = 0.$$

And from this it follows that

$$\mu_d = \mu\frac{N_n}{N} + \gamma\frac{N_{dam}}{N}. \tag{6.51}$$

Now Equations (6.48) to (6.51) can be used to consider the effect of a single instantaneous impact pulse on the system. To solve Equations (6.48) to (6.51), the characteristic variation time for coefficients α, β, and γ is assumed to be much more than 1/m, and in this way α, β, and γ can be considered constant. Note that it is comparatively easy to find a solution for the opposite case where the variation time for α, β, and γ is much less than $1/\mu$.

As a result, the following formula for μ_d is derived:

$$\mu_d = \frac{(\mu_0 - \phi_0)e^{\mu t} + \phi_0 e^{\beta t}}{1 + (\mu_0 - \phi_0)(e^{\mu t} - 1)/\mu + \phi_0 (e^{\beta t} - 1)/B}, \tag{6.52}$$

where

$$\varphi_0 = \frac{A}{\mu - B}y_0; \quad \mu_0 = \mu_d(t = 0); \quad \varphi_0 = \varphi(t = 0)$$
$$A = \gamma^2 - \gamma(\alpha + \beta + \mu) + \alpha\mu; \quad B = \gamma - \alpha - \beta.$$

The examination of limit cases demonstrates that the model is capable of describing the most typical processes that occur during UV damage.

The framework of the model can be used to describe other cases, and not only the case of UV impact. This model can also be used for the case of continuous impact when the appropriate alterations are made in the coefficients.

Continuous cultivation enables isolating cells that grow at a higher rate more quickly than other techniques. In densitostat cultivation, as in nature, selection follows the course of maximum growth rate.

These experimental and modeling results make it possible to estimate the stability of unicellular algal population as a life-support-system component in the presence of various impact effects (UV, ionizing radiation, etc.). The course of restoration processes after the impact of a damaging factor may be a source of information about the self-restoration processes that adhere in a biological life support system.

Concluding Remarks

As the reader could see in this chapter, the photosynthetic process as it is realized in a cultivator with algal cell population is predictable, stable, and controllable. Photosynthesis can be likened to chemical synthesis with autocatalysis, with only one difference: this biotechnical process exceeds any modern chemical manufacturing process in the diversity of the synthetic organic substances it produces, which opens up broad horizons for biotechnological applications of photosynthesis in space and on Earth.

CHAPTER 7

Controlled Continuous Cultivation of Higher Plants

7.1 SELECTING PLANT SPECIES AND CULTIVARS TO REGENERATE THE HABITAT

As is the case with algae, higher land plants can function as a polyfunctional link in a closed ecosystem. In principle, they are able to regenerate atmosphere, water, and food for humans by using human gas exchange and liquid wastes, and after mineralization, their solid wastes. The capability of unicellular algae and other lower plants to regenerate food is undetermined, despite prolonged and comprehensive experimentation on that point. Therefore many researchers consider higher plants as the most likely link for these regenerative functions in closed systems where substantial or near to complete food regeneration is envisioned.

This section and Chapter 7 as a whole present mostly Soviet and Russian investigations of higher plants, which in 1960–80s provided a basis for the creation and examination of different variants of experimental closed systems including humans and higher plants. Later, during 1980–90s, the major research along these lines was performed on a broad scale in the USA by B. Bugbee, F. Salisbury, T. Tibbits, R. Wheeler, and others. This research has proved our fundamental findings and developed further the ideas of the potentials of higher plants as regenerators of the human environment.

When selecting higher plants for closed systems, the most important criterion is that of satisfying human food requirements. As was shown in Chapter 5, there is no single plant, in nature nor under cultivation that is edible and which can fully satisfy human requirements for carbohydrates, vegetable proteins and fats, vitamins and minerals. Therefore human daily food requirements can be satisfied by vegetable matter only if a specially-concocted selection of vegetable products is devised and supported in the closed system as a multicomponent link.

Reasonable limitations on closed system values with respect to volume, area, weight, and expended energy determine another important criterion in plant selection: their high productivity level. The following must be kept in mind when evaluating plant productivity:

1. Plant biomass taken as the harvest gathered from a unit area contains a variable water quantity depending on the crop: from 12–15% in grain to 95–96% in cucumbers. Therefore, when comparing the harvests of different crops, the harvest must be converted into a “hard currency unit”—dry mass.

2. The harvest should be counted not only in terms of area yield, but yield over time since different vegetables have different growing times, and even different cultivars of one and the same crop can have longer or shorter growing periods depending on the type of cultivation regime used in each instance. The dried biomass of a harvest obtained for the growing period expressed in terms of area and time (dried biomass, $g/m^2 \cdot day$) is used as the productivity indicator in the narrow sense of the term.
3. It is important to evaluate the plant's: a) total biological productivity along with its gas and water exchange functions, and b) agronomic productivity in terms of its dry edible biomass expressed as $g/m^2 \cdot day$ or in terms of per cent of total biomass (harvest index).
4. The energy value of the yielded production should be adjusted for the radiant energy of the photosynthetically active radiation (PAR) spent on photosynthesis (380–710 nm)[1]. This characterizes the energy efficiency for planting. Energy efficiency should not be confused with the photosynthetic efficiency coefficient since the latter does not account for energy loss due to secondary biosynthesis, plant respiration, etc.

In general, the criterion of high productivity should include an optimal combination of high total productivity, a high harvest index, and a high energy efficiency coefficient. For any one of the crops none of these characteristics can be considered integral or constant. They can vary to a large degree dependent on the direction of the cultivation regime, and also dependent on the selection process used to create the culture.

A more strict productivity criterion must be applied for crops selected to provide the main portion of required human biomass. Crops having a small specific mass in the total mass transfer of the system (for instance, lettuce or herbs) can be evaluated less strictly with respect to this productivity criterion.

When selecting plant species and cultivars for inclusion in the experimental closed system, a series of additional criteria were taken into account:

1. Plant cultivation and product processing techniques that require a) the least number of essential operations and b) either the auxiliary devices for their performance or a reduction in time that humans spend on their execution.
2. Ecological compatibility of various crops and various growth stages of one and the same crop in the system with the purpose in mind to unify cultivation regimes and remove the necessity of creating a specific ecological niche for each of the various crops and growth stages.
3. Self-regeneration by seed propagation or cuttings. This avoids the need to have seed stores on hand (bulbs, etc.) in the closed system. Instead, they can be harvested fresh from the plants in the system.
4. Self-pollination in cases of seed propagation. This limits the amount of flower pollen that must be cast into the plant environs.

[1] We think that PAR, measured in W/m^2, can be more conveniently used to estimate processes of plant energy exchange in the closed system than PPF, often used by plant physiologists, measured in photon units that have to be converted into energy units, depending on the spectrum of the radiant flux.

However, the evaluation of any plant based on these and several other criteria makes sense only when the plant sufficiently satisfies the criteria examined earlier: the ability to satisfy human food requirements and to ensure high productivity levels.

Starting with human requirements, the vegetative portion of the daily ration should contain: 1) 45–50 g of protein; 2) 20–25 g of fats, of which 3–6 g are essential fatty acids; 3) 400–500 g of carbohydrates, of which about 80% are in the form of starchy polysaccharides; 4) the required quantity of vitamins and mineral components; and 5) a sufficiently and subjectively determined collection of components that can guarantee gustatory variety.

When taking into account these human food requirement quantities, it becomes clear that the higher plant link providing gas exchange, water exchange, and the vegetative portion of the ration must include starch crops. On Earth, grains and potatoes are the primary sources of starch in the civilized world. A diet that includes grain products (bread, grits, etc.) is preferred to the more voluminous and monotonous potato diet. A series of grains (wheat, rye, barley, oats) are long-day crops. This enables cultivating them under continuous illumination in smaller growing areas than are required for getting the same quantity of potato or short-day grain products such as rice and maize.

Among long-day grains, wheat was preferred as the main crop for the higher plant link because of high product quality and a simpler cultivation technology and product processing. Since rye is cross-pollinated, it casts a large quantity of pollen into the air during flowering and is therefore undesirable for use in a closed system. The usual sorts of barley and oats having a chaffy grain require complicated processing. Bare-grained forms of these crops are less productive, not very disease-resistant, and unless they are carefully selected, they cannot compete with wheat in terms of productivity per time and area. This does not mean that crops such as potato and bare-grained barley cannot participate in the higher plant link in small quantities. These crops can be used to increase diet variety if medical–biological indicators show this to be necessary.

From the enormous variety of wheat cultivars available, preference should be given to short-stemmed, beardless spring cultivars with spikes. Among other things, these cultivars should output a high fraction of grain per biomass yield and be tolerant of thick planting. During the first stages of closed system experiments, the early-maturing, medium height, beardless cultivar Skala was chosen. Later, cultivar Sonora-64 was chosen. It was lower-growing and more lodging-resistant than Skala, but it also had bearded spikes and smaller harvest yields. To dispose of Skala's medium height and Sonora's beardedness, G.M. Lisovsky created cultivar 232 by deriving a hybrid from these two cultivars (see Section 7.2). Cultivar 232 combined the desirable qualities of both parents and had no major defects. Beginning in 1972, it was precisely this cultivar of wheat that was used in all closed system experiments that were made on higher plants and humans together.[2]

Wheat and other grains produce almost no vegetable fats with essential fatty acids, and these are of the utmost importance in the diet. While striving to increase the production of vegetable food in the system, the problem of producing sufficient

---

[2] B. Bugbee and his colleagues have recently released a highly productive dwarf-wheat cultivar named "Apogee" (Bugbee *et al.*, 1996).

vegetable fats stands alongside the problem of producing enough protein and carbohydrates.

Many oil-producing (oleaginous) plants were investigated while keeping in mind the general criteria enumerated above for selecting closed-system crops. These oleaginous plants were cultivated under artificial light and were observed, and the numerous data in the literature concerning these plants were analyzed. Such widespread crops as sunflower, flax, and mustard yield an exceedingly small amount of edible parts for the large specific mass in the harvest of its stems, roots, and leaves. Also the oilcake which is left over after the oil has been removed from the seeds is neither usable nor unusable. Soybean and peanut are also of little use in a closed system due to specific developmental traits—a need for a short growing day, and for peanut, the complexity of inseminating the seed buds. A more promising crop is chufa, *Cyperus esculentus L.*, from the sedge family (Shilenko *et al.*, 1979). Chufa multiplies by vegetative propagation, forming numerous edible tubers on its root system that are 15–20 mm long and 5–10 mm wide (Figures 7.1a and 7.1b). Chufa responds positively to lengthening the daily light exposure even up to exposing it to illumination 24 h a day. Chufa also grows well in hydroponic conditions. Chufa can be harvested only 70–90 days after it has been set. Under artificial light of 120–150 W/m^2 PAR, the harvest of absolutely dry biomass is 3.5–4.5 kg/m^2, including tubercles which come to 2.0–2.5 kg/m^2 when the harvest index is 50–60%. Chufa productivity with respect to its edible part calculated for a unit of time is not less than, and is even greater than, that of wheat. Dried chufa tubercles contain 20–25% fats, about 60% carbohydrates, of which about a half are starch types, and around 8% proteins.

The unsaturated fatty acids of oleic (rapic) acid and linolenic acid predominate in the composition of lipids, comprising up to 76% of the total fatty acids. This attests to the high food value in chufa fats. Experiments that included 130 g/day of chufa in the human diet showed that the test subjects experienced no negative reaction to this product (Okladnikov *et al.*, 1977). A simple calculation shows that 100–150 g/day of chufa tubercles fully satisfies the human daily requirement for vegetable fats, including essential fatty acids. Besides, chufa can supplement wheat in providing carbohydrates, including polysaccharides. Selection cultivars of chufa are not known. In the experiments, a chufa sample was obtained from the Russian Institute of Oleaginous Cultures.

When wheat and chufa provide the main dry-mass portion of the human vegetable ration in a closed system, then 10–20% of the growing area is left for cultivating vegetables. Therefore, the vegetables should not be distinguished as much for their high productivity as for their ability to satisfy human requirements for vitamins, organic acids, and for gustatory, aromatic, mineral, and other substances. The main body of literature about the various vegetable crops was obtained under field or hothouse conditions. They form a relatively rich, preparatory material for evaluating the performance potential these crops might reach as part of the higher-plant link in a closed ecosystem. However, special explanations of certain characteristics or quantitative refinement of data necessary to calculate mass transfer rates in the system had to be developed, such as, for instance, 1) whether or not these crops can be grown under continuous illumination or other photoperiods and, if they can, their productivity levels under continuous cultivation or these other photoperiods; 2) soilless cultivation techniques; 3) product biochemical and mineral composition during cultivation under completely manmade conditions, and so on. With these purposes in mind, several vegetable crops were tested in the phytotron under regulated conditions: carrot, table beet, turnip, radish, green pea, tomato, cucumber, kohlrabi, kale, onion, dill, sorrel, and a series of others.

Figure 7.1a Vegetable part of chufa grown in a Bios-3 phytotron.

Figure 7.1b Tubers of chufa grown in phytotron.

As a result of these tests, beet, carrot, radish, cucumber, sorrel, dill, and onion were chosen as the major vegetable components included in the closed system higher plant link. This group of vegetables is able to provide sufficiently varied meal preparation, the necessary amount of vitamins, gustatory substances, and aromatic substances. Carrot, beet, onion, and cucumber are close to the main two crops of wheat and chufa in terms of productivity. The green crops of dill and kale could be grown as companion crops between the beet and carrot, so they do not need a separate growing area. More detailed productivity characteristics and the biochemical composition of these crops when grown in fully manmade conditions are listed in Section 7.6.

Crops such as turnip, tomato, potato, kohlrabi, sorrel, and a number of others can be included in the composition of the higher plant link. But because of their limited edibility (turnip, sorrel) or their excessively low productivity under continuous illumination in general habitat conditions not tailored to the needs of individual crops (tomato, potato, etc.), their participation in the closed system was deemed unnecessary.

It should be noted that many vegetable crops are represented by a large number of cultivars that are distinguished by biological, morphological, biochemical, and other traits. Often, one or another feature of a particular cultivar renders it unsuitable or ineffective for use in a closed system, while other cultivars of this crop turn out to be completely satisfactory. For instance, many early-maturing cultivars of radish—Saks, red with white tip, and others—flower before they have a chance to develop proper roots when continuously illuminated, while the Virovsky White, Red Giant, and other cultivars yield a wonderful root crop. Radish, carrot, and beet cultivars with an elongated root shape demand a deep layer of rooting-habitable substrate down to 20–25 cm, while the cultivars with stubby or rounded roots produce a normal harvest when the layer substrate is 7–10 cm. Seedless cucumber cultivars that do not require pollination are more productive and more technologically suitable than insect-pollinated cultivars, etc. Therefore it was considered best not to be limited by the question of choosing suitable cultivars based on their variety, but to examine the question of specific directions in plant selection to enable human life support in a closed system.

7.2 PERSPECTIVES FOR PERFECTING PLANT SELECTION FOR THE HIGHER PLANT LINK

Those plants included in the phytotron link that would be ideal with respect to mass transfer functions would be those that would yield a biomass that: 1) is fully edible by humans, 2) has a biochemical composition that completely satisfies human requirements, and 3) can have its requirements for carbon dioxide, water, and mineral nutrients completely satisfied by human wastes. Such plants do not exist in nature. The majority of higher plants suitable for food are not completely edible. And those few that can be eaten in their entirety cannot satisfy all human food requirements.

The most efficient, although not the quickest, path to bring the plants closer to fulfilling human requirements is their breeding. This necessity originates from a whole series of specific requirements presented to the plants under conditions of "space crop husbandry." In connection with this, it is expedient to concretize the main tasks in breeding work in this field for the direction taken.

From the point of view of perfecting the food value, two main directions in higher plant breeding for closed systems can be envisioned: 1) increasing the fraction of edible portions in the total biomass harvest, and 2) giving a qualitative "push" to the biochemistry of the edible portion to more fully meet human requirements. It must be noted that in the majority of cases, plant breeding in these directions will be just as useful for the advancement of agricultural purposes.

Increasing the edible biomass fraction can be achieved by two distinct paths. The first path involves increasing the edible portion harvest with a corresponding decrease in the harvest of the inedible parts while maintaining or increasing the total harvest. The advantages of such a breeding goal are clear. The second path involves decreasing the inedible portion of the biomass harvest while decreasing the total biomass harvest and maintaining the edible portion harvest constant. The advantage of the second method is that it increases the "closability" of the reference and performing elements by reducing the substances which require special processing or which are deadlock products.

Both of these methods assume breeding based on a fundamental restructuring of the plant architecture. For grain crops, for instance wheat, this means, first of all, 1) shortening the stems, even to the point of creating rosette forms having only leaves and spikes; 2) decreasing the weight of the root system—this is achievable thanks to the abundant mineral supply in the artificial nutrient medium; 3) eliminating spike beardedness, and so on.

For such crops as cucumber and tomato, this means not only a decrease in the total stem and root mass, but breeding to produce continued and repeated fruiting on one long-living leaf and stem apparatus. For root crops, this means decreasing the mass of buds on the leaves or increasing their food value, as has been done by B.S. Moshkov for a cabbage–radish hybrid (Moshkov, 1966). Resolving these tasks by the usual field breeding practice would require years or even decades. But, when cultivating under intense artificial illumination, which is the necessary backdrop for breeding plants for use in the closed system, the breeding process is sped up by 4–6 times, and these tasks can be resolved more quickly.

The tasks involved in selecting species and cultivars and in perfecting their quality by breeding are varied. In many cases it is not necessary to change the biochemical direction of biosynthesis via breeding if the biochemistry of the total product can be nudged in the required direction by introducing some other crop. For instance, it is scarcely expedient to conduct a prolonged and complicated breeding process to increase wheat grain fat content if the human requirement for vegetable fat can be satisfied by an oleaginous crop that does not detract from the system link structure. Therefore, the biochemical "push" of plants to fulfill human requirements should take place not so much on the level of separate species as on the level of the performing link in its entirety, which can consist of groups of synergistic species. Breeding to better the biochemical composition of various crop products can be limited to a series of partial tasks. For instance, to increase the amino acid completeness of proteins, to increase vitamin content, to decrease the quantity of hard-to-digest substances, to perfect technological and gustatory biomass properties, and so on.

Plant species and cultivar selection and breeding to ensure their compatibility with human gas exchange is a more general problem. Breeding plants so that their biochemical properties conform to human requirements is at the same time breeding them in order to bring the plant assimilatory quotient in line with the human respiratory quotient. Besides such "gross" compatibility, breeding must be directed at plant resistance to the numerous airborne exchange products from humans and other system links, as well as to its own gas exchange products.

By respiration alone humans emit up to 150 airborne substances (Kustov and Tiunov, 1969), and their effect on plants can be very significant. The plants themselves give off hundreds of substances like hydrocarbons, aldehydes, alcohols, ketones, ethers, and so on (Mokhnachev and Kuzmin, 1966). Various plant species produce various airborne substances in nonidentical quantities (Dadykin *et al.*, 1967; Dadykin, 1970).

All of these substances are far from being identified, but among those already known, there are a few toxic ones. Therefore several researchers suppose that plant allelopathic activity in a closed system will turn out to be one of the leading forms of their cooperation (Dadykin, 1970a; Dadykin and Nilovskaya, 1969). Finally, it is well known that plants not only excrete, but that they absorb a series of organic substances from human and plant gas exchange.

Therefore plant selection and breeding for closed systems to facilitate human gas exchange compatibility must be conducted taking into account plant capabilities to:

1. Produce a minimal quantity of substances toxic to humans, themselves, and other system links.
2. Evoke resistance to airborne substances excreted by humans and other system links, as well as by system devices.
3. Absorb substances toxic to humans that are excreted by themselves and other system links.

Plant selection and breeding based on these characteristics is still an extremely difficult and complicated affair. Human gas exchange chemistry has undergone little study, and even less is known about plant gas exchange. Methods have not been developed to research and evaluate plants with respect to their production and absorption of airborne substances. Work must be conducted in hermetic cultivation systems within targeted background ranges of airborne substances in the atmosphere.

A whole range of problems with regard to plant selection and breeding arises when using human liquid and solid wastes as a source of elements for plant mineral nutrients in a closed system. Since most plant mineral nutrient elements are found in human liquid excretions (Bazanova, 1969), it seems promising to directly apply them to the higher plants with a small adjustment made for salt. In this case, the plants should be able 1) to use urea as the main form of their nitrogen nutrition, 2) to be resistant to the medium being salted by NaCl, 3) to absorb NaCl up to 1–1.5% of its dry mass in order to provide for the removal of table salt entering the medium from humans, 4) to bear those organic substances in the nutrient medium that originate in liquid excretions such as ureic acid, creatinine, purine, etc., 5) to be compatible with the microflora that process these organic substances, and so on. Human solid wastes cannot be directly applied to higher plants. Therefore, if they are to be used, they must undergo preliminary and biological or physicochemical mineralization into compounds that can be used by plants.

Different higher plants are characterized by nonidentical ecological requirements. If the link will be made up of a few plant species, the most likely case, then creating a separate ecological niche for each will be extremely difficult. From this, the task arises to artificially bring plant requirements close to such general habitat requirements as, for instance, air temperature, humidity, illumination period duration each day, light spectral composition, and so on.

V.P. Dadykin (1968a) has proposed that it is desirable to use plants in a closed system that have air temperature and humidity requirements close to those of the human comfort zone. We start from the fact that humans will be among the plants for the minimum amount of time necessary to perform one or another kind of work and so it is scarcely necessary to maintain conditions commensurate with the human comfort zone in the compartment where the higher plants are growing. Taking into consideration intensified photosynthesis, plant development tempo, and heat removal among other factors, it seems more expedient to maintain the air temperature and humidity at the highest levels allowable for all plants participating in the system and yet still bearable by humans. Therefore, the selection and breeding of all plants should be undertaken on the basis of their thermophilic and hydrophilic nature.

An extremely desirable ecological trait of all plants included in the link is their adaptability to continuous illumination, or at least in those situations where continuous illumination is not excluded by special requirements, for instance, by a lunar regime of night and day. Overcoming photoperiodicity would go a long way to simplifying the governance of the plant light regime and to bettering the use of system working areas and volumes while increasing cultivation intensity. Selection can solve such a problem.

When studying ecological features of plants that form the phototrophic link in a manmade ecosystem, attention must be paid to increasing plant requirements with respect to their resistance to various emergency situations, as for instance, overheating, prolonged darkness, decompression, etc.

Finally, selection and breeding tasks that arise due to the peculiarities of plant cultivation can be mentioned. It is not possible to clearly formulate all tasks in this way at the present time because a plant cultivation technology that has proven acceptable in closed systems has not yet been worked out in detail. However, it is more than certain that plants should have traits such as, for instance, 1) low height combined with a compact leaf bunch or rosette that simplifies the culture conveyor technology, 2) a stubby root that is nearly round, and 3) nearly vertical foliage orientation, which favors the application of intensive irradiation from the top, and so on.

Proceeding from the plant selection and breeding tasks outlined above, various plant cultivars were evaluated under artificial illumination, and breeding research was conducted on one of the plants used in the model—spring wheat. From a large collection of over 800 samples of spring wheat used in the Krasnoyarsk Agricultural Research Institute, about 40 samples of mid- and early-maturing, medium- and low-growing wheat with a productive spike were chosen. A few samples of short-spiked wheat were obtained from the collection of the All-Union Institute of Plant Husbandry in Saint Petersburg. A comparative evaluation of these samples under photoculture conditions has shown that very few of these are worth considering for use in closed human life support systems. The final choice was to use the Skala and Sonora-64 wheat cultivars. These cultivars are close to one another in the duration of their growing period when subjected to photoculture conditions, extending from 60–70 days from sprouting to maturation. The Skala cultivar is one of the most productive amongst the tested samples that has a short growing period. However, it has a tall straw stem that measures up to 90–100 cm and it lodges easily in the grain formation period. The Sonora-64 cultivar is one of the low-growing cultivars, reaching only 60–80 cm under artificial lighting conditions, and is more resistant to lodging than Skala. At a young age it forms wide, impenetrable leaves. However, with respect to productivity, it defers to Sonora-64 by 20–30%, and its beardedness constitutes another deficiency.

The presence of both good qualities and deficiencies in each cultivar provided a basis for hoping that it is possible to develop new cultivars by breeding that would combine the good qualities of the Skala and Sonora cultivars, but which would lack their deficiencies.

With this purpose in mind, forced hybridization of Skala and Sonora-64 was conducted under photoculture conditions. Moreover, several hundred spontaneous hybrids of Sonora-64 were produced when it was freely pollinated with the pollen of various highly productive mid-growing cultivars.

Beginning with the third generation, all of these hybrids were used for single and repeated individual selection. The selection was conducted according to the following list of traits:

1) low-growing (in the Sonora-64 range)
2) quickly-developing, straight-standing leaves
3) spike beardlessness
4) spike high grain yield
5) grain size
6) high harvest index
7) high productivity calculated in terms of a unit growing time and other factors.

Cultivating parent forms and hybrid material, breeding material for several different generations, as well as subsequent sort testing was conducted against the background of hydroponics under continuous artificial light at 60–80 W/m^2 and at a temperature of 22–26°C. During 4 years of work in the phytotrons, from 6 to 12 generations of various hybrid lines were cultivated, and the final preference was given to Line 232. Line 232 successfully combined the low-growing quality and the close and vertical foliage arrangement of Sonora-64 with the lack of beard and

Table 7.1 "Competitive" tests on Line 232 in the phytotron in comparison with parent forms (0.36 m^2, 3 replicates, paired comparison).

Cultivar	Average plant height, cm	Yield, g/m^2		Grain yield, % of standard	Harvest index, %
		Aboveground biomass	Grain		
Sonora-64	58	908	313	100.0	34.5
Skala	89	1343	432	138.0	32.2
Line 232	59	879	427	136.7	48.6

high harvest yield of Skala with a growing period duration that is close to that of the original parents. The evaluation of this line in the phytotron during competitive testing in 3 repeated verifications (Table 7.1) have demonstrated its advantages over both its parents and has allowed using it under the name of "Cultivar 232" in the higher plant link in a whole series of experiments with closed systems (Lisovsky, 1972; Gitelson *et al.*, 1975).

It stands to reason that the work performed in breeding wheat did not resolve all tasks in this area as they were formulated earlier in this chapter. Notwithstanding, this work graphically demonstrated that wheat can be efficiently improved in a timely fashion under completely controlled conditions that to a certain degree approximate future conditions expected for wheat cultivation in closed systems. In recent years B. Bugbee and his co-workers have made considerable progress in this direction (Bugbee *et al.*, 1996).

7.3 LIGHT AND TEMPERATURE REGIMES FOR CULTIVATING PLANTS IN MANMADE CONDITIONS

The energy encapsulated by closed-system primary production—that is, the energy plants accumulate when they photosynthesize radiant sunlight energy or radiant artificial light energy—is the energy that propels the biological material exchange. Therefore, when there is a favorable temperature regime, the plant cultivation light regime is the main determinant value for both quantitative and qualitative plant productivity indicators. Three main characteristics of the plant regime can be singled out: 1) the power of the radiant stream falling on the planted area, measured in the photosynthetically active radiation (PAR) region and having a wavelength within the range of 380–710 nm; 2) the rhythm with which light and dark periods alternate, that is the rhythm of the photoperiod; and 3) the spectral composition of the radiant stream. It is not possible to use natural light (sunlight) to research different variants of directed higher plant cultivation since it is impossible to provide photoperiod stability and needed radiation intensity levels, which are bound to change in the course of a day from 0 to 400–450 W/m^2 PAR. Another reason why sunlight cannot be used is that it is impossible to create any and every kind of photoperiod, including 24-h-a-day continuous illumination and continuous illumination that spans multiple days. Therefore, practically all research on the development of plant cultivation light regimes was performed using artificial illumination in various special laboratory assemblies, which shall henceforth be called phytotrons.

To evaluate the characteristics of the entire higher plant population as a photosynthesizing system, it is enough to examine the photosynthetic rate for one or another experimental condition for individual plants or leaves. It is necessary to research the entire higher plant population as a whole. To this end, hermeticized phytotrons in which a large enough quantity of interacting plants could be cultivated for the whole growing period were employed. This large quantity of plants shall hereafter be referred to as a cenosis. Continuous gas exchange control—mainly, continuous control of the CO_2 concentration in the phytotron—allows expeditious evaluation of plant cenosis cumulative responses to other various changes in conditions without destroying plant integrity. Beginning in the 1960s, this method received widespread acceptance among researchers (Tranquillini, 1967; Lisovsky *et al.*, 1968; Laptev and Nilovskaya, 1968; and others).

To study cenosis gas exchange for various crops, several hermetic phytotrons were developed and built, which enabled the regulation and control of the plant growth conditions. The seeded area of laboratory phytotrons was 1.5–2.1 m^2, which permitted simultaneous cultivation of, for instance, a few hundred radish plants or several thousand wheat plants. The volume of assimilation chambers was from 2000 to 3100 l. A generalized description of the phytotrons can easily be drawn since all phytotrons employed in the experiments were uniform with respect to their main characteristics (Figure 7.2).

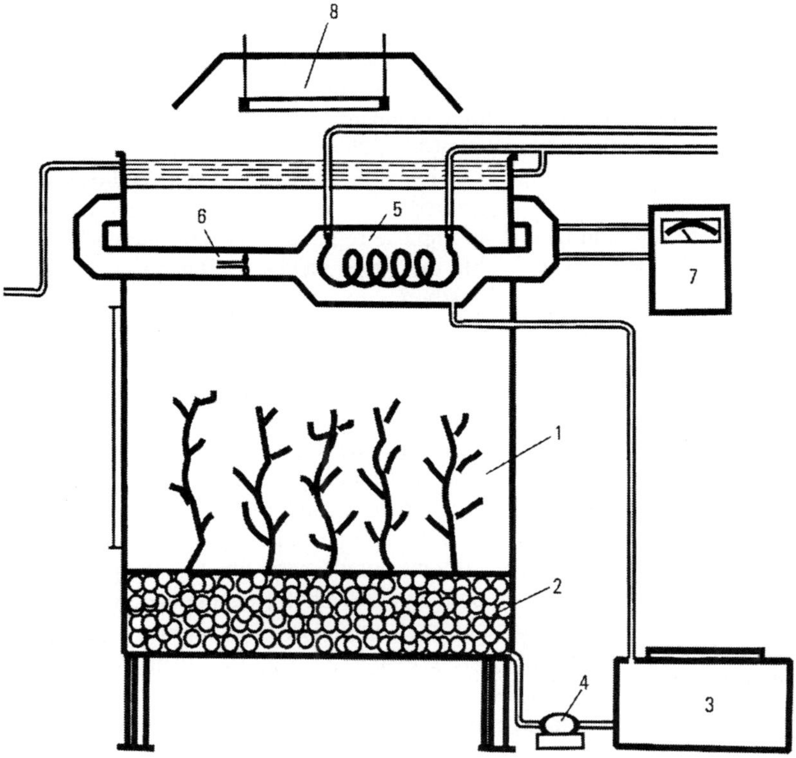

Figure 7.2 Flow diagram of the hermetically sealed phytotron. 1—phytotron chamber; 2—chamber section filled with substrate; 3—nutrient solution tank; 4—pump for injecting nutrient solution; 5—conditioning device with cooling water coil; 6—fan of conditioning device; 7—gas analyzer; 8—light source.

The phytotron assimilation chamber was either a cube or a parallelepiped made either of acrylic plastic sheets glued together or of stainless steel sheets welded together, with a glass ceiling and a hermetically sealed door in one of its walls. The light sources consisted of either 1) one or two 6-kilowatt xenon lamps or 2) four iodine loop lamps KI-220-1000, or 3) 2–4 metal halogen lamps of wide specifications, or finally 4) special (lithium, gallium, indium, etc.) lamps that emit radiation at various spectral regions.

There was a layer of silicate glass 6–8 mm in thickness and a layer of running water 35–50 mm in thickness between the light sources and the plants. The layer of running water absorbed a significant portion of the infrared radiation. The root medium was a layer of expanded clay aggregate up to 20 cm in thickness for the hydroponic phytotrons, and the root space was left unfilled for aeroponic phytotrons or for an air subirrigation culture. The nutrient solution is pumped into the root area periodically from a hermetic tank that is jointed to the assimilation chamber. The tank is outfitted with a device that automatically regulates the nutrient medium temperature. An air conditioning unit serves to condense transpirational water and to maintain the target air temperature. The air conditioning unit is made from PVC, it is joined to the assimilation chamber, and it is equipped with: 1) a fan for air circulation, 2) a coiled water-cooling pipe, and 3) a heating element. The transpirational water was returned to the nutrient solution.

An optical–acoustic gas analyzer (AO-2209 or Infralit) with a scale measuring CO_2 concentrations in the range of 0–1% or 0–2% was attached to the assimilation chamber, and it continuously registered the carbon dioxide concentration in the phytotron. With the help of a relay-impulse regulation system, the carbon dioxide concentration was maintained within the target values. A change in the phytotron air pressure during an influx of carbon dioxide was counterbalanced using an adjustable volume pressurizer. This pressurizer is not included in Figure 7.2. Surplus oxygen formed during the course of photosynthesis is removed once every few days by releasing the phytotron hermetic seal and airing out the phytotron. The illumination was regulated by changing the voltage to the xenon lamps, either by using a regulated ignition device or by changing the number of installed lamps. The intensity of the radiant stream was continuously registered by a Yanishevsky pyranometer, and its signal was diverted to an automatic recording device. The pyranometer readings were compared with 1) periodic illumination measurements taken on luxometer Yu-16 at 16 different points at the level of the plant tops, and with 2) PAR measurements taken on a Kozyrev phytopyranometer.When measured at 16 different points, the maximum and minimum illuminations differ from the average by ±20% for plants that have grown to a height of 30 cm. When the plants were 70 cm tall, the light distribution in phytotrons having two xenon lamps grew worse, deviating at individual points as much as –40 +65% from the average illumination. This does not happen in phytotrons that have 4 or more lamps of a lower power rating.

To maintain air temperature, a contact thermometer was set at the air output from the assimilation chamber to the air conditioning unit. When the air heats up past the targeted temperature, the contact thermometer signal opens a pinching device, upon which cold pipe water enters the water cooler of the air conditioner. A drop in temperature separates the thermometer contacts so that the pinching device stops supplying water to the cooler. When the illumination in the phytotron is turned off, heat is supplied by a constantly engaged electric air heater. Air temperature is regulated in the same way when the electric heater is turned on. The air temperature registration system is independent of the air temperature regulation system. It consists of 1) a thermistor installed in the path of air exiting from the phytotron and 2) a device that continuously records the thermistor signal.

Thermistor readings were periodically checked against the readings of the dry thermometer in the psychrometer located in the phytotron.

The temperature of the root system could be maintained within the target limits by heating or cooling the nutrient solution, which was being periodically pumped into the zone where the roots were growing. The temperature of the nutrient solution was regulated in the same way as was the air temperature. However, there was no need to specially regulate the root system temperature in the majority of experiments since the root system was shaded by the plants and free evaporation from the layer surface facilitated a sufficient (3–4°C) negative temperature gradient between the root system and the aboveground layers.

The humidity in all of the phytotrons depended on the air temperature, illumination, and transpiration capacity of the studied planting, except in the case of one particular phytotron that was supplied with air by a special regulation system. Humidity usually fluctuated between values of 60 to 85%. When the temperature and illumination regimes were stable, air moisture deviated from the average value by not more than 5%.

In the described phytotrons, various parameters were maintained constant or were set to enable the achievement of targeted values. Each of the following parameters was independently varied within the following range values: 1) the average intensity of the radiant stream was varied between 20 to 150 W/m^2 PAR ±10% from the targeted value; 2) illumination duration was from 0 to 24 h a day; 3) air temperature was varied between 20 to 30±1°C; 4) root zone temperature, from 15 to 25±2°C; 5) CO_2 concentration in the air was varied from 0 to 2%±0.1%.

Laboratory phytotrons permitted the study of a cenosis of various higher plant species as a single photosynthesizing system. It also enabled the gathering of the cumulative energy and mass transfer characteristics of the cenosis as an integral and functional unit so that these data could be used to perform calculations and build closed systems for higher plants.

Above all else, the productivity of plants cultivated in a closed system is determined by the light regime created for that particular system.

The intensity of the radiant stream recommended for production conditions using artificial light was usually set low (20–60 W/m^2) since, in the first place, this intensity is determined by considerations about the economic expenditure of electric power and is not determined by considerations about any physiological optimum (Leman, 1961; Van der Win and Myer, 1962; Latyshev, 1967; Kenkhem, 1967; and others). It is true that if the cenosis could have a sufficient amount of foliage surface at this illumination level so that it absorbed 80–90% of incident energy in the PAR region, and if the cenosis would provide for a high energy coefficient of harvest formation (10–12%), then its productivity would be sufficiently high—15–45 g/m^2 per day of dry biomass. However, when a cenosis has a large foliage surface value and a correspondingly high biomass value, then a relatively large amount of organic substance is spent on respiration. On the other hand, a cenosis with a small collective foliage surface uses light poorly. Therefore, when the radiant stream intensity is low, it is almost impossible to realize a highly productive cultivation practice.

To facilitate intense plant cultivation it is more expedient to physiologically define an optimal radiant stream intensity, for example, by considering the light curves of leaf photosynthesis.

It is well known that when irradiation is increased in the visible region, the photosynthetic rate in the leaf increases to a definite level, and then photosynthetic light saturation ensues. The level of leaf photosynthetic light saturation is different for plant species with different light adaptation mechanisms, and it also changes substantially for different plants to differing extents depending on the temperature, CO_2 concentration, and a number of other factors. At the atmospheric CO_2 concentration the photosynthetic rate for leaves on the majority of higher plants is close to its maximum as long as the PAR reaches 60–120 W/m^2. However as work on the structure of the planting's photosynthetic apparatus developed (Boysen-Iensen, 1932; Watson, 1952; Nichiprovich, 1955, 1956, 1966; and others) it led to the idea that, aside from foliage photosynthetic productivity, several other factors have a significant impact on how light usage figures in total harvest formation: 1) leaf area index; 2) leaf spatial orientation; and 3) the ratio of photosynthesis to plant respiration, including the respiration of stems, roots, and other plant parts. Therefore, when determining the optimal PAR level for the cenosis when its total gas exchange is realized within a closed system, it is not enough to consider only leaf photosynthesis light curves.

Leaf photosynthesis values for wheat increase with an increase in illumination to 95–120 W/m^2 (Friend, 1964; Stoy, 1965; Wardlow, 1967). When the CO_2 concentration grows to 0.3–0.4%, the photosynthetic rate of wheat increases (Stoy, 1965;Wardlow, 1967), as it does for other grains and vegetables (Ford and Thorne, 1967;Nilovskaya, 1968; and others). A temperature in the range of 20–30°C exerts little influence on the photosynthetic rate (Friend, 1964). Changes due to plant age also bear insignificantly on the wheat foliage photosynthetic rate (Stoy, 1965).

The data in the quoted literature demonstrates that when conditions are solidly optimal, a light level of 50–60 W/m^2 PAR must be regarded as the lowest permissible level for a plant cenosis. Experiments had to be performed to determine the optimal radiant stream intensity that would facilitate sufficiently high values with respect to the cultivation rate indicator and cultivation energy efficiency for cenoses of wheat and other plants.

In order to evaluate how wheat cenosis photosynthesis depends on the irradiation level, a series of experiments in a hermetic phytotron were conducted. The growing area in the phytotron was 2.14 m^2, and its volume was 3100 l. Spring wheat Sonora-64 was sown to a density of 660 plants per m^2. The cultivation method was that of hydroponics on expanded clay aggregate using the Knop nutrient medium. Air temperature in the phytotron was 22 ±1°C during hours of light, and 20 ±1°C during hours of darkness, with respiration maintained at a specific level. CO_2 concentration in the air was maintained at 0.4 to 1.4%, and air humidity was kept at 75–85%. The photosynthetic and respiratory rates were determined according to cenosis CO_2 absorption (and emission), assuming a 1 m^2 planted area during exposure to stepped and varied PAR illumination intensities within the range 120–0 W/m^2. For the remainder of the time, radiant stream intensity was held round the clock at 80–100 W/m^2. Once a week, 2 groups of 5 plants were taken as samples from the phytotron to determine the rate at which the biomass and foliage surface were increasing. Experimental results were processed using the method of regression analysis.

The rate of visible photosynthesis for a wheat cenosis calculated for a unit planting area changed under identical illumination depending on the plant age (Figure 7.3). The photosynthetic rate was steady during the period that spanned the beginning of leaf-tube formation (15–17 days) to the end of flowering (35–37 days). For this period, a high level of respiration in the dark was characteristic for the cenosis.

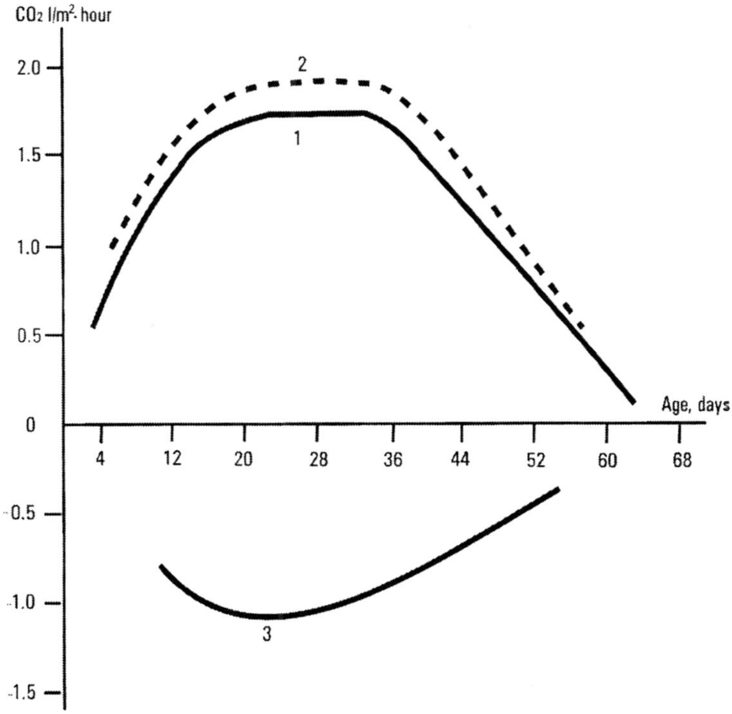

Figure 7.3 Rate of apparent photosynthesis of the wheat cenosis for the growth period, at 88–96 W/m^2 PAR (1) and 104–112 W/m^2 PAR (2), and dark respiration rate of the cenosis (3).

Any change in irradiation (it was kept within the range of 20–120 W/m^2) that was incident to a cenosis that had developed an extensive foliage surface was accompanied by an almost linear change in the photosynthetic rate (Figure 7.4). The photosynthetic compensation point was found to be from 25 to 41.6 W/m^2, as was calculated using both the linear regression equation and the plant age. The cenosis energy efficiency coefficient grew as the illumination was increased up to the studied limit of 120 W/m^2. With the illumination held steady at 120 W/m^2 during the period when the plant was forming leaf-tubes and flowering, the energy efficiency coefficient reached values of 10–11% (Figure 7.5). Although a PAR radiant stream intensity of 120 W/m^2 is the photosynthetic saturation intensity for individual leaves, this intensity was far from being the photosynthetic saturation intensity for a cenosis with a sufficiently developed foliage surface.

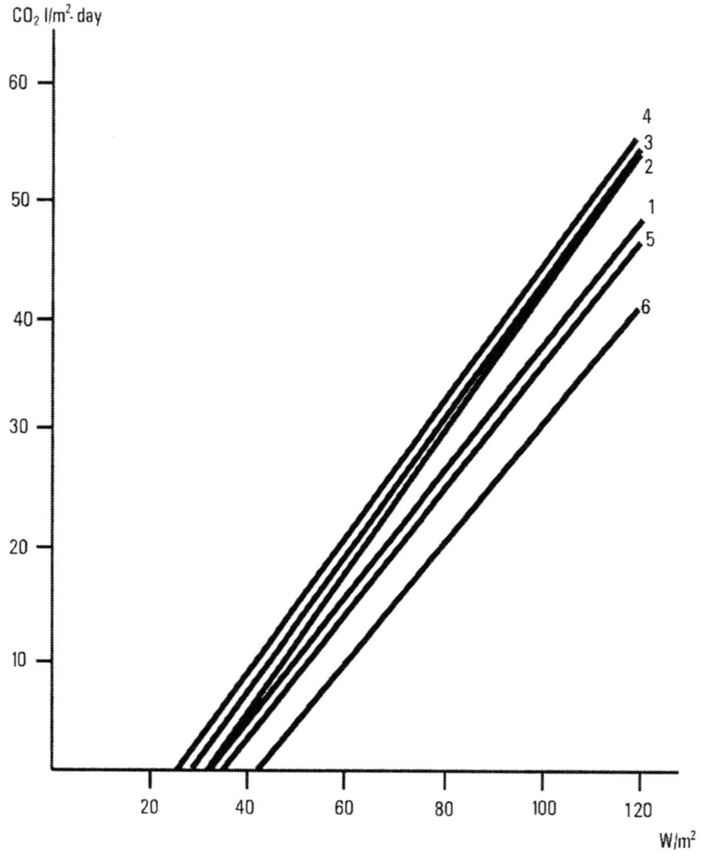

Figure 7.4 Rate of apparent photosynthesis of the wheat cenosis vs. Intensity of photosynthetically active radiation for plant age: 1—17–18 days; 2—22–24 days; 3—26–27 days; 4—33–34 days; 5—40–41 days; 6—44–45 days.

The high cenosis productivity for higher plants is due to the fact that higher plants produce a lot of biomass, and this in turn is due to the increased respiration for nonphotosynthesizing organs and to the photosynthetic compensation point value. The energy efficiency for a cenosis having a large biomass can reach high levels only when the light level is sufficiently far from the photosynthetic compensation point. When speaking of the photosynthesis of a single leaf, it is true that the lower the illumination, the higher the photosynthetic efficiency coefficient. But this same thing is not true of the visible photosynthesis of a highly productive cenosis. When photosynthetic activity is determined using the criteria of energy efficiency, data procured in experiments encourage the supposition that the optimum photosynthetic activity of a wheat cenosis lies at the upper limit or a little higher than the upper limit of the radiant stream intensity interval studied in these experiments, i.e. 120 W/m^2. At the same time, the wheat cenosis visible photosynthetic rate did not exhibit any signs of saturation at 120 W/m^2 (Figure 7.4). This means that any potential for increasing the wheat cenosis productivity at a radiant stream intensity of 120 W/m^2 remained unexplored (but see below).

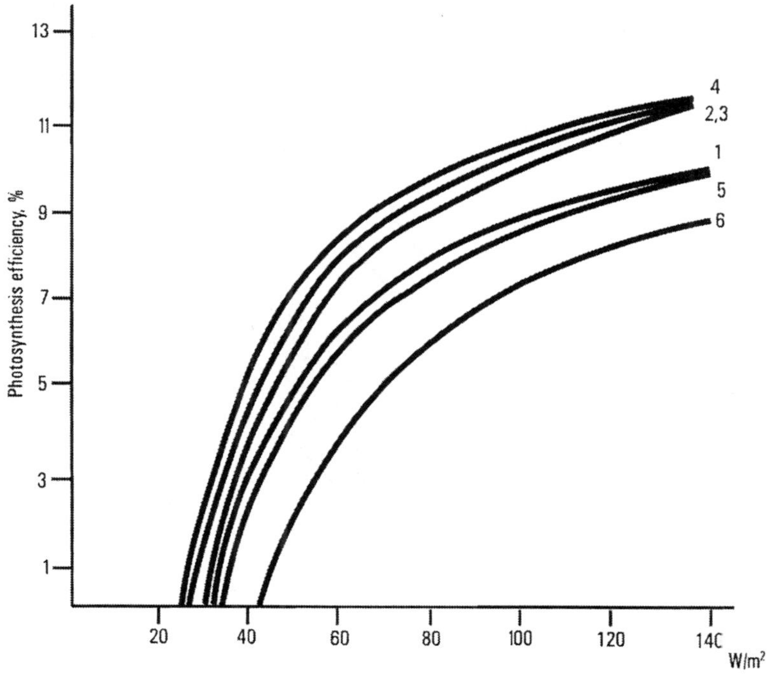

Figure 7.5 Efficiency of wheat cenosis photosynthesis vs. radiant flux rate for plant age: 1—17–18 days; 2—22–24 days; 3—26–27 days; 4—33–34 days; 5—44–45 days.

During the 1970s at the Institute of Biophysics of the Siberian Branch of the USSR Academy of Sciences a large series of experiments was performed with the stream intensity was increased to levels substantially surpassing those of noontime sunlight. A detailed description of these results can be found in the works of Polonsky *et al.* (1977, 1977a) and Polonsky and Lisovsky (1980). Generalizing these results, the authors come to the conclusion that the wheat cenosis is able to respond with an increase in total productivity and in grain productivity when the radiant stream measures from 700–1200 W/m^2 PAR, that is when it passes the PAR in noontime sunlight by double or more (Figure 7.6). But this productivity at heightened PAR levels was achieved only by substantially changing the structure of the wheat cenosis itself to significantly increase the plant stand density from 500–800 plants/m^2 to 2000–4000 plants/m^2. As a result, the optical structure of the cenosis changed: the foliage surface area in a unit planting area increased, and the leaf orientation was approximately vertical, enabling efficient light distribution within the cenosis.However, the increase of cenosis productivity experienced when

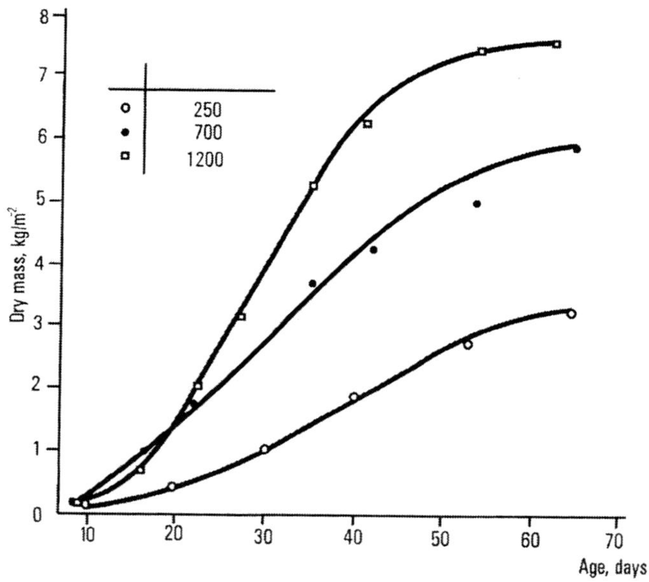

Figure 7.6 The age dependence of accumulation of dry biomass of shoots in wheat crops at various (250–1200 W/m^2) PAR irradiances (Polonsky and Lisovsky, 1980).

the radiant stream intensity was increased incurred an increase in the energy expended on a production unit (Figure 7.7), that is, cenosis energy efficiency was reduced.

An analogous study performed by B. Bugbee and F. Salisbury (1988, 1989) demonstrated that variety selection, temperature optimization, photoperiod optimization, and optimization of other elements of the cultivation regime can better the harvest index, the energy efficiency coefficient, grain formation, CO_2 optimization, and other indicators. But in general, this research confirmed the conclusions of previous authors, which asserted that a change in wheat cenosis structure allows an increase in its productivity when the radiant stream is increased at the cost of decreasing the energy efficiency coefficient of the production process. Analogous data were procured even on a radish cenosis (Tikhomirov *et al.*, 1976).

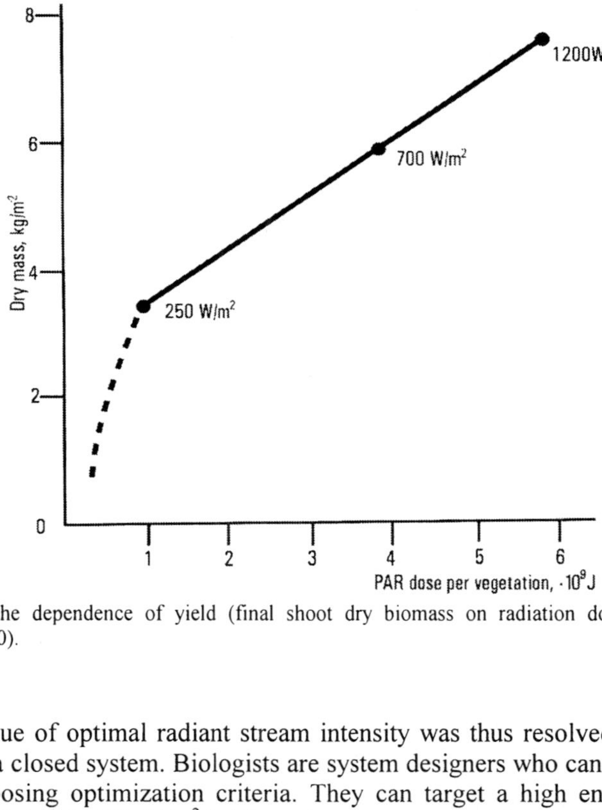

Figure 7.7 The dependence of yield (final shoot dry biomass on radiation dose (Polonsky and Lisovsky, 1980).

The issue of optimal radiant stream intensity was thus resolved for plant cultivation in a closed system. Biologists are system designers who can work from one of two opposing optimization criteria. They can target a high energy efficiency coefficient for 120–200 W/m^2 PAR or they can target a high productivity per unit area. The latter comes at the price of rising energy expenditures per production unit, but provides saving advantages with respect to the growing area size, system volume, equipment materials, and so on. Apparently, the real choice will be something of a compromise between these two types of system. The compromise solution would depend on the technical parameters of future life support systems. Biological parameters necessary in the making of such a decision have already been examined with sufficient precision. Spectral composition of the radiant flux from the Sun outside the Earth's atmosphere is similar to the equal-energy spectrum over the PAR range (from 380 to 710 nm). If solar radiant flux is used to illuminate plants by confining solar rays within the life support system with the help of light conductors, the main task will be to get rid of ultraviolet radiation of wavelength < 380 nm. The effect of ultraviolet radiation on plants gets more adverse at shorter wavelengths, and there seems to be no proof that plants need UV radiation. Elimination of the infrared solar radiation could lessen the load on thermal control systems in the LSS. Experiments with artificial lighting, however, demonstrated (Tikhomirov *et al.*, 1991) that reduction of the IR-radiation fraction in the total radiant flux below 40% (the flux rate in the PAR range being 150 W/m^2) leads to a longer growth period and lower daily average productivity of wheat (Figure 7.8). Similar results were obtained by the same authors in experiments with radishes. As IR-radiation makes up not more than 50% of solar radiation, it would hardly be worthwhile reducing this fraction of the integral radiant flux confined within the LSS.

Correction of solar light spectral composition within the PAR range by eliminating one or another spectral region would reduce the total intensity of the radiant flux confined within the LSS. Besides, there is no strong opinion as yet on the possibility of significantly increasing plant productivity or improving the quality of the produce by correcting the "white" solar light.

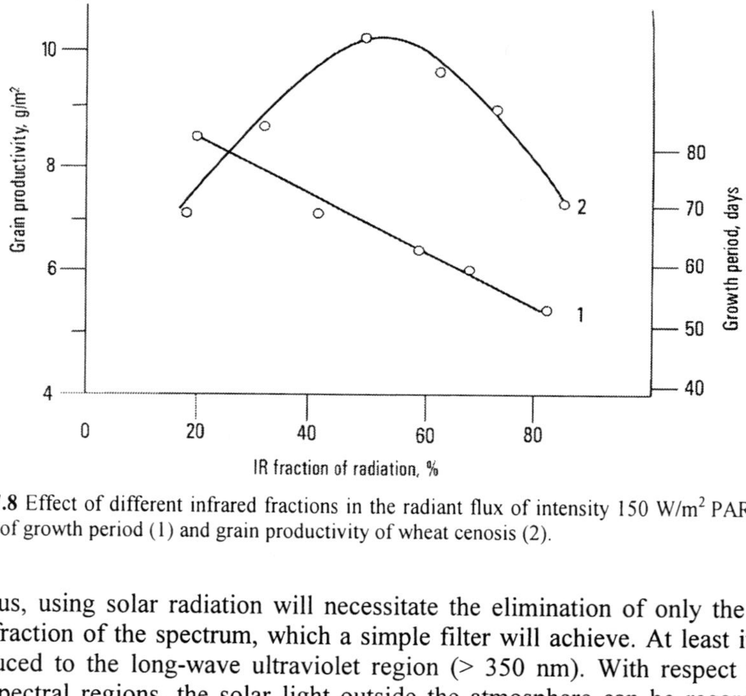

Figure 7.8 Effect of different infrared fractions in the radiant flux of intensity 150 W/m^2 PAR on the duration of growth period (1) and grain productivity of wheat cenosis (2).

Thus, using solar radiation will necessitate the elimination of only the ultraviolet fraction of the spectrum, which a simple filter will achieve. At least it must be reduced to the long-wave ultraviolet region (> 350 nm). With respect to the other spectral regions, the solar light outside the atmosphere can be regarded as quite suitable for the life support system, to be used for the illumination of plants. If artificial light sources are to be used in the life support system, the issue of the radiant flux spectral composition within the PAR range gets more complicated. First, artificial light sources that radiate in certain spectral regions (lithium, gallium–indium, sodium lamps) usually have higher coefficients of transformation of electrical energy into light than sources of "white" light, and this alone can make them more economical. Second, the efficiency at which plants use the light from some spectrum regions, estimated by the real productivity of the cenosis rather than by the true photosynthesis of leaves, varies greatly depending on the radiant flux intensity and plant species (Zolotukhin *et al.*, 1978; Tikhomirov *et al.*, 1991) For example, when wheat (Table 7.2) was illuminated through the growth period only with red light (600–700 nm, mercury–lithium lamps) or only with green light (500–600 nm, mercury–gallium lamps) at 100 W/m^2, the elements of the harvest structure were well developed, and productivity, both total and grain, was normal. However, the wheat grown at the same intensity of radiant flux under blue light (400–500 nm, mercury–indium lamps) differed from the previous variants by the very small number of grains on a plant and a lower mass per 1000 grains. As a result, the total productivity of the wheat grown in this variant was 2.5 times lower and the grain productivity 6 times lower than the productivities of the wheat grown under red and green lights. At the radiant flux of 600 W/m^2, the productivity of the plants illuminated with green and red lights increased 2–2.5-fold, and of those under blue light even more, particularly the grain productivity (8-fold); still the variant with blue light had much worse parameters than the other variants (Zolotukhin *et al.*, 1978).

Table 7.2 Productivity and harvest structure of the wheat cultivar 232 grown in certain regions of PAR at the radiant flux intensity of 100 and 600 W/m^2.

Radiant flux, W/m^2	PAR region, nm	Total number of		Mass of 1000 seeds, g	Growth period, days	Daily average productivity, g/m^2 · day (dry mass)	
		shoots	seeds			Total biomass	Seeds
	400–500	1.0	7.0	15.9	85	7.4	1.2
100	500–600	1.7	20.5	33.0	80	20.6	8.6
	600–700	1.1	20.8	27.0	80	19.3	8.0
	400–500	1.0	27.1	30.8	80	23.3	9.6
600	500–600	3.0	41.7	29.8	75	47.1	16.7
	600–700	3.2	55.3	30.4	75	51.8	22.8

Table 7.3 Total daily average productivity of the radish cultivar Virovskii White grown in certain PAR regions at different radiant fluxes (age: 24 days, g/m^2 · day; dry mass).

PAR region, nm	50 W/m^2	100 W/m^2	200 W/m^2
400–500	8.3	10.4	14.6
500–600	9.2	15.8	12.5
600–700	11.2	5.4	2.1

In experiments with radish, however, the situation was different (Table 7.3). At a low radiant flux (50 W/m^2) the total productivity of radish was the highest in red light and the lowest in blue. At 200 W/m^2 radish productivity in red light fell more than 5-fold, while in the blue light it increased more than 1.5-fold. Hence, the influence of PAR spectral regions on the productive process of wheat and radish proved to be nearly opposite (Lisovsky *et al.*, 1983). Similar experiments with some other crops (cucumber, tomato, maize, etc.) proved that intensively cultivated plants of different species exhibit specific responses to certain spectral regions (Tikhomirov *et al.*, 1987). It should be noted that none of the separate spectrum regions (blue, green, and red light) caused higher productivity than that obtained in "white" light. Nevertheless, various combinations of 2 or 3 components, with one or another radiation prevailing, may be more effective than the equal-energy "white" light with equal fractions of blue, green and red light.

Thus, by optimizing the spectral composition of radiant flux, it must be possible to choose a specific light regime beneficial for the crop at a definite radiant flux. If, however, light conditions are created for the entire higher plant link, white light, analogous to sunlight, is still the most universal. Xenon lamps can be used to produce artificial light of a similar spectrum.

The daily average productivity of plants strongly depends on the time during which they are subjected to photosynthetically active radiation, as the daily dose of radiant energy received by plants is defined as the radiation intensity times the duration of radiation a day. On Earth the period during which the plants are naturally illuminated is determined by the length of the day, which mostly depends on the geographical latitude of locality and the season. Near the equator the day lasts about 12 h, while in the North Polar Region the length of the daylight varies from 0 in winter to 24 h in summer. Outside of Earth, days and nights have more or less similar duration only on Mars, where the length of the day and the angle of gradient between the planet axis and the orbit plane, determining changes of seasons, are almost the same as on Earth. In other extraterrestrial conditions where it may be necessary to use closed ecological systems for long-duration human life support, using solar light (in near-Earth orbits and interplanetary flights, on the Moon, etc.), the Earth day cannot be reproduced.

Each individual LSS may have its own "local day:" e.g. in the low near-Earth orbit it will last only 1.5–2 h, on the Moon—29.5 Earth days, and in an interplanetary flight with definite attitude control the "local day" is endless, one side of the spacecraft being continuously sunlit, the other always dark. So, plants in manmade closed systems can be irradiated for as long as is dictated by the biological peculiarities of the plants and the day can be as long as necessary, irrespective of the length of the Earth day and the portion of the daylight in it.

It is known that by their response to the relative duration of day and night (photoperiods) all plants can be roughly divided into three groups: 1) short-day plants, requiring the alternation of 12–14-h daylight and 10–12-h continuous darkness to come into flower; 2) long-day plants, in which transition to flowering is hindered by daily continuous darkness during 10–12 h; and 3) day-neutral plants, which can come into flower independently of the length of the daylight. Although photoperiodic response is primarily manifested as more rapid or slower plant development, it may also determine the total plant productivity.

It should be remembered that the duration of the light period within a day, together with the radiant flux, determines the active part of the daily energy balance of plants as well as the ratio of the organic matter accumulated in course of photosynthesis during the day to that spent in the course of dark respiration during the night. So in principle it seems advantageous to make the lighting continuous, around the clock, so as to increase the plant yield per unit area per unit time. the situation was different (Table 7.3). At a low radiant flux (50 W/m^2) the total productivity of radish was the highest in red light and the lowest in blue. At 200 W/m^2 radish productivity in red light fell more than 5-fold, while in the blue light it increased more than 1.5-fold. Hence, the influence of PAR spectral regions on the productive process of wheat and radish proved to be nearly opposite (Lisovsky *et al.*, 1983). Similar experiments with some other crops (cucumber, tomato, maize, etc.) proved that intensively cultivated plants of different species exhibit specific responses to certain spectral regions (Tikhomirov *et al.*, 1987). It should be noted that none of the separate spectrum regions (blue, green, and red light) caused higher productivity than that obtained in "white" light. Nevertheless, various combinations of 2 or 3 components, with one or another radiation prevailing, may be more effective than the equal-energy "white" light with equal fractions of blue, green and red light.

Thus, by optimizing the spectral composition of radiant flux, it must be possible to choose a specific light regime beneficial for the crop at a definite radiant flux. If, however, light conditions are created for the entire higher plant link, white light, analogous to sunlight, is still the most universal. Xenon lamps can be used to produce artificial light of a similar spectrum.

The daily average productivity of plants strongly depends on the time during which they are subjected to photosynthetically active radiation, as the daily dose of radiant energy received by plants is defined as the radiation intensity times the duration of radiation a day. On Earth the period during which the plants are naturally illuminated is determined by the length of the day, which mostly depends on the geographical latitude of locality and the season. Near the equator the day lasts about 12 h, while in the North Polar Region the length of the daylight varies from 0 in winter to 24 h in summer. Outside of Earth, days and nights have more or less similar duration only on Mars, where the length of the day and the angle of gradient between the planet axis and the orbit plane, determining changes of seasons, are almost the same as on Earth. In other extraterrestrial conditions where it may be necessary to use closed ecological systems for long-duration human life support, using solar light (in near-Earth orbits and interplanetary flights, on the Moon, etc.), the Earth day cannot be reproduced.

Each individual LSS may have its own "local day:" e.g. in the low near-Earth orbit it will last only 1.5–2 h, on the Moon—29.5 Earth days, and in an interplanetary flight with definite attitude control the "local day" is endless, one side of the spacecraft being continuously sunlit, the other always dark. So, plants in manmade closed systems can be irradiated for as long as is dictated by the biological peculiarities of the plants and the day can be as long as necessary, irrespective of the length of the Earth day and the portion of the daylight in it.

It is known that by their response to the relative duration of day and night (photoperiods) all plants can be roughly divided into three groups: 1) short-day plants, requiring the alternation of 12–14-h daylight and 10–12-h continuous darkness to come into flower; 2) long-day plants, in which transition to flowering is hindered by daily continuous darkness during 10–12 h; and 3) day-neutral plants, which can come into flower independently of the length of the daylight. Although photoperiodic response is primarily manifested as more rapid or slower plant development, it may also determine the total plant productivity.

It should be remembered that the duration of the light period within a day, together with the radiant flux, determines the active part of the daily energy balance of plants as well as the ratio of the organic matter accumulated in course of photosynthesis during the day to that spent in the course of dark respiration during the night. So in principle it seems advantageous to make the lighting continuous, around the clock, so as to increase the plant yield per unit area per unit time.

It was recognized by Harvey in the early 1920s (1922) and, independently, by N. Maksimov (1925) that continuous artificial illumination can be used to grow plants "from seed to seed." N. Maksimov (1925), for instance, used continuous illumination to grow wheat, pea, and other long-day crops. Compared with 12-h illumination, continuous illumination accelerated the development of spring wheat and barley and enhanced the plant productivity in terms of dry biomass. In N. Maksimov's opinion, it could be accounted for by reducing the expenditures for respiration. The data of other authors show that dark respiration intensity is rather high. For instance, R. King and L. Evans report that dark respiration intensity in the wheat with well-developed leaf area is 14–15 mg CO_2/dm^2 planting area per hour while true photosynthesis at an illumination intensity of 2200 foot-candles (85 W/m^2) is not more than 30–35 mg CO_2/dm^2 planting area per hour (King and Evans, 1967). Five-week-old wheat plants illuminated for 12 h daily, the light intensity being 11.5 $c/cm^2 \cdot h$ (135 W/m^2), assimilated 11.8–13.5 mg CO_2 per hour per gram of dry biomass, respiring at night from 1.5 mg CO_2 (at 15°C) to 4.5 mg CO_2 (at 30°C) per hour per gram of biomass, i.e. expended for respiration 21–38% of organics synthesized during the day (Friend, 1964).

Our experiments with spring wheat (cvs Skala and Sonora-64) showed that the formation of a high-production cenosis with biomass exceeding 200–300 g dry matter/m^2 resulted in high levels of dark respiration, to 0.8 l/m^2 of planting area per hour, at a temperature of 20–22°C.

The net photosynthesis of the same cenosis at a light intensity of 100 W/m^2 was 1.6–1.8 l CO_2/m^2 planting area per hour. Simple calculations show that under continuous illumination daily CO_2 assimilation by the cenosis may amount to $1.5\ l \cdot 24\ h = 38.4$ l/day, whereas if the photoperiod is 18 h, assimilation minus dark respiration will be only $1.6\ l \cdot 18\ h - 0.8\ l \cdot 6\ h = 24.0$ l/day.

These calculations were verified in special-purpose experiments. The photosynthesis of two identical wheat cenoses was investigated in the sealed phytotron. One cenosis was continuously illuminated; the other was grown at a photoperiod of 16 h of light and 8 h of darkness. The light intensity in both variants was 80–90 W/m^2, the temperature during the light periods was maintained at 22±2.5°C, where

2.5°C is the single maximum deviation; in the dark periods of the variant with photoperiod the temperature was 16±3°C. Other conditions were the same for both variants.

As can be seen from Figure 7.9, when the plant growth rate was high, the net photosynthesis intensities in the two variants of the experiment were quite close. In the variant with a photoperiod, dark respiration intensity amounted to 30% of the net photosynthesis intensity. As a result, the total CO_2 assimilation by the cenosis (Figure 7.10) in the variant with continuous illumination was 33.8–35.6 l/m^2 of planting area a day, and in the variant with 8 dark hours it was only 19.6–21.0 l/m^2 per day. Thus, compared to continuous illumination, not only did the photoperiod of 16 h force the wheat photosynthetic system to be "idle" during dark hours, but it also caused unnecessary expenditures of previously synthesized substances for dark respiration.

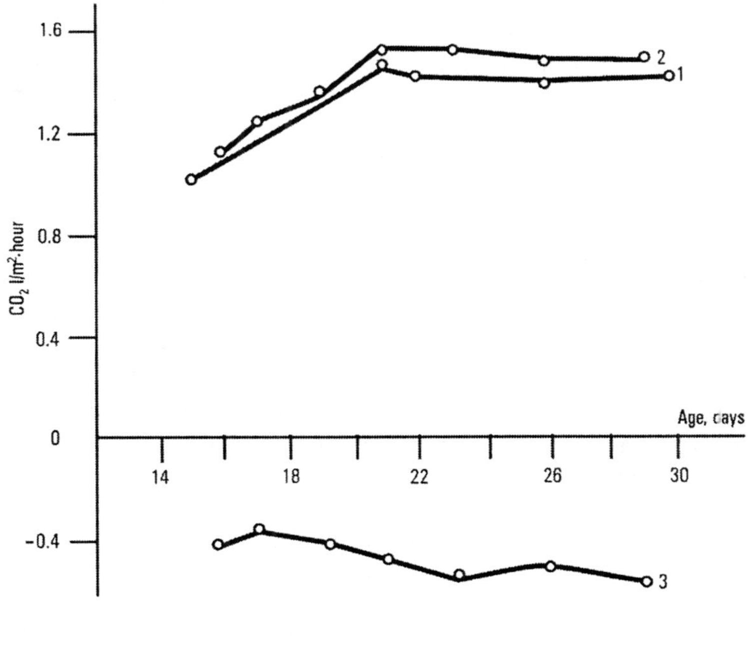

Figure 7.9 Photosynthetic rate of wheat cenosis in daytime. 1—continuous illumination; 2—16-h daylight; 3—intensity of dark respiration of the cenosis grown under a 16L:8D photoperiod.

Continuous illumination accelerates the development of wheat, barley and other long-day cereal crops even at a low radiant flux. This may lead to a disproportion between the reproductive organ developmental rate and the biomass accumulation rate, for the formation of low-production ephemeral plants. Coordination between growth and developmental rates can be reached by choosing a sufficiently high radiant flux.

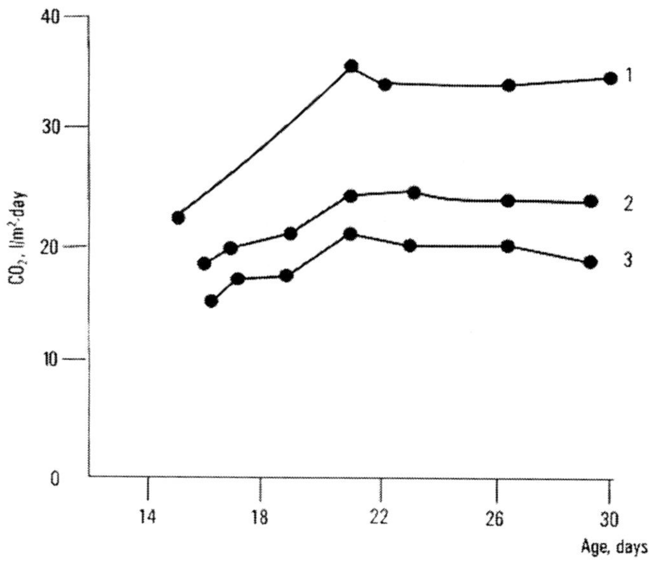

Figure 7.10 Daily CO_2 consumption by a wheat cenosis under continuous lighting and under a 16L:8D photoperiod. 1—continuous lighting; 2—CO_2 consumption by a cenosis for 16 light h; 3—16 h daylight (with dark respiration).

Our numerous experiments with various spring wheat and barley cultivars demonstrated that continuous illumination, together with a sufficiently high intensity of radiant flux (100–140 W/m^2), is perfectly suitable for growing these crops. Though plants developed at an accelerated rate and the productive shoots grew densely, the number of grains per ear was rather large and the weight of 1000 grains was normal, i.e. plants were by no means ephemeral. Biological productivity of the cenosis (as dry biomass) was rather high: 30–50 g /m^2 · day, including grain: 10–17 g/m^2 · day. For the tested early-maturing cultivars such productivity is indicative of the intensive growth of the plants under these light conditions.

These results suggest a conclusion that continuous illumination at a sufficiently high PAR intensity (80–140 W/m^2) is quite an adequate condition for intensive cultivation of spring wheat and barley.

Further enhancement of the radiant flux may be accompanied by an increase in cenosis productivity but at the cost of reducing energy efficiency (Polonsky *et al.*, 1977, 1977a; Bugbee and Salisbury, 1988). That is why what we are discussing here is not the optimal wheat or barley cultivation regime in terms of cenosis potentialities but rather the regime suited to solve the problems arising in the creation and experimental investigation of closed ecological systems.

An important parameter for assessing the suitability of one or another regime for the cultivation of plants reproduced within the closed system and not from the stored seeds of the field harvest must be the hereditability of cultivar qualities in the successive generations. For several years we worked with the wheat cultivars Skala and Sonora-64, which had 6 to 10 generations under the conditions of continuous intensive artificial illumination, and have not recorded any cultivar depression. Grown under similar light conditions, they had steady productivity, morphological and biological characteristics.

Hence, continuous illumination of wheat, together with a sufficiently high radiant flux to maintain active photosynthesis, ensures a high total productivity and harvest index of the cenosis as well as the preservation of the cultivar genotype in successive generations. Therefore, the investigated light conditions can be regarded as useable for the creation and investigation of closed ecological systems including wheat (or barley) as the component of the higher plant link.

Whereas accelerated development of wheat under continuous illumination essentially conforms with the aims of cultivation, it is undesirable, according to many authors, to grow another long-day crop, radish, not only under continuous illumination but also with a long photoperiod since under such light conditions plants pass to reproductive phases early, and often roots have no time to form.

On the other hand, B.S. Moshkov (1966) showed that continuous illumination with a sufficiently high radiant flux intensity can retard development of radish (cv. Pink-with-a-White-Tip) and effect the formation of roots weighing up to 30 g 12–14 days after emergence of plants.

Continuous photosynthesis of radish (and any other crop) is advisable in a closed system for the same reasons as were stated for wheat. It should also be noted that for practical, economic, purposes it is more desirable to cultivate radish intensively, under continuous artificial illumination as it is a more rational way of using the planting area and equipment than, e.g., under 14–18-h illumination

Our investigations aimed at choosing the light regime for growing radish in closed systems (Lisovsky and Shilenko, 1970) demonstrated that this crop can be efficiently grown under continuous illumination. In this case the choice of radish cultivar determines the yield. Of the many cultivars tested, some medium- and late-ripe cultivars, such as Duganskii, Virovskii White, effectively utilize radiant energy under continuous illumination, forming roots within 22–24 days with the yield of 4–6 kg of roots per 1 m^2 and biological productivity to 25–30 g/m^2 of dry biomass.

Our investigations show that continuous illumination can be effectively used in growing many other crops, such as pea, carrot, table and sugar beets, chufa, turnip, cucumber, onion, sorrel, etc. Some of these crops after testing were included in the structure of the higher plant link. Under continuous illumination, such a multispecies link coped with all tasks of atmosphere, water and vegetable food regeneration. With some crops, though (potato, tomato, soybean, rice, etc.), we failed to achieve satisfactory results by growing them under continuous illumination. If it is considered worthwhile to include these crops in the closed system, they will require a separate "ecological niche" with a 12–16-h light period and an 8–12-h dark period.

It should be emphasized that plants exhibit broad plasticity for light conditions if they deviate from the above-described variants in case of emergency or as a result of purposeful changes. Take for example plants growing well under continuous illumination. If, in the event of an emergency illumination is discontinued for intervals from 1–2 h to 1–2 days, plant photosynthesis will be arrested for this period and the development of the plants will be somewhat slower, but no irreversible changes will occur. However, longer (3 and more days) periods of continuous darkness may have detrimental effects on plants, particularly on the reproduction. To neutralize such effects, special measures should be taken, e.g. the temperature of the environment can be decreased. For short-day crops to grow and develop well, regular alternations of day (12–14 h) and night (10–12 h) are needed, depending on the crop. In short-day crops illuminated continuously for many days, the following changes are recorded: leaves turn yellow and ovaries fall in tomato; seed formation in maize and soybean as well as tuber formation in potato start later; etc. Our attempts to grow the short-day potato together with long-day crops under continuous illumination in the Bios-3 experimental complex have been unsuccessful, the edible yield being very low (Zamknutaya sistema…, 1979).

On the whole, the tolerance of plants for various deviations of light conditions from natural rhythms is rather high. To confirm this statement, we can present the results obtained in a series of experiments by plant physiologists. The growth of the cabbage *Brassica abissinica* under 30-, 60- and 120-min light/darkness alternation (Moshkov, 1973) demonstrated that plants can grow and come into bud even under such conditions, but their developmental rates slow down against those obtained under continuous illumination.

In Prof. V.P. Dadykin's experiments (1976) radish and kale were grown throughout the life cycle under "the photoperiod of a satellite in low orbit" (60 min L:30 min D). The results were essentially the same as when the crops received the same quantity of radiant energy but were grown under a usual Earth 16L:8D photoperiod. In the experiments with our colleagues we developed procedures for growing a number of crops (wheat, barley, beet, carrot, etc.) under the "lunar" photoperiod with alternating 15-day continuous light and continuous darkness (Mizrakh *et al.*, 1973; Terskov *et al.*, 1978).

No plants can be expected to grow and develop under the "lunar" photoperiod unless special conditions are created to keep them alive during the long dark period of the lunar night. It has been shown that the lowering of the air temperature to 3°C in the long dark period can significantly slow down biological processes in many plant species, causing no damage to them. Plants in the state of hypobiosis can stay without light for a long time. For instance, experiments with spring wheat grown under the "lunar" photoperiod demonstrated that in spite of some decrease in the pigment and carbohydrate content of leaves after the "lunar" night, the processes of growth and development were actually not disrupted. The growth period from germination to maturity (taking into account only light period) was the same as for plants grown under continuous illumination.

The overall length of the growth period was larger only due to the length of the lunar night. The life cycle of the spring wheat grown under continuous illumination (control) was 60 days, and the life cycle of the wheat grown under the lunar photoperiod was 4 lunar days (15 days × 4) and 3 lunar nights between them (15 days × 3), i.e. 105 days. Grain of the wheat grown under such conditions can be well used as food and seed material. It should be noted that wheat plants were more sensitive to prolonged darkness at the early stages of development, during the first and second "lunar" nights: elevation of temperature even for a brief time during the first "lunar" night resulted in significant disruption in the formation of reproductive organs.

Besides wheat, we attempted to grow some cold-resistant and heat-requiring crops under the "lunar" photoperiod. Experiments with heat-requiring crops (tomato, bean) have given no satisfactory results yet. Cold-resistant crops (barley, carrot, beet, turnip, radish, dill) can be successfully grown under the "lunar" photoperiod from germination to harvest maturity, producing the edible biomass comparable in nutritional quality to the produce of plants grown under continuous illumination (Lisovsky *et al.*, 1979).

These examples would suffice to prove that many higher plants are potentially adaptable to changes in the relative duration of day and night. We can assert that the cultivation of plants in artificial closed ecological systems will not be gravely impaired by various emergencies, when the light suddenly goes out or, vice versa, when it does not go out at a proper time. We can also state with assurance that "orbital" and "lunar" photoperiods will be quite appropriate for plant growth and development when natural sunlight can be used.

7.4 CONTINUOUS CULTIVATION OF HIGHER PLANTS

In the course of vegetation the photosynthetic rate varies, which is inconsistent with the relatively stable gas exchange of the main link of the closed system—humans. For this reason alone, traditional even-aged ("synchronous") culture over the whole planting area should be rejected, if it is to perform gas exchange functions in the closed system. Besides, in most crops the greater part of edible biomass is formed at the end of the growth period, and the food will have to be stored "from harvest to harvest." The time interval between two labor-consuming operations, harvesting and planting, will be too long and temporal distribution of the human workload will be uneven.

These difficulties can be overcome in the closed system by creating a continuous (conveyer) culture represented by plants of several age groups simultaneously. Practicability of higher plant continuous cultivation has been studied since the 1960s both for the goals of industrial hydroponics and for closed systems (Dadykin *et al.*, 1967; Chuchkin, 1967; Lisovsky *et al.*, 1969).

The main functions of continuous plant culture in the closed system are 1) to maintain the total gas exchange of the phototrophic link at a relatively steady level to effect direct gas exchange with humans; 2) to ensure uniform temporal distribution of energy and labor expenditures for maintaining functioning of the link; 3) to regularly supply humans in the system with fresh vegetable produce. (For cereal products this requirement is not obligatory.)

The optimal number of simultaneously present crop ages, determining the age interval of the conveyer, depends on the peculiarities of the crop, demands on the rhythm and amount of labor expenditures, harvesting practice, storing and processing of the grown produce, permissible limits of the variations in the composition of the gaseous environments in the system, buffer properties of the system determined by its gas volume, etc. Organizational and technological principles directing the choice of the conveyer age interval may change depending on the nature of the processes and the degree to which they are evolved as well as on the chosen variant of arranging the system as a whole, i.e. on the conditions that cannot be accurately defined now. Therefore, it is premature to decide what number of age groups will be optimal in the continuous culture of the higher phototrophs for one or another variant of the system.

Even now, though, basing our conclusions on the available experimental data, we can specify the gas exchange characteristics of the multi-age continuous culture, depending on the number of age groups. Knowing such characteristics for different higher plant species, one can calculate the dynamics of the atmospheric composition in life support systems with any area occupied by the crops studied, if the temporal gas-exchange characteristics of other gas exchange links of the life support system, e.g. crew, are also known.

Variations in the composition of the system's gaseous environment are primarily limited by changes in the carbon dioxide concentration of the gaseous phase, since nitrogen or inert gas concentrations are presumed to be constant, and humans tolerate much wider variations in the oxygen concentration than variations in carbon dioxide concentration. Of course the two concentrations are strongly related to each other. That is why the calculations that follow concern only the amount of atmospheric carbon dioxide in the system. To find the minimum number of conveyer plant age groups needed to keep variations in the CO_2 concentration within the bounds permissible for humans, we can take one crop (e.g. wheat) as a higher plant link and a human as its metabolic counterbalance in the system. It is a requisite to system closure that the daily CO_2 assimilation by the even-aged wheat cenosis averaged over the growth period should be equal to the daily CO_2 production by the human, which can be regarded as nearly uniform. However, at different stages of the growth period the actual CO_2 consumption by an even-aged ($n = 1$) wheat cenosis will differ from the required daily average consumption (Figure 7.11, $n = 1$). At the beginning and end of the growth period it will be lower than the required daily average, while in the middle, in the phase of intensive growth, actual daily CO_2 consumption by the cenosis will be higher than the daily average. Let us denote daily deviations in one or the other direction by $+\alpha$ and $-\alpha$. Then the overall deviation of the CO_2 amount consumed by plants in the system in the phase of intensive photosynthesis will be $\sum (+\alpha)$ and the amount of CO_2 deficiency for plants in the phase of intensive photosynthesis will be $\sum (-\alpha)$. Absolute values of these deviations are equal, making ±6584 liters for $n = 1$ (Figure 7.11), when calculated graphically. It is clear that even in the system with 100 m^3 of atmosphere for one human such changes in the atmospheric CO_2 volume content will constitute ±6.58%, which is absolutely impermissible.

With two plant age groups in the cenosis ($n = 2$) "gaps" in the photosynthesis of one age group will be partially compensated for by an increased photosynthetic rate of the other age group (Figure 7.11, $n = 2$), and the magnitude of deviations $\sum (+\alpha)$ and $\sum (-\alpha)$ will not exceed ±888 l. However, such deviations, even in a system with a large atmospheric volume, are also intolerable for humans. About the same is true for the version with $n = 4$. Of the versions under consideration, only in the system containing 8 age groups these deviations will not exceed ±11 liters (Figure 7.11, n = 8), which is quite admissible for a system with even a small (10–30 m^3) atmospheric volume for one human.

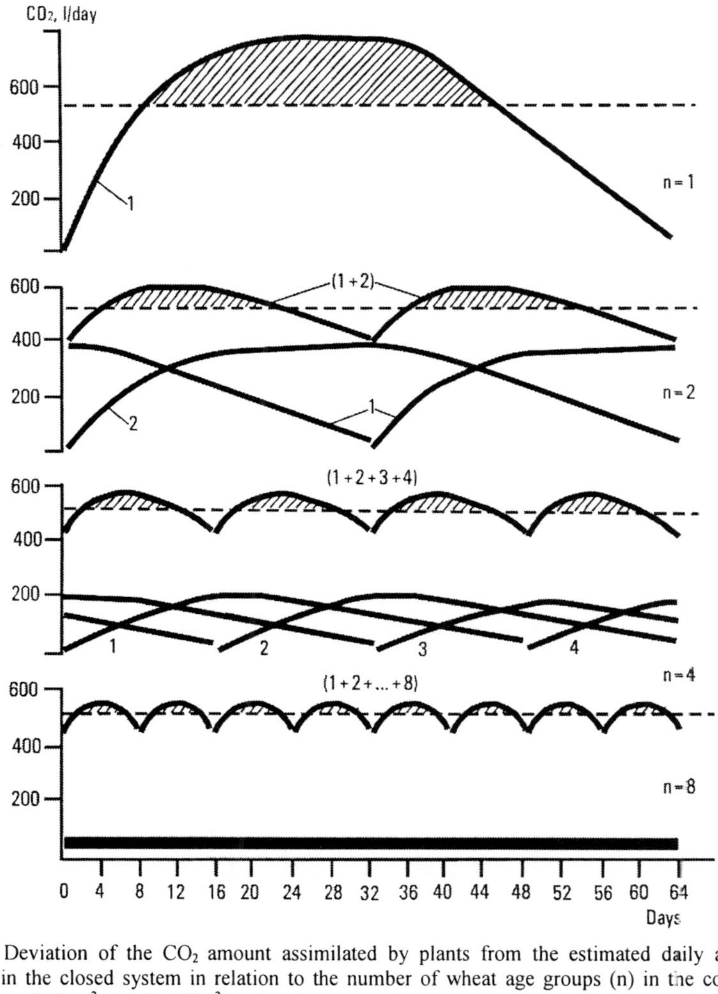

Figure 7.11 Deviation of the CO_2 amount assimilated by plants from the estimated daily average assimilation in the closed system in relation to the number of wheat age groups (n) in the conveyer (planting area of 20 m^2, 88–96 W/m^2 PAR). The dotted line is daily average CO_2 production by the human and assimilation by plants (525 l/day). 1,2,3... are individual assimilation curves for each wheat age group in the conveyer; (1+2), (1+2+3+4), (1+2+...+8) are net assimilation curves for all age groups in uneven-aged cenoses.

Values Σ $(+\alpha)$ and Σ $(-\alpha)$ depend not only on the chosen number of plant age groups in the conveyer but also on peculiarities of the photosynthesis age curve for each crop species and cultivar, as well as on peculiarities of the cultivation conditions. For example, Figure 7.12 (n = 1) shows daily deviations of actual CO_2 consumption by the even-aged radish cenosis. As the number of age groups in the radish conveyer gradually increases (Figure 7.12; n = 2, n = 4, n = 8), these deviations are smoothed out. It is safe to say that a conveyer of at least eight age groups of any crop can maintain essentially uniform CO_2 levels acceptable for humans and plant photosynthesis in the closed system.

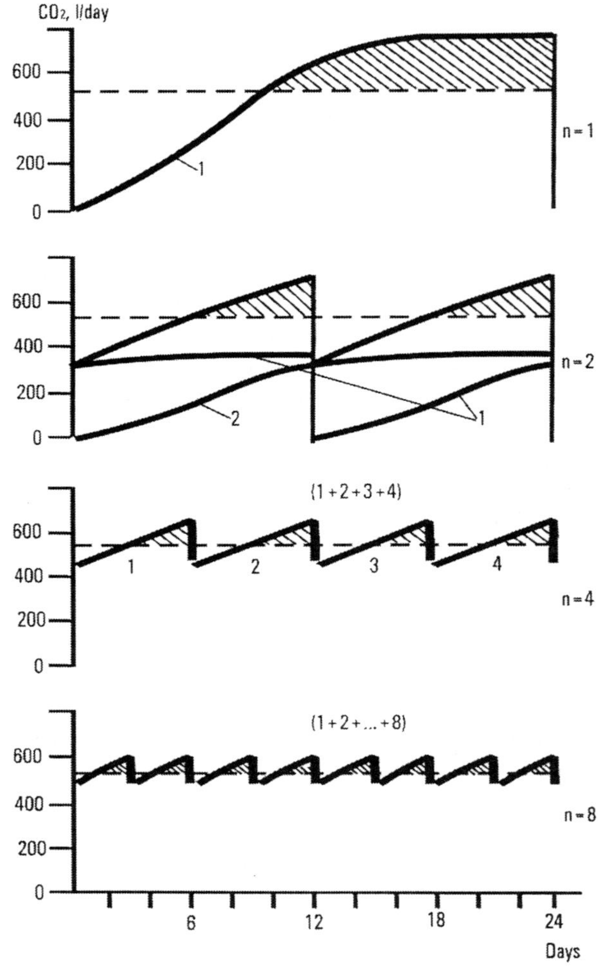

Figure 7.12 The same as in Figure 7.11, but for radish cenoses.

Each experimentally obtained age curve of photosynthesis for different crops and different cultivation modes can be empirically fitted with the equation describing it, which will allow computation of the gas exchange of conveyer plant cultures in the system. Such a computation was done by B. Kovrov and G. Lisovsky (1972) for the above age curves of wheat and radish photosynthesis; the result, to be sure, was the same.

When the closed system includes several conveyer plant cultures simultaneously, the integrated variation in carbon dioxide concentration in the system will be intermediate between variations caused by separate cultures. Thus, the use of higher plant continuous cultures for air regeneration can stabilize the rate of carbon dioxide uptake from the atmosphere of the life support system even if few (6 to 8, depending on the crop) different age groups are planted.

Continuity of the culture with several plant age groups grown simultaneously can be attained through a kind of age "crop rotation," when each individual plot has its own dates of planting and harvesting, different from the other plots of this crop. This cultivation mode is good for growing plants both on substrate (soil cultures, hydroponics on expanded mineral, polymer granules) and without it (water cultures, aeroponics).

The other way of attaining culture continuity is to create the plant conveyer proper, in which plants entering the next age group are transferred in a certain order, by maturity reaching the end of the conveyer, where they are harvested. Simultaneously with the harvest, new plantings are made at the start of the conveyer.

The second way is primarily applicable in cultivating plants without substrate. Though more complicated technically, it shows more promise than the first, allowing for a better use of the light area by changing the distance between plants when transferring them.

Both procedures of continuous cultivation have been intensively developed and investigated mostly for vegetable crops (Dadykin *et al.*, 1967; Chuchkin, 1967; Dadykin, 1968; Lebedeva *et al.*, 1969).

The method and procedure for the cultivation of wheat and other cereal crops had some specific features due to their biological and economic peculiarities, such as the absence of secondary stem thickening, capability for long tillering, high crop stand density in a highly productive cenosis, etc. We presumed that continuous cultivation of cereal crops would be most rationally effected in air subirrigated culture. The idea of such culture was suggested by V. Artsikhovsky (1915). The essence of the idea is that the roots of somehow fixed plants hang freely in the air and are periodically soaked in nutrient solution, which is the source of regular mineral nutrition and water for them. Unfortunately, this idea remained undeveloped.

We have evolved the following method of cereal continuous cultivation in the air subirrigated culture. Seeds of wheat (barley, naked oak) are preliminarily germinated in darkness, in a moist chamber, on filter paper. Then they are placed into sockets made in special plastic plates, where they are kept firmly through the life cycle. Along the underside of the plates there are slits for the roots. Socket and slit dimensions as well as plate height have to fit the following requirements: 1) seedlings must not fall out of the sockets; 2) embryo and nodal roots must grow freely; and 3) adult plants must be held in vertical position. A planting plate with plants is shown in Figure 7.13.

After the germinated seeds are placed into planting plates, they are held for about 2 days in a dark moist chamber until coleoptiles of seedlings are high enough (in the mean, as high as the top edge of the planting plate) and embryo roots emerge from lateral slits. Then the plates with seedlings are taken to light, in the phytotron. This is the starting point of the plant growth period.

To use the phytotron light area more effectively, the planting plates are first positioned at short intervals, being fitted into slots of special stands. As the plants grow, the plates are transferred, and the distance between them increases. The position may be changed many times until stem elongation. After that the nutrition area of the plant remains constant until maturity. The planting density was 1600–3200 plants/m^2 during the first days of their life cycle and 800–1100 plants/m^2 from stem elongation to maturity.

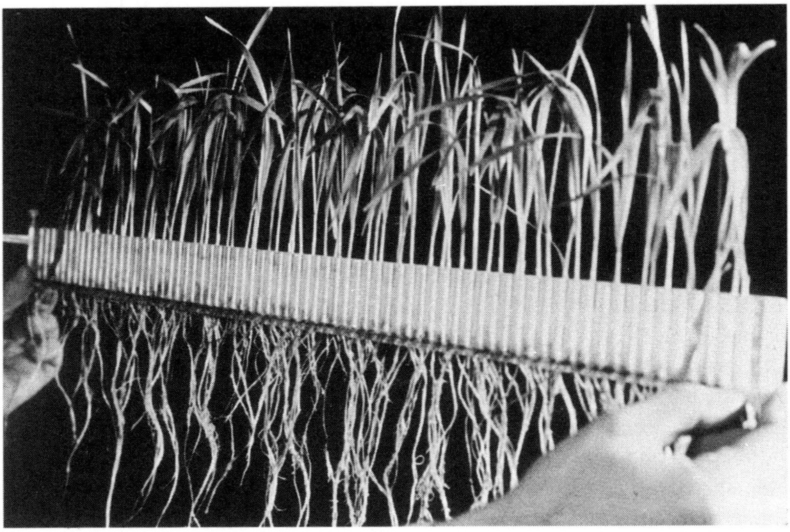

Figure 7.13 Sowing plate with wheat plants.

We chose a greater planting density than in the field based on the results of our experiments with wheat grown hydroponically. These results are in good agreement with the evidence that if the plants are fully supplied with nutrients and light, the greater crop density leads to greater crop productivity.

The roots of the plants fixed in the sockets of the planting plates, grew through slits made in plates, hung loosely and in the later phases of growth spread over the phytotron bottom. Experiments show that to avoid the interlacement of roots that could hamper the rearrangement of planting plates, the plates need not be fixed higher than 40–60 mm from the phytotron bottom. Then the most active part of the roots is spread over the phytotron bottom in a thin layer retaining some nutrient solution, so intervals between the supplies of nutrient solution can last for up to half an hour. Only the youngest plants aged 1 to 9 days, with few short roots, may need additional trays retaining some nutrient medium immediately under the planting plates.

To illustrate, in one of the experiments the nutrient medium was supplied to the roots every 27 min. The phytotron bottom was divided into 12 equal-sized sections. Every section was connected with a rubber hose to one of the 12 conduits of the hydraulic control valve linked to the pump working continuously. It took about 2.3 min to pump nutrient solution into each section. The same time was required for the medium to gravitate into the pump. The interval between supplies of nutrient solution lasted for about 22 min. Following this procedure, we did not need any tank for the nutrient medium.

The quantity of the required nutrient medium was only 12–15 l for each square meter of the wheat planting area.

In our experiments a unit load on the nutrient medium was 80 wheat plants per 1 liter medium. We calculated that intensive plant biomass growth usual for the phytotron must result in the complete removal of some elements from the nutrient medium within a day or two. Unwanted changes in the pH level may occur even sooner as the plants preferentially consume anions. Therefore, the nutrient medium must be often (up to 4 times during 24 h) corrected by adding the calculated amounts of salts and acids (see the next section).

Plants in the continuous air subirrigated culture intensively developed their leaf area, had the cultivar-specific number of embryo roots, and quickly and actively formed the nodal root system. Acceleration of plant developmental rates induced by continuous illumination did not affect the formation of normal plant habitus. Ripe wheat plants had sufficient numbers of shoots (total and productive) though the crop stand was rather dense (1000–1100 plants/m^2). As a result, the harvested yield amounted to 1380 ears/m^2. The mass of 1000 grains was typical of the tested wheat cultivar. The productivity of the wheat in continuous culture was 43.8 g/$m^2 \cdot$ day of ovendry biomass, including 11.2 g of grains, i.e. it was not lower than that of the same wheat cultivar grown hydroponically under the same illumination conditions.

More than ten different experiments lasting from two months to a year demonstrated that the described cultivation method and procedure are acceptable and reliable for growing wheat and barley in phytotrons.

Continuous air subirrigated culture of cereals in the phytotron has certain advantages over both substrate-based hydroponics and aeroponics.

Compared to hydroponics based on expanded clay aggregate, pearlite or any other filler, continuous air subirrigated culture uses the phytotron light area more effectively as the young plants can be placed closer to each other. It has been calculated that to grow on expanded clay aggregate with the same number of plants as in the experiment described above, the phytotron light area would have to be 16.7% larger. In the experiments where not only young but also ripening plants were closely spaced, 33.3% of the light area was saved. As plants are not fixed on constant nutrition areas, during the early stages of their life cycle the distance between them can be altered depending on age changes in their habitus. Thus the cenosis leaf area index influencing light utilization can be significantly improved. In this situation, during the early stages of the plant life cycle the amount of biomass per unit light area was much larger and the cenosis leaf area index much higher than when the plant nutrition area did not change.

The mass of planting plates and stands supporting them is about 10 kg for 1 m^2 of wheat planting area, which is about 5 times less than the weight of the expanded clay aggregate for the same planting area. All equipment can be reused.

Air culture is very convenient for the harvesting and measuring of ripe plant roots, whereas it is a tedious procedure to extract all roots from various solid substrates in hydroponics.

A considerable merit of the described plant cultivation method for application in experimental systems is that materials used to fix plants can hardly absorb or desorb any substances, such as nutrients or toxicants. Besides, investigations of mass exchange processes in the link can be performed much more easily and the results will be much more accurate than in any solid substrate, when different fillers with extended surface are used.

Unlike the various methods of spraying freely hanging roots with nutrient solution (aeroponics) the method we have described does not require any intricate pipe system to supply solution to each plant (which is particularly difficult to attain in cultures of high planting density) or numerous sprayers with sophisticated and troublesome mechanisms, etc.

Roots issuing from the planting plates and spreading over the phytotron bottom can receive incident light. This can result in algal growth on them, and as the nutrient medium is not changed in the water-closed system, the algae can be a source of metabolites harmful for higher plants. Algal growth on roots can be suppressed by darkening the gaps between the planting plates and by making the plates from opaque material.

Successful use of both continuous and usual even-aged air subirrigated wheat cultures in many different experiments suggests that this method and the cultivation procedure could be employed to good advantage not only in creating closed systems, but also in experimental investigations in plant physiology, agrochemistry, etc.

This method, however, can be used only under the conditions of natural gravity, e.g. on the Moon, Mars, or with sufficient artificial gravitation aboard spacecraft. For microgravity conditions experiments are still conducted with plants grown on porous or fibrous substrates, which can hardly be used over a long period because they get fouled with roots and contaminated with metabolites. Methods and cultivation procedures are proposed to grow plants in microgravity by creating artificial gravitation only within the cultivation installation, for instance, by employing aeroponics on a spinning disk with plants, by using pipes with porous filler through which nutrient medium is pumped (Dreschel and Sager, 1989), etc. None of these methods has been proven in real microgravity. It is hoped that these, mostly technical, difficulties should be overcome by the time when higher plants will be grown in the spaceflight microgravity environment.

7.5 GAS AND WATER EXCHANGE, NUTRIENT MEDIA FOR PLANTS

Direct gas exchange between plants and humans necessitates creating such conditions for plants that will be acceptable for humans. Thanks to evolution, the principal components of human respiration and plant photosynthesis, O_2 and CO_2, are contained in Earth's atmosphere in the quantities equally optimal for humans and plants. At normal atmospheric pressure, the usual O_2 volume concentration, about 21%, can be regarded as suitable for humans and plants alike. Elevation of O_2 concentration to 24–26% decreases apparent photosynthesis and increases dark respiration in many plants, prospective components of a closed system (Nilovskaya, 1973). Likewise, plants are adversely affected by low (1.5–4.0%) atmospheric O_2 concentration. In the experiments of Gerband and Andre wheat grown at 4% of atmospheric O_2 started earing 10 days later than the control and no grains set in the ears.

The normal atmospheric CO_2 concentration, about 0.03%, is quite sufficient for plants although this level is not saturating for photosynthesis. Elevation of CO_2 concentration to 0.1–0.3%, and sometimes to 1.0%, somewhat increases photosynthesis (Nilovskaya, 1973; Salisbury *et al.*, 1995) but further elevation starts suppressing photosynthesis. The CO_2 concentration of 0.01–0.008% can be considered the lower limit for apparent photosynthesis. At optimal illumination and air temperature the CO_2 concentration saturating photosynthesis is in the range from 0.15–0.30 to 1.0%. CO_2 concentration outside this range usually decreases photosynthesis intensity, but this is a reversible decrease. In most physicians' opinion, humans can stay and work in an atmosphere containing the amount of CO_2 that saturates plant photosynthesis.

The optimal water–vapor content of the atmosphere, often expressed as relative air humidity, is different for different plant species. For instance, according to standards accepted in hothouse vegetable-growing, tomato requires rather low relative air humidity, 45–70%, while the optimal value for cucumber is 85–95%. Investigations of S. Mizrakh *et al.* (1969) demonstrated that photosynthesis intensity in spring wheat remained approximately the same at 40 to 70% but decreased at 85%. In the closed human life support system, even with direct gas exchange, human habitat and plant growth chambers are separated from each other, so air humidity can be easily optimized in each section. However, we see little reason for optimizing air humidity for every plant species in the multi-species plant community in experimental closed systems. In our long-duration experiments (Gitelson *et al.*, 1975; Zamknutaya sistema..., 1979) a relative humidity of 65–75% maintained in the phytotrons, together with sufficient moisture supply to roots, made for quite satisfactory productivity of all plants within the system.

* * *

To optimize water supply to plants artificially cultivated by different methods, a particular frequency of nutrient medium delivery (soaking, spraying of roots, etc.) is selected. This frequency is determined by species and age peculiarities of the culture, the nature of substrate in the root zone, temperature, humidity, etc. Continuous contact of plant roots with the nutrient solution leads to root damage, so too frequent solution supplies would be undesirable.

It has been found empirically that even during the heat of summer in open hydroponics (expanded clay aggregate, vermiculite) plants do not need more than 3–4 supplies of nutrient solution a day (Bently, 1965; Davtyan, 1967).

Under the conditions of continuous artificial illumination it was quite sufficient to supply nutrient solution to various crops grown hydroponically on expanded clay aggregate twice a day in the early age and at the ripening phase for cereals, and 4 times a day at the phase of active growth.

Clearly, continuous air subirrigated culture must be supplied with water more frequently. It has been recorded that to attain an average dry biomass increment of 50 g/m^2 · day, 0.3–0.4 l of water an h is used. The mass of plant roots per m^2 of the planting area is relatively small. Water cannot be stored on them for several hours. Uneven distribution of dropping liquid water over the surface of roots can cause their drying in some zones, which is particularly harmful for young plants. There is also a danger that the salt concentration on the root surface can increase if the liquid water is completely absorbed from it. That is why to fully satisfy plant requirements for water and to prevent the roots from drying, the nutrient solution should be fed to the air wheat culture 2–3 times per h.

The amount of transpirational moisture from plants and water partially vaporized from the substrate surface is easy to measure in the tightly sealed phytotron. Repeated measurements show that under the conditions described above, continuous wheat culture produces 6.2–7.1 l/m^2 · day of condensate in an illuminated area. Approximately the same quantity of condensate was produced in our experiments by a multi-component uneven-aged cenosis of vegetable crops (tomato, radish, cucumber, table beet, carrot, etc.) grown hydroponically under an 18L:6D photoperiod. Hence it follows that the daily requirement of one person for drinking and hygiene water can be quantitatively provided by higher plants occupying 2–3 m^2. R.I. Kuzmina analyzed the quality of the wheat transpirational moisture condensate obtained in our experiments. Low salt and organic contents, pH and other characteristics, except the total nitrogen content, fell within the accepted standards. The total nitrogen was mostly in the form of ammonia. After refinement and mineralization the condensate can be safely used as drinking water, and as hygiene water after simple boiling.

To satisfy human requirements for oxygen and food, the planting area must be several times larger than that needed to satisfy drinking and hygiene water requirements. Thus, in a system with full gas exchange between humans and higher plants, there will be surplus condensate of transpirational moisture, useless for humans, that must return to the nutrient solution. A simple calculation shows that with plant transpirational water amounting to 6–8 l/m^2 · day and with the process necessary to have 12–15 l of nutrient solution for 1 m^2 of air subirrigated culture, the water inside the phytotron must pass through a complete cycle every two days. We conducted several experiments in tightly sealed phytotrons in which over a period of 4 to 6 months the condensate of transpirational moisture from unispecies and multispecies cenoses was used as a source of water for cultivation. In all experiments quite satisfactory results were obtained.

* * *

In a closed system the basic problems of supplying plants with mineral nutrients and water stem from the condition of nutrient medium unchangeability. The great diversity of nutrient mediums proposed by different researchers for water, gravel and air cultures of plants, as G.S. Davtyan pointed out, is far beyond our current scientific comprehension (Davtyan, 1967). In most cases researchers sought to attain long-term use of such media at steady pH levels, etc., without correction (Hewitt, 1952; Zhurbitsky, 1968). In intensive plant culture growing on nutrient media, the volume of which is limited in closed systems, correction of medium composition is inevitable and must be done more often than in some research experiments or industrial hydroponics. In our experiments with continuous wheat culture there were only 12–15 l/m^2 of nutrient medium of the cenosis area. An increment of 50 g/m$^2 \cdot$ day of ovendry biomass was accompanied by active uptake of mineral elements. As can be seen from Table 7.4, the relationship between plant productivity and the volume of nutrient medium allowing for this productivity being what it was, most elements, e.g. from the Knop medium, were to be completely consumed within 1.7–3.1 days. Unwanted changes in medium pH occurred even earlier, due to an imbalance between cation and anion uptakes.

Table 7.4 Rate of uptake of mineral elements by plants in continuous culture.

Element	Content in daily dry biomass increment (50 g)	Content in 15 l of Knop solution	Rated time of complete element uptake, days
Potassium	1.52	2.52	1.66
Calcium	0.29	3.67	12.70
Magnesium	0.12	0.37	3.07
Phosphorus	0.34	0.85	2.51
Sulfur	0.17	0.49	2.87
Nitrogen	1.05	3.08	3.00

Most of the other standard media contain smaller amounts of mineral nutrients and, therefore, will be depleted even sooner.

As plant requirements for mineral nutrients vary with their developmental stage, some authors suggest altering the medium composition, depending on plant age (Zhurbitsky, 1963, 1965; Aliev, 1966; and others). For continuous culture, however, in which many age groups are present simultaneously, it would be an extremely complicated job to introduce a separate medium for every age group and correct each of them. Thus, first of all we considered the feasibility of growing continuous plant culture on the medium of a more or less constant composition, often replenished with proper additions.

Given the stabilization of all growth parameters, the amounts of elements added to the nutrient medium for higher plant continuous culture can be determined by their amounts recorded in biomass. Although plants of each age consume different elements at a different intensity, the integral parameter of element uptake by all ages in unit time will be the elemental composition of the harvested yield produced in the same time.

Hence, we can calculate the correction of nutrient media for continuous higher plant culture by the same method as for continuous algal culture (see Chapter 6):

$$m = C_b P + CV,$$

where m is the daily dose of the element in the corrective addition, mg; C_b is the element concentration of the harvested biomass, mg/g; C is the element concentration of the nutrient medium, mg/l; P is the ovendry yield, g/day; V is the volume of nutrient medium, l/day, removed from the phytotron (samples for analyses, technical losses, etc.). If $V = 0$, then the corrective addition of each element is equal to the removal of this element with the yield: $m = C_b P$.

In this case, to determine the corrective dose of any element added to the nutrient medium per unit time is to measure the amount of the element removed with crop yield produced in the same time.

How often the corrective addition must be introduced will be determined by the permissible deviation of the concentration of the most quickly absorbed element from the selected level. Under intensive subirrigated cultivation of wheat, when, for instance, potassium is completely taken up from the Knop medium for 1.66 days, the nutrient solution has to be corrected every 6–10 h.

Optimal levels of trace elements in the nutrient medium can be maintained in a similar way. It should be noted that standard trace element solutions of Hoagland, Bertlo, and others prove to be inadequate for the occasion. Relative trace element concentrations in these solutions are in no way optimal (Hewitt, 1960). Much less are they fit for the correction of the nutrient medium under continuous cultivation, as trace element contents of standard solutions do not match trace element contents of plant biomass. So we had to prepare a nutrient solution containing trace elements at optimal concentrations and to calculate the amount of each trace element to be included into the corrective addition by the above equation, on the basis of trace element contents of plant biomass.

In most experiments with continuous subirrigated wheat culture we assumed that normal trace element concentrations of the nutrient medium were those of the standard Knop nutrient solution. Trace element concentrations were as follows:

Fe—3.5 mg/l B—1 mg/l Mn—0.5 mg/l
Cu, Zn—0.2 mg/l Mo—0.1 mg/l

With wheat productivity of 50 $g/m^2 \cdot day$ of dry biomass, daily corrective addition per 1 m^2 of the cenosis contained the following substances (g):

KNO_3—3.95
$Ca(NO_3)_2 \cdot 4H_2O$—1.58
HNO_3—1.9
H_3PO_4—1.08
$MgSO_4 \cdot 7H_2O$—1.26
$FeC_6H_5 \cdot H_2O$—0.10

Corrective additions for other crops can be calculated in a similar way, with regard to the peculiarities of their mineral composition.

To add the necessary amounts of elements in adequate proportions in salts only turned out impossible. Therefore, the corrective addition also contained nitric and phosphoric acids.

Besides the salts and acids listed here, salts of trace elements (B, Cu, Mn, Zn, Mo) were added, their total weight less than 1 g/kg of biomass. A corrective solution of such a composition not only keeps elements at prescribed concentrations but also maintains the pH level in the range between 5.8 and 6.5.

If the productivity of the higher plant link in the system is at a steady level, the properly calculated correction can be done without checking the chemical composition of the nutrient media. In real experiments, however, there is a need to make regular control analyses of the nutrient media because 1) plant productivity and yield structure may vary; 2) one of the aims of the experiment is to obtain data on the dynamics and coordination of different processes. In most long-duration experiments samples of nutrient media were analyzed for macronutrients once a week and for trace elements once a month. If required, based on the results of analysis, "the correction was corrected."

Growing plants in continuous culture from germination to maturity on a medium of constant composition suggested that the uptake of elements by plants could differ from the true requirements for them. Mineral composition of wheat plants grown on the Knop medium in the air subirrigated culture is not the same as that of soil grown wheat plants. The main distinction is the elevated amounts of mineral nutrients in the roots and shoots of plants grown in subirrigated culture. It can be assumed that the reason for this is the passive uptake of elements by plants during the final growth stages, whereas wheat plants growing in the field stop consuming mineral nutrients once they reach milky ripeness. The assumption of passive element uptake can be inferred from the following: in our experiments, in contrast to field conditions (Olifer, 1965), the shoots of ripe plants contained elevated amounts of elements in the form of mobile ions, K^+, NO_3^-, etc. To illustrate, nitrates, which were not recorded in the shoots of field-grown plants, were found in the shoots of subirrigated culture plants, amounting to 0.4–0.5% of the dry biomass in terms of nitrogen. In our experiments too much phosphorus and calcium were removed with the root biomass.

It was proven long ago by a number of researchers that the withdrawal of some elements from the nutrient solution in the last stages of a wheat life cycle does not decrease the yield if the plants received adequate nutrition in earlier periods (Ovechkin, 1940; Podvalkova, 1959; Pavlov, 1969; and others).

Progressive removal of elements from the nutrient medium could seriously complicate the method of cultivation of air subirrigation we used in our experiments. So we were confronted with the task of finding the plant age at which we could simultaneously and completely remove all nutrient elements from the nutrient medium and avoid a drop in plant productivity. For this purpose we conducted experiments with continuous air subirrigated wheat culture containing 21 plant age groups separated by a 3-day age interval. The plants were grown on the full Knop nutrient solution, changed twice a week. After harvesting the yield of several age groups that were grown from germination to maturity on the full Knop solution, the whole "conveyer" with 21 age groups began to receive tap water. Further irrigation with water instead of nutrient solution allowed the estimation of the productivity and chemical composition of plants undernourished for 3 to 63 days (counted from the end of growth period).

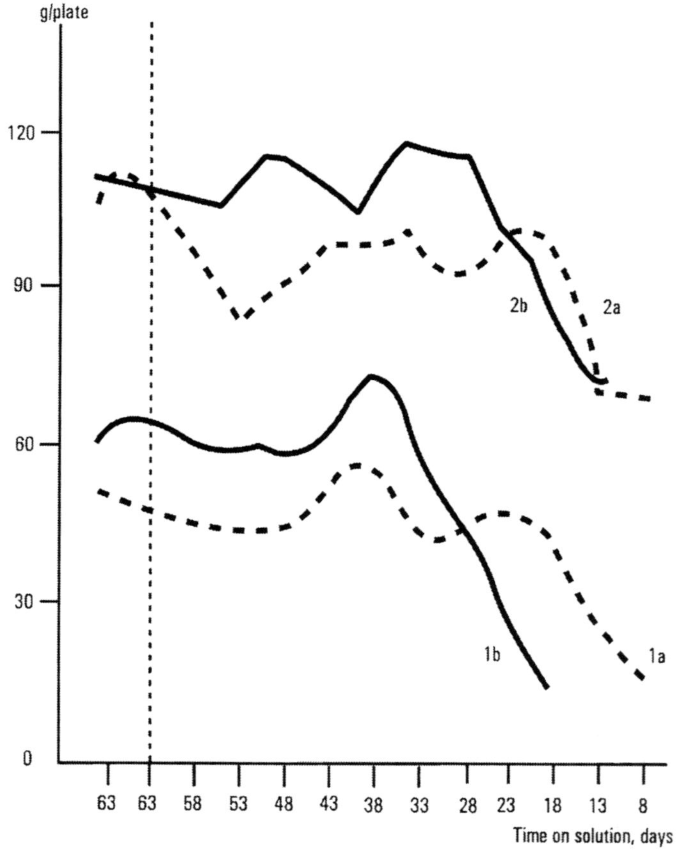

Figure 7.14 Harvested grain (1a and 1b) and vegetative fraction (stems, chaff, roots, 2a and 2b) in two experiments (a and b), in which wheat was first grown on the Knop solution and then on water.

Two replicate experiments were performed, with similar results. As illustrated in Figure 7.14, growing plants on tap water for 25 to 27 days at the end of their growth period (after 36–39 days of growing on the nutrient medium) did not cause any decrease in the total biomass or the wheat grain yield. The yield of the plants irrigated with water for a longer period decreased: the shorter the period of growing on the nutrient solution the lower the yield. In the plants grown on water for the last 24–27 days of the life cycle, the grain fraction of the total plant biomass increased appreciably, by 39.7% averaged for the two replicates against 33.4% in the control. Longer irrigation with water reduced this parameter dramatically. The proportion of mineral nutrients (except nitrogen) in the biomass of plants grown for 36–38 days on the nutrient solution and then for 24–27 days on water was one quarter smaller than in the control for grain and twice smaller for shoots and roots (Table 7.5).

Table 7.5 Element contents in the biomass of wheat grown on the Knop medium throughout the growth period (control) and that grown on water during the last 24–27 days of the life cycle.

Element	Variant of experiment	Element content, % of dry matter		
		grain	shoots	roots
P	Control	0.62	0.68	2.8
	Water during 24–27 days	0.48	0.43	1.00
K	Control	0.55	5.22	1.11
	Water during 24–27 days	0.43	3.33	1.37
Ca	Control	0.08	1.43	5.53
	Water during 24–27 days	0.04	0.80	2.12
Mg	Control	0.25	0.36	0.53
	Water during 24–27 days	0.19	0.22	0.21
S	Control	0.23	0.41	0.61
	Water during 24–27 days	0.22	0.23	0.58
P+K+Ca+Mg+S	Control	1.73	8.10	10.57
	Water during 24–27 days	1.35	4.01	5.28
Total nitrogen	Control	2.50	1.69	4.45
	Water during 24–27 days	2.57	1.25	3.90

The experiments demonstrated that wheat in the air subirrigated culture need not be grown on the full nutrient medium through the life cycle. Moreover, in case it is done, the total yield structure deteriorates to some extent and plants take up excess amounts of some mineral elements accumulating predominantly in shoots and roots, which is extremely detrimental to the closed system. It may seem that excessive salting of ripening plants in the continuous culture could be precluded by decreasing the number of elements in corrective additions. This method, however, proved unusable because adult plants with a well-developed root system “intercept” nutrients from young plants with higher nutrient requirements but with small absorbing root surface. The experiments prove that a more rational approach is to grow wheat plants for 38–40 days on the full nutrient medium and then on pure water.

Cultivation of chufa and vegetable crops does not seem to create a problem of passive uptake of surplus amounts of elements because they are usually harvested at picking maturity, long before biological ripeness. The mineral composition of radish roots and some other crops hydroponically grown on Knop medium is similar to the mineral composition of these crop species and cultivars grown in the field (Trubachev *et al.*, 1975).

Ideally, in the closed life support system plant requirements for mineral nutrients must be fully satisfied by mineral components of wastes from the heterotrophic link—humans. If a human is to be provided with oxygen only due to plant photosynthesis, 0.7–0.8 kg of dry biomass must be synthesized in the system daily. Comparing the spectrum of daily human mineral exchange with the mineral composition of 0.75 kg of dry biomass of an “average” plant (Table 7.6), we can see that the human requirements for basic mineral elements, except Na and Cl, can be completely satisfied by using plant biomass as food.

Table 7.6 Comparison of human and plant mineral exchange.

Element	Mineral composition of 0.75 kg of “average plant” dry biomass, g	Human daily mineral exchange, g	Human daily liquid waste mineral composition, g	Minimal imbalance (g) in plant supply due to	
				liquid wastes	all wastes
N	12–15	12–16	10–14	0	0
P	2.5–4	1–1.5	0.7–1	−1.5	−1
K	8–20	2.5–5	2.3–4.5	−3.5	−3
Ca	4–6	0.8–1	0.3–0.4	−3.6	−3
Mg	1.5–2	0.3–0.5	0.1–0.2	−1.3	−1
S	2	1	0.8	−1.2	−1
Na	1	4–6	4–6	+3.0	+3
Cl	2	5–7	5–7	+3.0	+3
Fe	0.5	0.02	–	−0.5	−0.5

Even though one person in a gas-balanced system with plants can consume daily only 0.3–0.4 kg of plant biomass (dry mass) and the mineral composition of some products (grain, roots) is usually somewhat poorer than that of an “average” plant, plants are capable of supplying a human with the required amounts of phosphorus, potassium, calcium, magnesium, sulfur, iron, and nitrogen of the vegetable fraction of the protein ration.

Plants cannot be fully provided for with mineral nutrition from human wastes. The closed system would, however, benefit greatly from maximal waste utilization by plants. It is quite clear that having undergone preliminary physicochemical or biological mineralization, human wastes can be successfully used by higher plants. This, however, would make the system more complex, with additional units for thermal, thermocatalytic or some other kind of oxidation, or with a special link of biological mineralization, like an aeration tank (Varlamov *et al.*, 1967).

Direct utilization of human solid wastes by higher plants cannot be permitted in the closed system for hygienic reasons. As for the idea of direct utilization of liquid wastes, containing the major part of mineral elements released by humans, it seemed worth studying to us. The ability of plants to use urea as a source of nitrogen in root nutrition (Kalinkevich, 1964; Curuc, 1967; and others), to grow on media with moderate chloride salting (Genkel, 1954; Petra, 1968; Strogonov *et al.*, 1970), and to take up Na and Cl from the medium either actively or passively allowed us to be quite optimistic.

The main problems associated with the direct utilization of human urine are that plants need little if any Na^+ and Cl^-, and that organic nitrogen excreted by humans cannot be the only source of nitrogen for a number of plants. Besides, for hygienic and/or psychological reasons it would be better to avoid using unprocessed urine in the nutrient medium for the plants that have their edible parts (e.g. tubers, bulbs, roots) in direct contact with it.

In the 1970s we used wheat conveyer culture to investigate the possibility of the direct utilization of liquid wastes from human activity (simulated domestic sewage water from the kitchen, shower and sink, and native urine). Use of domestic sewage water by plants (potash soap was used instead of soda soap) created no difficulties. Attempts to use native urine failed at that time (Zamknutaya sistema ..., 1979). It was only when we added the daily urine output of one person to the nutrient medium in two or three steps, for 16–18 m^2 of wheat planting, with the total volume of the nutrient medium of 200–220 l, and when all age groups were supplied with the common nutrient solution, that we obtained a wheat yield close to the reference yield. The biochemical composition of the wheat grain grown on the nutrient medium containing human urine was similar to that of the reference wheat grain. Sodium chloride taken up from the solution by ripening plants constituted 1.16% of the dry mass in roots, less (0.54%) in shoots, and was recorded only in trace amounts (0.003%) in grain (Shilenko *et al.*, 1985). The addition of urine did not cause a rapid NaCl increase in the culture medium due to uptake by wheat roots and shoots. Gradually the NaCl concentration of the medium reached the steady level of 1.5 g/l, which had little if any effect on wheat productivity. So, the study directed the way to immediate utilization of human urine in experiments with the "human–higher plants" closed system. In one of such experiments it was actually done (see Chapter 9, Section 4).

7.6 PLANTS GROWN UNDER ARTIFICIAL CONDITIONS: EVALUATION OF PRODUCTIVITY AND QUALITY OF BIOMASS

To judge whether we have selected an adequate combination of artificial lighting regimes, cultivation methods, and regimes of gas, water and mineral supplies for plants, we can use data on the productivity and quality of the produce of the principal crops grown as an integrated multispecies cenosis in control experiments before the system was closed with humans.

In several sets of control experiments plants were grown in the phytotron under continuous illumination with xenon lamps providing a radiant flux intensity of

130–145 W/m^2. Wheat was grown in air subirrigated culture; chufa, carrot, beet, and other crops were grown hydroponically on granulated expanded clay aggregate. The Knop nutrient solution was not changed, with its composition being

Table 7.7 Productivity of different crops in the Bios-3 phytotrons.

Crop, cultivar	Growth period, days	Yield for the growth period, g/m^2			
		Total		Edible	
		fresh	dry	fresh	dry
Wheat cv. 232	63	–	3182	–	1118
Chufa	90	14500	4556	4464	2476
Carrot cv. Chantane	80	24219	3491	12322	1709
Beet cv. Bordo	80	17149	1813	7125	927
Radish cv. Virovsky White	27	8705	624	5330	342

Crop, cultivar	Growth period, days	Productivity, g/m^2 · day				% of edible fraction (dry), harvest index
		Total		Edible		
		fresh	dry	fresh	dry	
Wheat cv. 232	63	–	50.5	–	17.7	35.1
Chufa	90	161.1	50.6	49.6	27.6	54.3
Carrot cv. Chantane	80	302.7	43.6	154.1	21.3	49.0
Beet cv. Bordo	80	214.4	22.9	89.1	11.6	50.6
Radish cv. Virovsky White	27	322.4	23.1	197.4	12.6	54.7

regularly corrected. The level of CO_2 was either atmospheric or elevated to 0.2%. The air temperature was 23–25°C, the air relative humidity 65–75%. Representative yields of the crops for a life cycle and crop productivity per day are listed in Table 7.7. It is evident from the table that the most productive crops are wheat, chufa, carrot (with dill as a companion crop), i.e. these are the crops that must predominate in the higher plant link.

It should be pointed out that the production process of the leading crops is very energy efficient. During the life cycle their biomass utilizes an average of 5–7% of the total incident radiation, and in the period of the most active growth this value amounts to 10–11% (Chuchalin, 1980).

In addition to the crops listed in the table, we also investigated cucumber, welsh onion, bulb onion, and some other vegetable crops serving as gustatory, spicy, and vitamin supplements. These crops can neither occupy large areas in the closed system nor produce any significant effect on the total productivity of the higher plant link. That is why they are not discussed here.

The quality of produce from plants grown under artificial illumination was examined by a number of authors (Moshkov, 1966; Dadykin *et al.*, 1967; Rozov, 1973). In most papers, however, no comparison was made between the quality of the experimental produce and field produce of the same crops, and if such a comparison was made, it concerned only limited groups of substances under study. To base our research, even tentatively, on the abundance of literature data on the biochemistry of field-grown crops, we had to find out how much the plants grown under the conditions we had chosen were similar in their biochemical composition to those grown in the field.

With this object in view I.N. Trubachev, G.M. Lisovsky and colleagues (1970; 1975) made a large number of analyses of wheat biomass grown under continuous artificial illumination in the air subirrigated culture and in hydroponics. Some vegetable crops hydroponically grown under 18-h and continuous artificial illumination were also investigated. Concurrently the same crop species and cultivars grown during the usual growing season (1967 and 1971) under field conditions in leached chernozem soil in the vicinity of Krasnoyarsk (forest–steppe zone) were investigated.

The total biochemical composition of wheat grain grown in the phytotron does not differ much from that of field-grown grain (Table 7.8). A somewhat higher crude protein content of the phytotron-grown grain is accounted for by quicker grain filling due to the steadily elevated, compared to field conditions, air temperature. These differences, however, are within the fluctuation range of field-grown wheat cultivar Skala in the years differing by humidity and temperature conditions.

Grain of the wheat grown both in the field and in the phytotron contained the same variety of amino acids, including the essential ones. The amounts of each individual amino acid (except methionine) in different grain samples varied little. The proportion of essential amino acids was nearly the same (Table 7.9), but the absolute quantity of essential amino acids was somewhat larger in the phytotron-grown grain containing a larger amount of crude protein. Hence it follows that in terms of biochemical characteristics the phytotron-grown grain was at least as good as that grown in the field.

The overall biochemical characterization of vegetables grown in the field and in the phytotron is presented in Table 7.10. Most of the phytotron-grown vegetables contained somewhat lower levels of sugars and elevated levels of cellulose. Carrot, beet, and especially cucumber also contained increased nitrogen levels. The reason is probably that these plants grew quicker and reached harvest maturity earlier than field-grown plants, therefore the harvest index was lower. In our opinion, however, the recorded differences are not significant, and we can assert that the compositions of field- and phytotron-grown vegetables are qualitatively similar. Table 7.11 confirms this, showing that phytotron-grown vegetables contained all the essential amino acids we determined although their total amount was somewhat smaller. It is noteworthy that phytotron-grown vegetables contain an elevated level of glutamine, which is indicative of active amino acid and protein synthesis in the plants harvested in the period of intensive growth.

Table 7.8 Biochemical composition of the grain of wheat cv. Skala grown under different conditions, %.

Growth conditions	Crude protein (total No. · 5.7)	Carbohydrates without cellulose	Cellulose	Fats	Ash
Field	12.5	80.3	2.8	2.5	1.7
Phytotron, hydroponics, 1st generation	15.4	80.0	2.2	1.7	2.2
Phytotron, hydroponics, 2nd generation	15.7	77.2	2.7	2.0	–
Phytotron, air subirrigation culture (average of 3 samples)	14.4	76.1	2.0	1.3	2.7

Table 7.9 Essential amino acid content of the wheat cv. Skala grown under different conditions (% of dry mass).

Amino acid content	Field	Phytotron		
		Hydroponics, 1st generation	Hydroponics, 2nd generation	Air subirrigation culture
Total amino acids	8.64	9.55	9.45	10.22
Total essential amino acids	2.92	3.15	3.31	3.24
Essential amino acids as % of total amino acids	34.5	33.0	35.0	31.7

Phytotron-grown vegetables compare well with those grown in the field in ascorbic acid and carotene content (Table 7.12). This means that vegetable crops grown under artificial conditions can potentially serve as a “vitamin planting” in closed ecological systems.

Table 7.10 Biochemical composition of the vegetable edible fraction (% of dry matter).

Vegetable crops and their growth conditions	Total No.	Crude protein	Nitrogen-free extractants		Cellulose	Fats	Ash
			sugars	other			
Carrot							
field	1.4	8.7	51.5	25.27	7.3	0.63	6.6
phytotron	1.8	11.2	42.3	31.23	8.8	1.47	5.6
Beet							
field	2.3	14.4	63.0	6.88	6.4	0.22	9.1
phytotron	2.8	17.5	53.0	12.98	7.6	0.62	8.3
Radish							
field	2.5	15.6	24.2	38.57	8.0	1.13	12.5
phytotron	2.2	13.7	28.7	34.25	8.8	0.25	14.3
Turnip							
field	2.6	16.2	37.7	28.89	6.8	1.31	9.1
phytotron	2.2	13.7	40.0	28.60	7.8	2.20	7.7
Tomato							
field	3.3	20.6	29.4	31.60	8.4	–	10.0
phytotron	2.9	18.1	20.8	40.10	9.9	–	11.1
Cucumber							
field	2.5	15.6	28.3	35.99	8.1	0.91	11.1
phytotron	3.9	24.4	16.8	37.58	9.0	1.22	11.0
Onion							
field	4.0	25.0	36.5	12.25	7.6	1.95	16.7
phytotron	3.5	21.9	34.3	21.30	6.5	–	16.0

The trace element content of radish roots and other crops hydroponically grown in our experiments in expanded clay aggregate on the Knop medium is similar to that of field-grown crops and cultivars (Table 7.13).

As reported by I.V. Tsvetkova and her colleagues (1970), expanded clay aggregate used over the long term for growing vegetables, they can accumulate aluminum at concentrations exceeding the permissible levels for foodstuffs. It is not improbable that vegetables may take up other harmful elements leached from material used to manufacture pumps, hoses, trays, etc. From investigation of the trace-element composition of vegetables grown in the field on leached chernozem soils and in the phytotron in local (Krasnoyarsk) expanded clay aggregate that had been used many times, it became obvious (Table 7.14) that such trace elements included in the nutrient solution as manganese, boron, copper, and zinc accumulate in roots of phytotron-grown vegetables at the same concentrations as in field-grown vegetables. There is no evidence to suggest accumulation of aluminum from expanded clay aggregate in roots, which may be accounted for by the greater durability of local expanded clay aggregate against that examined by I.V. Tsvetkova. It would hardly be justified if we, on the strength of those data (Tsvetkova *et al.*, 1970), completely refuse to use expanded clay aggregate in creating experimental closed systems.

Table 7.11 Amino acid composition of vegetables grown in the field (1) and phytotron (2) (% of dry matter).

Amino acids	Carrot		Beet		Cucumber	
	1	2	1	2	1	2
Cystine	0.02	Traces	0.47	–	0.13	0.14
Lysine+histidine	0.20	0.24	0.49	0.43	0.55	0.45
Arginine	0.22	0.58	0.36	0.40	0.65	0.86
Asparagine	0.46	0.21	1.70	0.64	0.93	1.14
Serine	0.27	0.37	0.38	0.35	0.52	0.43
Glycine	0.43	0.54	0.52	0.46	0.84	0.74
Glutamine	1.28	2.22	3.20	3.80	2.54	4.27
Threonine	0.46	0.52	0.48	0.37	0.66	0.68
Alanine	0.69	0.70	0.57	0.63	0.61	0.70
Proline	0.33 }	0.75 }	0.43 }	0.62	0.74	0.58
Tyrosine	0.15 }				0.28	0.27
Methionine	0.13	0.25	0.25 }	0.48	0.18	0.14
Valine	0.58 }	0.45	0.45 }		0.87	0.75
Phenylalanine	0.46 }		0.26	0.24	0.78	0.77
Leucines	1.07	1.12	0.93	0.69	1.19	2.01
Total amino acids	6.75	8.95	9.59	9.11	11.47	13.93
Total essential amino acids	2.90	2.58	2.86	2.21	4.23	4.80
Essential amino acid % of total amino acids	42.90	28.8	29.8	24.2	36.87	41.68

Table 7.12 Ascorbic acid and carotene content (mg/100 g product) of the edible fraction of vegetables grown in the field and in the phytotron.

Crop	Ascorbic acid		Carotene	
	Field	Phytotron	Field	Phytotron
Radish	21.3	37.6	–	–
Turnip	34.0	26.2	–	–
Beet	9.6	18.8	–	–
Carrot	9.7	8.8	2.6	3.4
Tomato	14.0	18.2	1.9	1.6
Cucumber	6.5	9.9	–	–
Dill	–	121.9	–	–
Onion	46.5	57.8	–	–

Table 7.13 Mineral composition of vegetable crop roots grown in the field (1) and in the phytotron hydroponically (2), % of ovendry matter.

Element	Radish		Turnip		Carrot		Beet	
	1	2	1	2	1	2	1	2
P	0.60	0.68	0.58	0.58	0.52	0.44	0.44	0.48
K	6.00	6.10	4.40	2.65	2.60	1.70	3.66	3.50
Ca	0.56	0.60	0.46	0.46	0.36	0.41	0.24	0.37
Mg	0.26	0.36	0.23	0.23	0.22	0.20	0.37	0.36
S	0.47	0.51	0.69	0.59	0.19	0.22	0.22	0.26
ΣP,K, Ca,Mg,S	7.89	8.25	6.36	4.51	3.89	2.97	4.93	4.97

Table 7.14 Trace element content (mg/kg of dry mass) in roots grown in the field (1) and in the phytotron (2).

Element	Radish		Carrot		Beet	
	1	2	1	2	1	2
Mn	18.8	7.4	20.6	17.5	34.5	37.4
B	53.8	58.6	53.3	42.5	23.6	27.5
Cu	1.3	2.3	2.8	3.0	8.0	8.4
Zn	15.0	2.9	10.0	19.0	15.4	17.5
Mo	0.1	1.0	0.1	0.6	0.1	0.8
Al	375.0	58.6	106.0	75.0	50.9	91.7
Cr	1.4	2.4	0.6	1.6	0.9	2.4
Ni	0.9	2.4	0.6	1.5	0.9	1.8

Interestingly, phytotron-grown vegetable crops contained 2–10 times more molybdenum, chromium and nickel than field-grown ones. As for molybdenum, it was a component of the nutrient medium and could be taken up from it by phytotron-grown vegetables, while neither chromium nor nickel were specially included in the nutrient solution. They could be contributed both by admixtures of salts used to prepare nutrient mediums (chemically pure and analytically pure salts were used) and by leaches from construction materials, primarily from stainless steel containing 18% of chromium and 10% of nickel. As doses of nickel and chromium contained in the daily vegetable ration (300–600 g of fresh mass) were within the limits permissible for humans, it seems that the presence of these elements can be neglected when creating experimental closed systems.

On the whole, the analytical comparison of phytotron- and field-grown crops suggests that the edible fractions of the two plant groups are closely similar in macronutrient and trace element content; in accumulation of carbohydrates, lipids, and nitrogenous substances; in amino acid and vitamin composition. This means that vegetable food generated under artificial conditions is of full value and that, generally, scientists can lean upon qualitative characteristics of vegetable products grown in the field in evaluating them as candidates for closed ecological systems. However, the actual creation of an experimental closed human system would require purposeful detailed analyses to obtain biochemical and chemical characterization of the chosen crops and cultivars with reference to definite cultivation conditions.

Giving a qualitative evaluation of the total biomass of candidate crops, not only the food value of the edible biomass should be considered but also the proportion and composition of the inedible fraction. As it has been reported from Russian cultivar-testing stations, the grain fraction in the harvested aboveground biomass of spring wheat varies from 20.6 to 42.0% (Nosatovsky, 1957). In our experiments the grain fraction of the aboveground biomass varied depending on light and, particularly, temperature cultivation conditions: from 24.7 to 39.0% for the cultivar Skala and from 35.1 to 48% for the short-stalked cultivar 232. Similarly, for the field-grown radish the ratio of leaf mass to root mass varied from 6.1 to 0.5, depending on the stage of the life cycle, cultivar, etc. (Kalacheva, 1971), i.e. the root fraction of the fresh biomass yield ranged from 14 to 67%. In our experiments it was 39 to 65%, depending on cultivation conditions.

Inedible biomass of higher plants can be a sort of a dead end in the closed system mass exchange. It must be either removed from the system in exchange for incoming substances, stockpiled in the system with appropriate expenditure of substances from the system's storage, or it may undergo chemical or biological oxidation within the system. The amount of oxygen required for these processes would equal to that evolved by plants in the synthesis of this biomass. It is usually undesirable to decrease the system closure by using stores instead of the edible biomass regenerated in the system and to make the system larger and more complex by biomass oxidation. The likely avenue to the ideal solution of this problem will be to increase the edible fraction of the total biomass in every possible way, bringing the quality of the biomass produced in the system into a maximum agreement with human requirements. At the same time, this could bring the human respiratory quotient and the plant assimilatory quotient closer to each other.

CHAPTER 8

Bios-1 and Bios-2 Experimental Facilities

8.1 TWO-LINK ECOLOGICAL LIFE SUPPORT SYSTEM "HUMAN–MICROALGAE"

The investigation of continuous microalgal culture as a link of an ecological system, described in Chapter 6, the accurate definition of the ecological niche for it, the mathematical simulation of the cultivation process, and finally, the realization of the stable prolonged process with a specified productivity of microalgal culture created a foundation for designing an experimental system for prolonged operation, including humans.

The creation of a closed system called for the technological development of all mass exchange processes in the system, continuous microalgal cultivation being the central process but not the only one.

Basic Technological Scheme of the Two-Link System with Closed Gas and Water Exchange

In the "human–microalgae" system with closed gas and water exchange, humans receive bound energy as food from the system stores and dissipate it as heat given up to the outer environment. The microalgal link receives free energy from the outside as radiant energy, transferring it in the bound state to the system store of biomass and, partly, converting it to heat. In the course of the system operation, dynamic equilibrium is established among inflow of free energy, binding of it in photosynthesis, release of the energy in the oxidation of food by humans, and dissipation of the depreciated energy as heat.

If food chains are closed within the system, the storage of unregenerated food products becomes unnecessary (they are all regenerated), and bound energy is linked directly from the producer to consumers. Only temporary storage will always be needed as a buffer because rates of production and consumption will not always be equal (Figure 8.1).

Like energy, system substances continuously move between links; and with the steady-state process the inflow rate of any element entering each link of the system is equal to its outflow rate. The more accurately the condition of equal rates is fulfilled, the nearer the closed system is to the ideal case and the longer it can maintain self-contained existence.

Creation of the system presupposes the fulfillment of the following tasks:

– to satisfy human requirements for oxygen and for drinking, cooking and hygiene water;

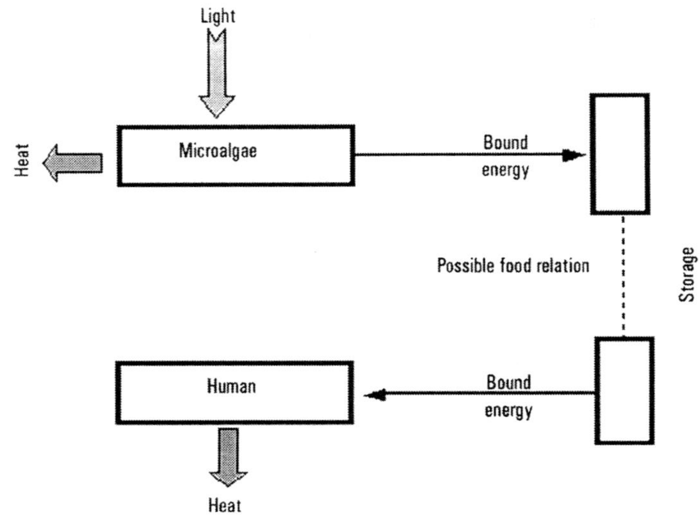

Figure 8.1 Energy exchange of the "human–microalgae" system.

– to provide light energy, water, components of the nutrient medium, and carbon dioxide to the microalgal link.

After a number of variants had been analyzed, the scheme shown in Figure 8.2 was accepted.

Human respiratory requirements are satisfied by supplying gas emerging from the algal cultivator (2) into the sealed cabin (1), where the test subject lives for the duration of the experiment. On its way the gas is dehumidified in the heat exchanger (3), where it gives up to 2–4 l of condensate a day; then it passes through the filter with activated charcoal (4) and finally enters the sealed cabin. The air is taken up from the sealed cabin with the compressor (5), which pumps it into the algal cultivator, where the carbon dioxide respired by the human is consumed by algae.

The human requirements for drinking and cooking water are satisfied by the condensate from the heat exchanger (3). The condensate passes through the measuring device (6), 1.5 l a day is taken for filtering (7), then salts and trace elements are added to the water, to improve its taste.

Hygiene water is provided from several sources. One of them is the condensate from the heat exchanger (3) that has passed through the measuring device (6). The second source is the condensate of the sealed cabin atmosphere, formed on conditioning devices working in the cabin (not shown in the scheme). The third source is vacuum distillate of the culture centrifugate supplied from the unit (8). The fourth, additional, source is moisture supplied from the outside, to replace the losses resulting from the removal of water-containing products for analyses.

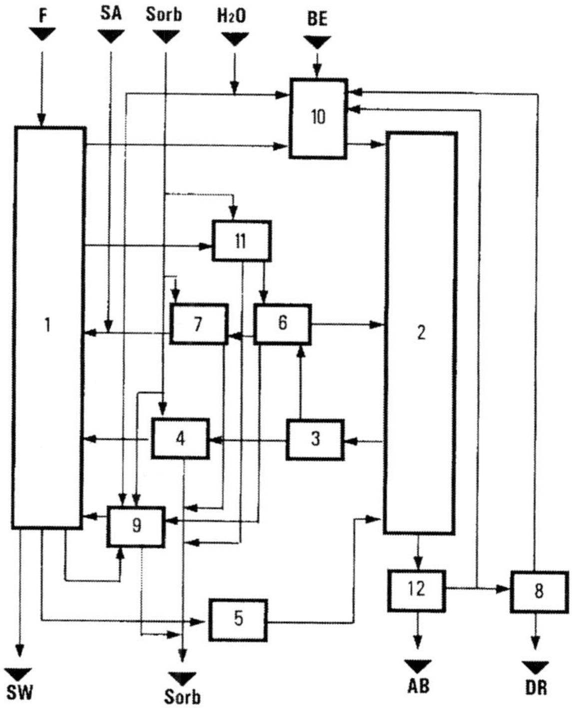

Figure 8.2 Schematic diagram showing mass exchanges in the "human–microalgae" system. 1—sealed cabin; 2—algal cultivator; 3—heat exchanger; 4—charcoal filter; 5—compressor; 6—metering pump for condensate and wastewater; 7—filter for drinking water; 8—distillation unit; 9—filter for hygiene water; 10—nutrient solution tank; 11—filter for wastewater; 12—centrifuge.
F—food; **SA**—salt additives to drinking water; **Sorb**—sorbents; **H_2O**—water added to the system to compensate for process losses and water expenditure in algal photosynthesis; **BE**—biogenic elements; **SW**—human solid wastes; **AB**—algal biomass; **DR**—dry residue.

The availability of several sources of hygiene water in the system allowed the regulation of water exchange in the course of the experiments, necessitated by the changing requirements for such water. This considerably increased on the days of general washing, cleaning and laundering.

In the experiments with the two-link system human food requirements were provided completely by outer sources. The foodstuffs entered the system dehydrated.

Human liquid waste is removed from the sealed cabin and supplied untreated to the algal nutrient vessel (10).

The used hygiene water is supplied from the sealed cabin to the filter (11), then through the measuring device (6) to the cultivator. The measuring device (6) coordinates the rate at which the used hygiene water is supplied to the reactor with the rate of condensate removal. These flows must be coordinated to keep the suspension volume in the reactor constant.

Human solid waste was removed from the system. The water removed with fecal mass amounts to 50–70 g/day, which approximately corresponds to the difference between the amount of metabolic water produced in the human organism and that of the photolysis water disintegrated in the algal link.

The requirements of the microalgae in the two-link system are satisfied in the following way. Carbon dioxide needed for photosynthesis enters the cultivator as a component of the cabin atmosphere.

Part of the biogenic elements are supplied to the algal nutrient vessel (10) with human urine. However, this amount of biogenic elements, in the case of nitrogen in particular, is insufficient to synthesize such a quantity of biomass that could assimilate all the human-produced carbon dioxide. The nutrient medium is complemented with biogenic elements from the source outside the system (store). It is a mixture of salts or alkalis and acids, of balanced elemental composition.

Sources of water, which constitute the bulk of the nutrient medium, are human urine, the centrifugate of microalgal suspension (12), and the vacuum distillate of the centrifugate (8).

Products of photosynthesis are removed in the following ways:

The oxygen-enriched atmosphere from the cultivator returns through the filter to the cabin for the human.

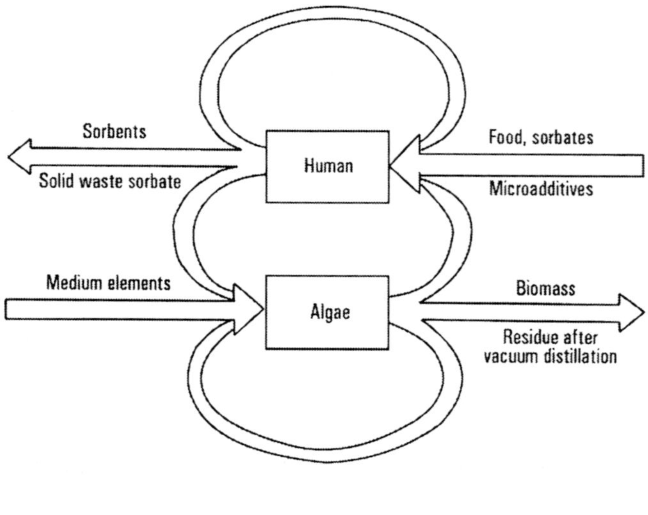

Figure 8.3 Qualitative pattern of mass exchange in the two-link system "human–microalgae."

Part of the algal suspension, on command of the automatic sensor for monitoring the suspension optical density, is supplied to the separator (12). Ninety percent of the centrifugate is directed to the nutrient vessel; the remaining part is subjected to vacuum distillation (8). The distillate is used to prepare hygiene water and the nutrient medium, and the dry residue, consisting for the most part of NaCl, is removed from the system.

Wet biomass from the separator undergoes lyophilic drying; the condensate returns to the water cycle; and the dried biomass is removed from the system a second time, to be replaced by lyophilized foodstuffs for humans. Figure 8.3 shows a qualitative pattern of mass exchange in the "human–microalgae" system.

Biotechnological and Physicochemical Processes in the Two-Link Life Support System

The material cycle in the closed "human–microalgae" system takes effect with the help of a whole set of processes. These are the densitostat cultivation of microalgae in a community with bacteria, water and gas purification, conditioning of the atmosphere in the sealed cabin, preparation of the nutrient medium, etc.

Chapter 6 describes microalgal cultivation in considerable detail. In this section we deal with the process of the preparation and purification of drinking, cooking, and hygiene water in the system "human–microalgae" as well as with atmosphere purification and some other processes. Certain processes will not be described since they are widely known and have long been used in laboratory experiments. These are the centrifugation of the microalgal suspension and the vacuum distillation of the suspension centrifugate.

The technology of obtaining drinking water in life support systems has been worked out quite well. Water suitable for human use can be produced both directly from human urine and from the condensate that forms inevitably in the systems from moisture exuded by the human (Chizhov, 1973).

The proposed methods include purification with ion exchange resins, permeable membranes, vacuum distillation and sublimation, simple drying in the air flow, evaporation with catalytic oxidation of volatile organics, and biological methods of purification (Humphries *et al.*, 1993).

Of these methods the ion exchange method is the simplest to perform. This is the most acceptable method for our system as water can be obtained from condensates or the vacuum distillate of the centrifugate—the liquids containing low levels of impurities and, therefore, requiring insignificant amounts of ion exchange resins to remove them.

Three sources of drinking water were investigated in the two-link system: the condensate from the conditioning unit in the sealed system, the vacuum distillate of the algal suspension centrifugate, and the condensate resulting from the cooling of the gas issuing from the reactor. Water from each source was passed through columns with activated charcoal, a cationic exchanger charged with the H+ and a+

forms, and a marble–sand mixture, so as to filter it and increase its hardness. After treatment, the following salts were added to the water to improve its taste: 50–80 mg/l NaCl, 100 mg/l $MgSO_4 \cdot 5H_2O$, 61 mg/l $NaHCO_3$, 0.1 mg/l NaI, and 1.5 mg/l NaF.

By this method, drinking water complying with safety standards can be obtained from all the three sources, but the amounts of sorbents expended will be different because the water from different sources contains different quantities of impurities (Table 8.1). Table 8.2 shows the composition of the purified water obtained from two sources. State standards and the composition of tap water in the city of Krasnoyarsk are added for comparison. Both tables demonstrate that in the two-link system the best source of drinking water is the condensate of the gas issuing from the microalgal cultivator. The applicability of the water obtained in the way described was tested on mice in a six-month experiment by S.V. Chizhov and A.A. Pak at the Institute of Biomedical Problems, where cosmonauts are trained.

Table 8.1 Water quality from the possible sources in the "human–microalgae" system.

Parameter	Reactor condensate	Cabin condensate	Vacuum distillate of centrifugate
pH	7–8	7.4–7.8	9–10.4
Nitrogen content, mg/l:			
NH_4^+	6–11	12–20	12–19
NO_3^-	0.08	0.07–0.14	0.07
NO_2^-	0.04	0.18–0.25	0.7–2.0
albumins	0.88–2.8	5.4–11.1	0.74–4.6
General alkalinity, mg · eq/l	2.4	4.2–5.2	0.7–2.0
General hardness	0.5	0.5	0.05
Oxidizability, permanganate, mg O_2 (mg/l)	6.9–11.6	35.2–48.2	5.69

The amount of condensate reached 4 l/day, which is enough to satisfy the water requirements of a human.

The sources of hygiene water in the system were the same: reactor and cabin condensates, and the vacuum distillate of the culture centrifugate. The three sources were all used simultaneously. The procedure for preparing hygiene water was simpler than that used to prepare drinking water from these liquids, as the requirements placed upon the composition of hygiene water are much less rigid. All hygiene water was passed only through the column with activated coal. After this treatment the water was transparent, colorless and odorless, which made it acceptable for hygiene purposes. The used (waste) water was supplied to the microalgal reactor after being passed through an anion exchanger, which adsorbed residual fatty acids that were brought into the wastewater with soap. The amount of water expended from the anion exchanger was 6 mg per liter of treated sewage water.

Table 8.2 Comparison of drinking water from different sources.

Parameter	State standards	Tap water	Reactor condensate	Vacuum distillate
pH	6–9	7.9	7.7	7.7
Nitrogen content, mg/l:				
organic	6–8	–	2.0	12.0
NO_3^-	2–40	0.85	traces	0.12
NO_2^-	0.01	0.0002	0.03	0.003
NH_4^+	0.2–1.0	0.16	0.16	0 26
Oxidizability, permanganate, mg O_2 (mg/l)	2–6	1.6	2.08	3.3
Hardness, mg · eq/l	7	4	2.96	2.50
Concentration, mg/l:				
Cl	to 300	4	2.72	130
SO_4	to 100	15	44.3	78
Na+K	to 300	7	47	72
P	0.0001–0.1	0.000	0.000	0.000

In the mass exchange schematic diagram (see Figure 8.2), (9) designates the charcoal column through which hygiene water from three sources is passed; (11) designates the anion-exchange filter for wastewater treatment; (6) designates the metering pump for introducing the treated sewage water into the reactor to replace the outflowing condensate.

Observations for over two months showed no reduction in algal cultivator productivity with the introduction of the wastewater passed through the anion exchanger.

Process Water and Preparation of the Medium for Algae

Process water is the water that circulates in the system, bypassing the human link but is involved in the functioning of the other links of the system. In the two-link variant of the system, process water is the water of the nutrient solution used for algal cultivation.

In our experiments the required quantity of this water was 30–40 l/day.

Process water in the system was obtained by centrifugation or the separation of the microalgal suspension continually removed from the cultivator.

The nutrient solution, 30–40 l/day of which is needed, was prepared by the following procedure. The untreated centrifugate of the microalgal suspension was mixed with the test subject's urine and the portion of the vacuum distillate that was left after the preparation of hygiene water. To this mixture was added outer water, replacing the water removed from the system with the biomass of the cultivated algae, the solid waste of the test subject, and samples of liquids for chemical analysis. Then, the resulting mixture was analyzed to determine each biogenic element concentration, and calculations were made to estimate the supplementary quantities of every element to be added to the mixture so that a balanced nutrient solution could be obtained. Calculation of the balanced medium for algae is described in Chapter 6.

The main function of the microalgal cultivator in the system is air purification, primarily of carbon dioxide but also of other pollutants. The source of pollution is a human, evolving carbon monoxide and a host of other components. The algal culture also supplies pollutants to the atmosphere, evolving carbon monoxide and other gaseous impurities that are toxic for humans (Korotaev *et al.*, 1964; Nefedov and Zaloguev, 1967). Another source of pollution may be the materials from which the system equipment is made.

In the closed atmosphere, volatiles of both links, human and microalgal, will accumulate. Some researchers propose removal of gas microimpurities by methods that are based essentially on catalytic oxidation (Clemedson, 1959). Atmosphere purification by such methods involves the introduction of a rather elaborate unit into the system design as some catalysts can act only at high temperatures. Moreover, in the course of catalytic oxidation molecular nitrogen can be released, and the binding of it will be a very complicated independent problem (Gusarov *et al.*, 1967).

Yet it was likely that at least part of the impurities, readily soluble in water (e.g. nitrogen and sulfur oxides, ammonia), would be taken up by the microalgal culture, and one of the most hazardous components, carbon monoxide, could be used by the cells of the algal–bacterial cenosis as a source of carbon. Some bacteria consume hydrogen intensively (Voitovich *et al.*, 1971 and others).

These assumptions served as a basis for our attempt to close the gas exchange of a human and microalgal culture without involving the catalytic oxidation of gaseous impurities. If necessary, the catalytic unit could be introduced into the system in the course of the experiment.

The gas flue by which gas flows from the cultivator to the sealed cabin was provided with a charcoal filter, with a layer 20-cm thick. The charcoal filter cleared the air of the specific odor of microalgae. The filter was renewed with fresh charcoal before every experiment. The capacity of the filter was 900 g of charcoal. The longest period of service was 90 days, but no deterioration of its absorbing capacity was recorded. Therefore, there is reason to hope that the expenditure of charcoal per unit time could be reduced.

Design of the Experimental Installation

As illustrated in Figure 8.2, the experimental installation includes two major units—a microalgal cultivator for growing *Chlorella* and a sealed cabin for the human test subject.

Other units of the installation, described above, are auxiliary units, serving to maintain physicochemical processes.

Chapter 6 contains equations of microalgal culture productivity as a function of two principal parameters: the surface irradiance of the culture and the carbon dioxide concentration in the gas phase. These equations form the foundation for calculating the cultivator design parameters.

The starting value for the calculation is the maximum design productivity of the cultivator. This value can be expressed in terms of the liters of consumed carbon dioxide per day or liters of the evolved oxygen per day, or grams of the produced biomass per day. These values are related to each other.

In our calculations the value of the productivity will be estimated as the dry mass of the biomass produced within a day.

The maximum productivity of the cultivator must be higher than the productivity of the human, when converted to the same units. This condition ensures the automatic regulation of the carbon dioxide concentration in the atmosphere.

If the carbon dioxide productivity of the human is known in advance, it can be used to calculate the maximum productivity of the cultivator. With the help of this calculation, the carbon dioxide concentration in the system can be set at any specified level. Here is the calculation routine:

First, a carbon dioxide concentration should be selected that could be maintained in the system's atmosphere for a long time, causing no harm to human health.

At one time, many investigations were carried out to estimate the effect of an enhanced carbon dioxide concentration on humans (Shaefer, 1964; Genin, 1964). There is evidence that CO_2 concentrations of up to 5% may be tolerated for a short time and up to 1–2% for a long time (Haldine, 1955; Zharov *et al.*, 1966). In NASA life support systems the CO_2 concentration has to be kept below 1%. This is feasible for physicochemical systems, but in a biological system we must take into account plant requirements, providing them with CO_2 in concentrations that does not limit their photosynthesis rate. Thus, in a biological system a compromise between human and plant requirements has to be attained. Of course, the physiological requirements of humans will be of absolute priority. Our first experiments on closing the gas exchange of the "human–microalgae" system showed that a permissible CO_2 concentration was 1.3% (Kirensky *et al.*, 1967).

To calculate the maximum productivity of the cultivator, we took the 1% CO_2 concentration, the permissibility of which is undisputed, and we used the equations and numerical coefficients obtained in the experiments described in Chapter 6.

Based on these calculations, we took the maximal design productivity of the cultivator to be 600 g/day in terms of the dry algal biomass. Such must be the productivity of the cultivator if microalgal growth is not limited by anything but irradiance.

The next step of the calculation is to determine the necessary area of the light-receiving surface of the culture. Here two variants are possible—to save on the light energy or to save on the area of the light-receiving surface of the cultivator. In the first case the cultivator must work at low illumination levels and have a large light-receiving area; in the second case, in contrast, the light-receiving area is smaller and the intensity of the light flux higher, which, of course, decreases the radiation-use efficiency.

First, we will determine the area of the light-receiving surface for the first variant. From Chapter 6 it follows that the maximum radiation-use efficiency, 12%, is at the surface illumination intensity of the culture of about 50 W/m^2. In this case the culture productivity is 25 $g/m^2 \cdot$ day. The light-receiving area will be 24 m^2.

With the light-receiving surface of such an area, the cultivator would be bulky and heavy, ensuring a high radiation-use efficiency.

To diminish the size of the cultivator, the illumination intensity should be increased. We took 270 W/m^2 as the upper limit of the illumination intensity because at higher levels the acrylic plastic of which photoreactor walls are made quickly ages and becomes unfit for use.

The productivity of a unit illuminated area of a culture layer at a given surface irradiation is derived from the equations of the mathematical model (Chapter 6). The calculation yields the productivity of the unit area $P = 89$ $g/m^2 \cdot$ day. Then the area of the light-receiving surface of the reactor is 6.7 m^2.

In the constructed installation the area of the light-receiving wall was 8 m^2 as 10–15% of the photoreactor volume is always occupied by air.

Furthermore, the physical thickness of the culture layer in the reactor has to be determined. It is evident from the theoretical reasoning in Chapter 6 that this value has little effect on the culture productivity, and the choice depends on the reasons associated with the process: increased thickness of the layer leads to an increase in the amount of water in the reactor; decreased thickness of the layer, to 1 mm, causes technical problems. In our work we chose a culture layer thickness of 5 mm, which corresponds to a reactor volume of 20 l.

Two sides of the reactor were illuminated, so the quantity of the culture could be halved and each of the two illuminated sides of the reactor was 4 m^2 in area. The productivity of the cultivator grew by about 9%. It is inconvenient to construct and keep in operation a reactor made as one cuvette of such dimensions, so it was sectioned into 8 separate cuvettes, each of them with an area on one side of 0.5 m^2, the area of both sides being 1 m^2. The cuvettes were positioned vertically, their edges facing the common light source—a set of xenon lamps of total power up to 36 kW. The light struck the reactor surfaces only after being reflected from glass mirrors. Figure 8.4 shows a schematic diagram of the cultivator.

Devices for the temperature control and stabilization of the biomass cell concentration were the same as those described in Chapter 6.

The capacities and dimensions of the gas–liquid separator, in which foam is separated from gas at the reactor outlet, and of the cooler of the issuing gas are different from those described in Chapter 6.

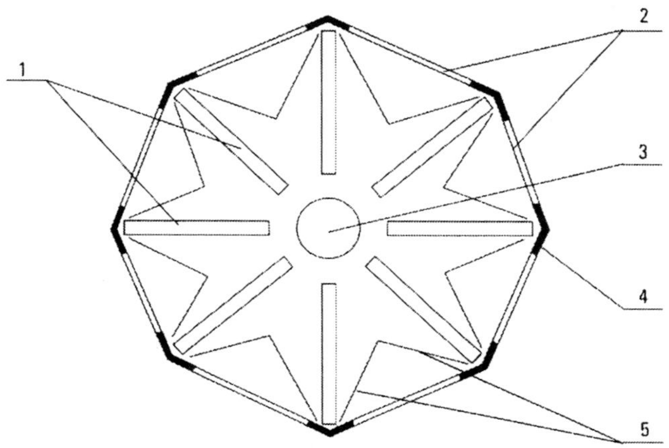

Figure 8.4 The plan view of the reactor with eight flat-parallel cuvettes. 1—cuvettes; 2—doors; 3—light source; 4—case; 5—reflectors.

The volume of the sealed cabin for the human must keep the system sufficiently stable with respect to carbon dioxide content. In our Bios-1 system the cabin measured as follows: 2 × 3 m in floor area, 2 m high, 12000 l in volume. The floor area is large enough to place a bed, a table, a chair, and a toilet, so that any part of the cabin surface could be within reach for cleaning. The height of 2 m gives an opportunity to move freely while doing physical exercises and, on the other hand, allows the test subject to clean the cabin without using any gadgets.

Due to the considerable capacity of the cabin, the cultivator could be potentially taken out of service for a long time. If, for instance, the CO_2 atmospheric concentration of the system is 1%, the permissible limit for a short period being 2%, the human produces about 20 l CO_2/h, and the microalgal link does not function, then the CO_2 concentration will reach the limiting value in 6 h:

$$\Delta t = \frac{(2\% - 1\%)\ 12000}{20\ \text{l/h}} = 6\ \text{h}.$$

This means that the human can stay in the hermetically sealed cabin for 6 h at most, if the cultivator is out of service. This characteristic of the system is of great significance, as 6 h may be enough to repair or replace any unit of the system.

The cabin is provided with two locks—one for passing food and other things to the cabin and removing them from it, the other for the removal of untreated solid wastes. Gas ducts connect the cabin and the cultivator. The air temperature in the cabin is regulated with a conditioning device.

The system atmosphere is completely separated from the outer atmosphere, which is why the difference in temperature inside and outside the system is subject to considerable fluctuations. To avoid the mechanical loading of the hull, an elastic receptacle connected to the sealed cabin with a short pipe was added to the system. It serves as a compensator for variations in the atmospheric pressure. When the atmospheric pressure decreases, part of the cabin air enters the compensator; when the outer pressure increases, the gas flows out of the compensator into the cabin. The volume of the compensator is about 8% of the volume of the sealed cabin. With the compensator, in our experiments the difference in pressure inside and outside did not exceed ±3 mm of water.

Experiments with the Two-Link System "Human–Microalgae"

The installation described above—Bios-1—made possible experiments in the "human–microalgae" system upon the closing of, first, gas exchange, then process and hygiene water exchange, and, last, drinking and cooking water exchange.

The key task of the work was to study the possibility of an experimental system of biological regeneration of the atmosphere with the help of microalgae. A great deal of work was done by scientists in a number of countries to study microalgae in the hope of using them as a means of gas exchange in spacecrafts (see Chapter 3).

To find out whether microalgae are suitable for gas exchange in an inhabited spacecraft it was necessary first of all to define their biological compatibility with humans and then to study the quantitative side of gas exchange between human respiration and microalgal photosynthesis in a closed atmosphere.

A number of experiments on direct gas exchange between humans and a microalgal cultivator were conducted. The duration of these experiments increased gradually from several hours to 90 days, and the accumulated data prove that the regenerated atmosphere is not toxic for humans.

The first experiments with humans in the Bios-1 were performed in 1964. Their duration was 12 h (I. Gitelson was the test subject) and then 24 h (the test subject was Yu. Gurevich). No reasons against long-duration experiments were found. During 1965–66 we carried out a series of experiments of increasing duration: 5, 14, 30 and 90 days. The test subjects were researchers of the Department of Biophysics (the Institute of Physics): G. Dralyuk (5 days), V. Pushkova (12 days), M. Bazanova (14 days), G. Tereshkova (30 days), G. Mazurkina (45 and 90 days in the Bios-2 system "human–*Chlorella*–higher plants") (Kirensky *et al.*, 1967, 1967a).

Healthy men and women of 20 to 33 years of age took part in the experiments.

Four weeks before the beginning of the experiment physiological parameters of the test subjects were determined by clinical and laboratory tests, and these data helped to control the reactions of humans during the experiment and the observation period following it.

During the course of the experiment the participants kept a regime of maximum activity; physical exercises were done according to the schedule, and active movements in the cabin were encouraged.

In observing the state of test subjects the following physiological parameters were monitored: the temperature and mass of the body, blood pressure, pulse and respiration rates, duration of respiratory delay, lung capacity, minute respiratory capacity, CO_2 concentration in the alveolar air, visual and motor reactions, state of mechanical memory, attention capacity, dynamometry, electrocardiogram, and oxygenometry.

Special attention was paid to possible signs of sensitivity to substances of algal origin.

The gas composition of the closed atmosphere was investigated: continuously for CO_2 and O_2 content and once a day for CO, ammonia, hydrogen sulfide, mercaptans, nitric oxide, indole, and skatole. The cabin bacterioflora was sampled daily.

In the course of the experiments no accumulation of deleterious contaminants in dangerous quantities was recorded in the cabin atmosphere; their concentration in the system was kept at a safe, stable level. No penetration of the cultivator bacterioflora into the cabin was detected either, though the cultivation was not sterile.

As an example we describe the results of the 45-day experiment.

Performance of the microalgal link in the 45-day experiment with the two-link system was characterized by the following values: irradiance of the reactor surface was 255±22 W/m^2 PAR; CO_2 concentration in the gas at the inlet was 1.02±0.23%; the cell biomass concentration of the culture (dry mass) was 12.2 g/l; culture productivity was 384 g/day; radiation-use efficiency was 5.6%. The dynamics of the basic parameters of this link is shown in Figure 8.5.

The productivity of the microalgal link in the experiment was much lower than the designed one because the microalgal culture was limited by the amount of carbon dioxide coming from the human. Thus, the cultivator had a reserve of oxygen-producing capacity that could be used in situations when the human requirement for it increased for a while, e.g. in case of more vigorous human activity.

The respiratory quotient (RQ) of a human with usual mixed diet = 0.82–0.85. The microalgal assimilator quotient (AQ) when using our cultivation method proved to be 0.89.

If the CO_2 balance is reached by adequate cultivator productivity, in order to stabilize O_2 in the system it is necessary to equalize RQ and AQ. For more details see Chapter 9.

Table 8.3 shows the system imbalance between the inputs and outputs of hydrogen and oxygen. However, this seems to be an imbalance; it is associated with the difference between the amount of metabolic water excreted by the human in the course of food oxidation and the amount of photolytic water utilized in algal photosynthesis. The difference is about 70 g/day and is exhibited as the imbalance between hydrogen and oxygen inputs and outputs. This imbalance means that oxygen and hydrogen enter the water exchange scheme, where the inverse imbalance of 70 g/day is recorded. As a result, the overall oxygen and hydrogen balance remains undisturbed.

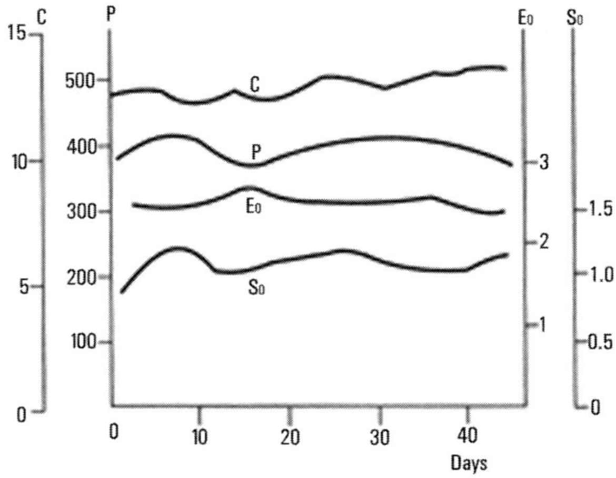

Figure 8.5 Dynamics of the basic parameters of the algal link in the 45-day experiment. C—cell biomass concentration in the culture (g/l); P—cultivator productivity (g/l); E_0—reactor surface irradiance (10^2 W/m^2); S_0—CO_2 concentration in the system atmosphere (volume %).

The exchange of some trace elements (iron, zinc, boron, manganese and molybdenum) in the system was studied in the experiment, too. The total imbalance for these elements was about 7%, the value three times smaller than that resulting from possible errors of a single measurement of the concentrations of these elements.

Based on investigations of the mass exchange in the two-link "human–microalgae" system we could compare the amounts of substances expended on human life support inside and outside the system. The results of this comparison are listed in Table 8.4. As can be seen from the table, the amount of substances expended in the two-link biological life support system (calculated without process losses and amounts removed for analysis) is 11 times smaller than that consumed by the human outside the system.

The results of the experiments suggest the following conclusions:

1. It has been experimentally proven that the atmosphere for human respiration can be maintained with the help of continuous algal cultivation by our method, and the atmospheric O_2 and CO_2 concentrations can be kept constant.
2. Contrary to our original fears, humans and microalgae (*Chlorella vulgaris*) prove to be biologically compatible with regard to gas exchange; their off-gas products are not toxic to each other.

3. The discrepancy between microalgal AQ and human RQ was eliminated by correction of the diet. The composition of the food ration needed for complete gas balance of the system is within the range of the physiological optimum.

Table 8.3 Balance of biogenic elements in the "human–microalgae" system (g/day).

Element	Inflow		
	food	mineral additives	total
Carbon	202.00	12.60	214.60
Oxygen	157.00	17.00	174.00
Hydrogen	32.00	4.24	36.24
Nitrogen	12.22	29.30	41.52
Phosphorus	0.94	0.37	1.31
Magnesium	0.44	1.22	1.65
Sulfur	0.65	1.20	1.84
Potassium	2.02	2.70	4.73
Sodium	4.23	–	4.23
Calcium	0.85	–	0.85
Total	412.35	68.633	480.98

Element	Outflow			Imbalance, g
	biomass	solid wastes and losses	total	
Carbon	206.60	8.00	241.60	
Oxygen	107.00	5.00	112.00	–62
Hydrogen	26.24	2.00	28.24	–8
Nitrogen	38.30	3.22	41.52	
Phosphorus	0.98	0.33	1.31	
Magnesium	1.31	0.35	1.65	
Sulfur	1.54	0.30	1.84	
Potassium	4.26	0.46	4.73	
Sodium	1.87	0.43	2.30	–1.93
Calcium	0.23	0.49	0.73	–0.12
Total	388.33	20.59	408.92	72.05

Having considered the two-link life support "human–microalgae" system we can conclude that it turned out to be a surprisingly convenient and reliable system for the regeneration of atmosphere and water.

Table 8.4 Minimal requirements of the "human–microalgae" system for stored substances vs. human requirements out of the system.

Substance	Daily, g		Annual, kg	
	in the system	out of the system	in the system	out of the system
Food (dehydrated)	412	421	150.4	150.4
Water	–	4840	–	1766.6
Oxygen	–	586	–	213.9
Soap, tooth-powder	10	10	3.65	3.65
Carbamide	62.70	–	22.90	–
K_2HPO_4	9.46	–	3.45	–
H_3PO_4	0.48	–	0.18	–
$MgSO_4 \cdot 7H_2O$	9.20	–	3.36	–
MgO	0.39	–	0.14	–
Compounds containing trace elements	0.21	–	0.08	–
Sorbents	20.04	–	7.32	–
Total	524.5	5848	191.5	2134.6

During the 45-day experiment over 100 generations of algal cells succeeded each other in the cultivator, and there was no tendency towards the deterioration of photosynthesis. The characteristics of the gas composition of the atmosphere, and the density and productivity of the microalgal suspension in the cultivator all remained constant.

Note that we use nonsterile microalgal culture, an algal–bacterial symbiosis. Dynamic equilibrium between phototrophs and heterotrophs is attained ecologically; no specific measures are needed to maintain sterility.

This is far more reliable and incomparably simpler technically than to use sterile culture.

This proves that continuous microalgal culture was in the state of stable equilibrium and that there were no biological obstacles to unlimited continuation of the cultivation, this process being restricted only by the safety margin of the technical devices performing it.

No toxic or allergenic gases were detected in the regenerated atmosphere in our experiments. However, the problem of biological compatibility is very complicated and must be studied further.

The specific hay odor emitted by algae is easily removed with the activated charcoal filter, which need not be changed for at least 3 months.

The 20-l cultivator continuously produces condensate of the moisture evaporating from the algal suspension. This moisture is subjected to simple procedures of physicochemical purification and conditioning. After that it can be used as drinking and hygiene water, satisfying human requirements almost immediately.

If the cultivator surface constantly receives the incident light flux of intensity 225 W/m^2, the light-receiving surface 8 m^2 in area is sufficient to satisfy the requirements of one human for fresh water and breathable atmosphere for an indefinitely long time.

A substantial advantage of a biological life support system over physicochemical ones, in which oxygen production and carbon dioxide consumption are not related, is automatic positive feedback from the human requirement for oxygen to the oxygen productivity of the cultivator. This feedback is effected due to the limitation of the algae by carbon dioxide. No directed regulation of the system is needed. A precise daily rhythm of variations in the carbon dioxide concentration in the cabin atmosphere demonstrates the automatic action of this feedback: each diurnal increase in the carbon dioxide content, caused by the excess production of it by the active human, alternates with a nocturnal decrease in CO_2, when algae consume the excess of carbon dioxide using the reserve capacity of the cultivator, while the CO_2 production by the sleeping human is reduced.

Besides oxygen and clear water, the cultivator produces algal biomass. The dry weight of the daily produce is approximately equal to the weight of the daily dehydrated food ration of the human. Thus, by replacing the consumed food from the storage with dehumidified algal biomass, we keep the mass of the system constant.

Algal biomass may serve as an added source of fuel for engines or other power-generating installations.

We probably must dampen the pioneer enthusiasm about the possibilities of using algae as food somewhat. Algal biomass does not meet human food requirements either biochemically (too high protein and nucleic acids contents and carbohydrate deficiency) or psychologically. In the future, proteins, amino acids, and vitamins contained in algae may be used in food, but one should not give too much credence to the unchecked advertising of "health-giving" algal pills (mostly of blue-green algae *Spirulina* or the green algae *Chlorella*) that must have started in Japan and have in recent years become widespread in America and Europe.

The merits of the algal cultivator as a regenerator of atmosphere and water, immediately utilizing solar radiation, became obvious as early as the 1960s, after the above-described experiments had been performed. To use microalgae in space is particularly easy because no special devices for overcoming the effect of microgravity are needed: even on Earth they exist suspended in water, i.e. actually under microgravity.

Hence, solar radiation in the visible spectrum region collected from the 5-m^2 surface in Earth's orbit outside of the atmosphere would be sufficient for the microalgal cultivator operating round the clock. If the flight were directed towards Mars, it would suffice to enlarge the light-receiving area 4 times.

The cultivator shaped as a tube directed towards the Sun, 1-m high and about 20 cm in radius, provided with inner and outer reflectors can fulfill the requirements of one human. The idea of the proposed construction is illustrated in picture in Figure 8.6.

It seems that such cultivators should have been used long ago in real space flights, at least as small models.

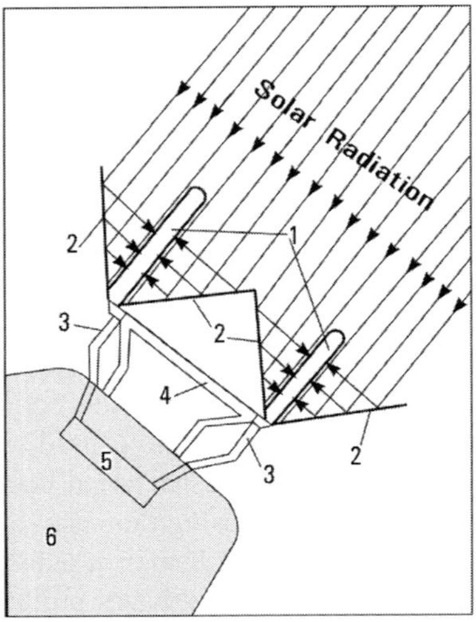

Figure 8.6 "Space sunflower"—microalgal cultivator on the outside of spacecraft, directly using solar radiation. 1—cuvettes both sides of which receive light from reflectors; 2—reflectors; 3—hoses for pumping through algal suspension; 4—bracket; 5—control system for algal continuous cultivation; 6—spacecraft hull.

Unfortunately, neither Soviet cosmonautics nor the astronautics of the USA has managed to overcome the justified conservatism of engineer thinking and the unjustified prejudice against "incomprehensible" biological technology. It only remains for us to hope that the next generation of spacecraft designers must appreciate the advantages of biotechnological installations for human life support during long-distance and long-duration space flights in the twenty-first century.

Our results are meant for them.

8.2 THREE-LINK SYSTEM IN BIOS-2 "HUMAN–MICROALGAE–HIGHER PLANTS"

The two-link "human–microalgae" system, as was predicted by calculations and proved by experiments satisfies the human requirements for atmosphere and water. However, this simple and low energy-consuming system did not seem promising for complete food regeneration. Apparently, human food requirements cannot be satisfied in the system unless it is more complex structurally. Even when impediments to using high protein algal products as food are overcome, the main problem will remain unsolved—regeneration of the vegetable carbohydrate fraction of food, constituting a significant portion of the human diet.

Basing our plans on these statements substantiated in detail in Chapter 5, we made the next step on the road to system closing and introduced higher agricultural plants, traditional human food.

This step involved the development of a conveyer culture conforming to the technological conditions of the closed system, described in Chapter 7.

Experiments with the two-link system provided a basis for introducing higher plants as the third link. The main aims of introducing higher plants into the already existing two-link system were to test their compatibility with the two links, to develop and test the mass exchange pattern, to go into technological issues and, finally, to achieve partial closing of the system with respect to food. However important this task may seem, it was of least importance for investigation because there was no doubt that traditional produce of higher plants could be used as food.

To illustrate, we will describe one of the experiments with the three-link system. It is shown schematically in Figure 8.7.

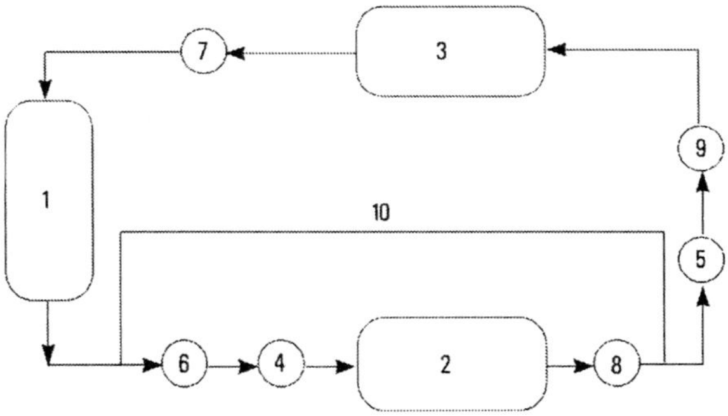

Figure 8.7 Schematic diagram of gas exchange in the three-link system. 1—hermetically sealed cabin with test subject; 2—algal cultivator; 3—phytotron; 4,5—compressors; 6,7,8—heat exchangers; 9—carbon filter; 10—bypass pipe.

A thirty-day experiment with a test subject was held in such a system. The hermetically sealed cabin and the algal cultivator were the same as in the experiments with the two-link system. Besides, the system included a phytotron in which vegetable crops were grown hydroponically on expanded clay aggregate. The planting area of the phytotron was 4.5 m^2. The work related to growing plants in the phytotron was all done by the test subject, which is why the hermetically sealed cabin was connected to the phytotron with a sealed hatch.

The frame of the phytotron was made of stainless steel; it measured $2 \times 2.5 \times 1.7$ m. Four 6-kW xenon lamps provided with a running water filter were used as light sources; they were located over the glass ceiling of the phytotron. The average illumination rate was 90 W/m^2 PAR, the duration of the light period was 24 h for 4 days and 18 h for 2 days, the average duration was 22 h a day. The air temperature was 21–23°C in the light hours and 16–17°C in the dark hours. The relative humidity in the light hours was 70%.

Plants were grown on the Knop nutrient solution with iron citrate and a mixture of trace elements. The solution composition was corrected every day, the solution was changed once a week and fed to the root zone 4 times a day.

As the main goal of the experiment was to study the compatibility of a human with the new link of the life support system—higher plants—under direct gas exchange of the links, we did not propose to study mass exchange in detail during this experiment. So, while the gas exchange in the system was closed entirely, the water exchange was closed only partially. At the same time in this experiment, we made an attempt to close the food chains of the life support system — the phytotron produce, vegetables, was used by the test subject as food.

The main characteristics of the algal link in the experiment were as follows: the production rate of the culture was 308 g/day of dry biomass, the cell biomass concentration of the culture was 9.26 g/l, the atmospheric carbon dioxide concentration at the cultivator inlet was 1.4% by volume. (The values of the parameters are averaged over the experiment.)

Parameters of the higher plant link in the thirty-day experiment are listed in Table 8.5.

For a number of crops the inedible biomass increment was not determined, as this biomass was not harvested during the experiment. The first plantings in the phytotron were done 70 days before the start of the experiment with a human occupant. By the time when complete gas recycling was achieved within the system, some crops were represented by plants of 3–4 different ages, other crops, such as tomato, were steadily fruiting. The total increment of inedible biomass calculated by the system gas exchange was 58 g/day of dry biomass.

Human oxygen requirements were satisfied completely through the photosynthesis of the autotrophic links. The mean daily oxygen consumption by the test subject was 462 l; of them 339 l were supplied by algae and 123 l by higher plants. The requirements for drinking water (2.155 l/day) were satisfied by the condensate of the algal link purified by the methods applied in the two-link system. For hygiene water the test subject used the condensate of higher plant transpirational moisture (3.689 l/day) with no additional purification but pre-boiled.

Human food requirements were basically satisfied by the lyophylized foodstuffs (370 g/day) and bread (200 g/day of wet mass) entered into the system and, in part, by the phytotron higher plant produce, which was added to the human ration at a rate of 44 g/day of dry mass, or 730 g/day of fresh vegetables. Some of the vegetables grown in the system (36.6 g/day of dry mass) were withdrawn from the system so as not to overload the diet with these components.

Human wastes were utilized in different ways. The respired carbon dioxide (415 l/day) was consumed by algae (302 l/day) and higher plants (113 l/day). Liquid waste (1.8 l/day) entered the algal cultivator; the wastewater preliminarily filtered through a gauze filter entered the algal cultivator (3.392 l/day), too. Solid excrements were removed from the system untreated.

Water exchange of the algal link was closed on the test subject (exchange between drinking and sewage water); it was also closed within the algal link (nutrient medium was used many times).

Water exchange of higher plants was hardly closed since every 7 days the nutrient medium was completely removed from the system and replaced by a fresh tap-water medium.

Table 8.5 Productivity of the higher plant link in the thirty-day experiment.

Crop	Area, m^2		Edible mass yield, g/day		Productivity by dry food biomass, $g/m^2 \cdot day$
	Start of experiment	End of experiment	fresh	dry	
Cucumber	0.87	0.87	487.0	16.5	19.0
Tomato	0.75	0.75	184.0	9.8	13.1
Radish	0.49	0.36	59.0	3.5	8.2
Beet	1.12	0.98	161.0	26.0	24.8
Carrot	0.72	0.86	143.0	16.3	20.6
Turnip	0.18	0.31	22.0	2.0	8.2
Kale	0.13	0.13	43.0	3.0	25.4
Dill	0.10	0.10	13.5	1.2	12.0
Welsh onion	0.14	0.14	17.7	2.0	14.3
Total	4.5	4.5	1130	80.6	

Water exchange between the human and the phytotron was not significant and involved only "pure" water: the human received transpirational moisture condensate from the phytotron and the phytotron received back the condensate of the cabin conditioning device.

The results of the experiment demonstrated that the links of the system did not affect each other depressively through the common atmosphere. Algae and vegetable crops investigated in the experiment retained their production rates at steady levels. The health and body mass of the test subject remained the same throughout the experiment.

The experiment also showed that to completely satisfy the requirements of one person for fresh vegetables, it would be enough to have 2.5 m^2 of continuous vegetable culture under the chosen growth conditions.

8.3 BIOS-2 FOUR-LINK SYSTEM "HUMAN–MICROALGAE–HIGHER PLANTS–MICROBIAL CULTIVATOR"

The experiment with the three-link ecological system briefly described in the previous section proved the compatibility of the investigated crop species with a human in direct gas exchange. The same experiment confirmed that vegetable crops can function as the autotrophic link of the system only to a limited degree. Even when the planting area was 4.5 m^2 and the illumination rate rather low (90 W/m^2), some of the vegetables were not used as food but removed from the system as excess.

To increase the closing of system food chains with the help of higher plants, other crops had to be introduced into the system, e.g. cereal crops, the produce of which can be used as food in far larger quantities than vegetables. Therefore, spring wheat, a traditional producer of vegetable food, was used as the higher plant link at the next stage of investigations.

The closing of human food chains on food sources within the system increases the closure of the system material cycle, but even if all human food requirements are satisfied by the system links, the material cycle will not be closed entirely. To achieve complete closure, human wastes must all be utilized within the system. In the two-link and three-link versions of the system described above gaseous and liquid human wastes were utilized completely, while all human solid wastes were removed from the system.

An attempt to utilize these wastes biologically and reintroduce the elements contained in them into the material cycle necessitated the addition of the fourth link—a microbial cultivator.

Biological utilization of solid wastes closely resembles the biological treatment of sewage water. Their aims, however, are different. The aim of the biological treatment of sewage water is to obtain pure water for the most part rid of insoluble and soluble substances. The aim of human solid waste utilization in the biological system is to oxidize organic substances to simple organic compounds—gaseous or dissolved—which may be contained in water in large quantities. This difference in aims can account for some peculiarities of the utilization procedure.

The most-used installation for the biological treatment of domestic and industrial sewage in the laboratory and on a pilot scale is the aeration tank, various modifications of which are described in the literature. The aeration tank has the following important characteristics: small quantities of air are used in aeration; there is a settling tank for microorganisms; and the treated substrate is supplied continuously.

The apparatus used in the experiment with the four-link system was fundamentally different from conventional aeration tanks: the aeration rate is higher—up to 5 liters of air per 1 liter of liquid a min; and the process loading–treatment–emptying was done in batches. Owing to the periodicity of the process, the whole portion of human solid wastes could be loaded for treatment immediately, and there was no need to store the wastes. Besides, the apparatus was not provided with a feeder for continuous supply of the substrate, which made it much simpler. Due to all these peculiarities of the apparatus it cannot be referred to as an "aeration tank," so, in what follows it will be termed a microbial cultivator.

The experiment in the four-link system including a human lasted 73 days. The gas exchange of the links was entirely closed, the water exchange was largely closed (water was removed from the system mostly with samples for chemical analyses), and the human food exchange was partially closed on the higher plant link.

The experiment with the four-link system involved the hermetically sealed cabin, the algal cultivator and the phytotron, described above, as well as the microbial cultivator first added to the system.

The microbial cultivator was a flat-walled cuvette in an upright position, 350 mm wide, 900 mm high, with a layer 20 mm thick (Figure 8.8). In the lower part of the cuvette there were aeration openings, in the upper part there was a gas and foam outlet to the centrifugal separator that separated the liquid phase from the gaseous phase and returned it to the cuvette. The separator was of the same design as that used in algal cultivation. Thermostabilization was performed with a water thermostat, the water from which was run through the water jacket located on one of the walls of the microbial cultivator cuvette.

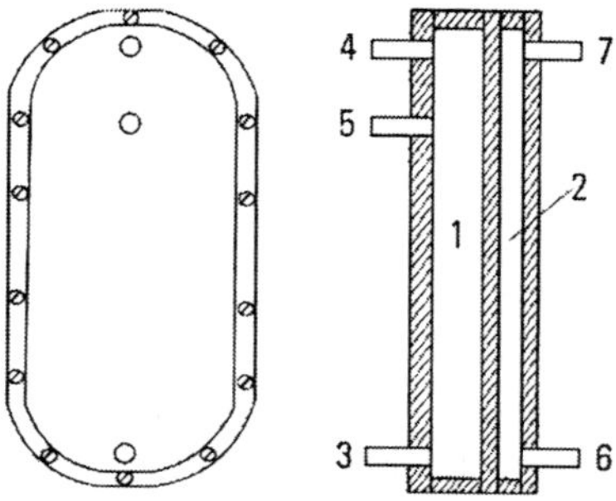

Figure 8.8 Microbial cultivator. 1—microbial culture cuvette; 2—water jacket; 3—air inlet; 4—gas and foam outlet to separator; 5—liquid phase return from separator; 6,7—thermostat-connected inlet and outlet.

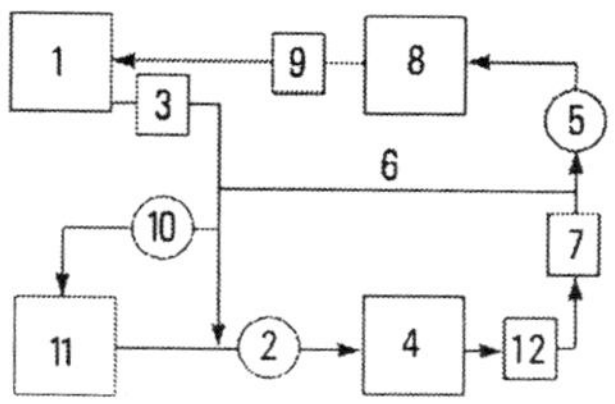

Figure 8.9 Schematic diagram of gas exchanger in four-link system. 1—hermetically sealed cabin; 2,5,10—compressors; 3,9,12—heat exchangers; 4—algal cultivator; 6—bypass pipe; 7—carbon filter; 8—phytotron; 11—microbial cultivator.

The gas exchange of the links in the 73-day experiment is presented schematically in Figure 8.9. The air enriched with carbon dioxide (to 0.93% by volume on the average) from the hermetically sealed cabin with the human (1), through the cooler–dehumidifier (3), was taken with the compressor (10) to the microbial cultivator (11) and with the compressor (2) to the algal cultivator. The algal cultivator also received the air from the microbial cultivator.

The air leaving the algal cultivator, depleted of carbon dioxide to 0.66% on the average, was dehumidified on the cooler (12), passed through the carbon filter (7), and pumped into the phytotron (8) with the compressor (5); in the phytotron it was depleted of carbon dioxide to 0.54% and through the cooler (9) entered the hermetically sealed cabin. The flow rate of the air in the chain of the links cabin–algal cultivator–phytotron–cabin was about 64 l/min. Such a sequence of links for air circulation was determined by the following factors. In any event the highest CO_2 concentration is in the atmosphere of the hermetically sealed cabin. Algae have the highest requirements for this gas, therefore they must receive air from the cabin.

To elevate the CO_2 concentration at the algal link inlet, part of the air from the hermetically sealed cabin was previously run through the microbial cultivator. The CO_2 requirements of higher plants are lower, and they can grow in an atmosphere somewhat depleted of this gas in the algal link.

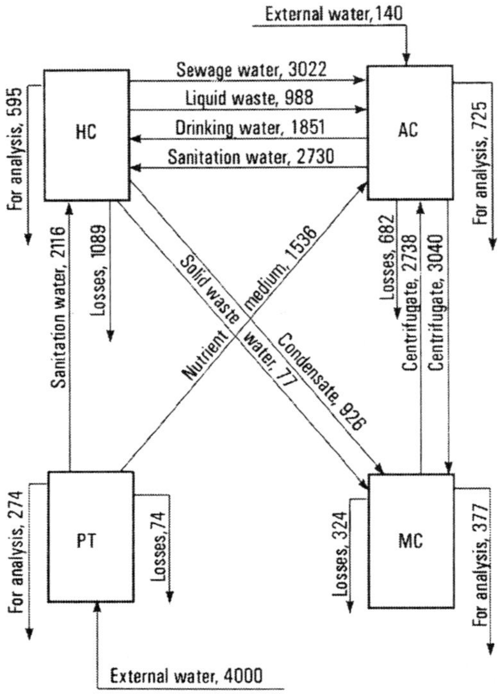

Figure 8.10 Scheme of water exchange in the four-link system: **HC**—cabin with a human; **AC**—algal cultivator; **PT**—phytotron; **MC**—microbial cultivator. Figures denote mean daily water expenditure, g.

Water exchange in the four-link system is schematically presented in Figure 8.10, showing the main water flows during the experiment. The numbers denote the daily average expenditure of water in grams.

As can be seen from the diagram of water exchange, significant quantities of water, 1971 g/day, were taken as samples for chemical analyses from different sections of the system. They were mostly the samples of the nutrient medium, biomass, drinking water, hygiene water and sewage, human liquid and solid wastes, etc. Process water losses in the experiment were about 2170 g/day; most of them were due to evaporation during processes outside the system: centrifugation, medium preparation, conditioning of drinking and hygiene water, sewage water filtering, etc.

Water outputs were all recorded and replaced with equal amounts of external distilled water. Water was brought into the system via two links: about 140 g/day entered the algal link (mostly as distilled water). The higher-plant link was the main site for replacing the water removed from the system.

The water treatment procedure in the experiment with the four-link system was hardly altered when compared to the experiments described above.

In this experiment the processes in the algal link were run in the same way as in the experiments with two- and three-link systems, but here an installation for the lyophilic drying of algal biomass was used. The biomass was dried at a temperature from –40 to –50°C, at a pressure of 10–3 mm of mercury. The resulting moisture was returned to the system. Water losses in drying were 3%.

The productivity of the algal culture in the experiment averaged 296 l/day, culture illumination rate 143 W/m^2, CO_2 concentration of gas at the cultivator inlet 0.93% by volume, CO_2 consumption 260 l/day, oxygen production 297 l/day.

The higher-plant link included continuous culture of the spring wheat cultivar Skala (of Siberian selection) and the Mexican semi-dwarf cultivar Sonora-64. Simultaneously the culture contained plants of 28 ages, with an age difference of 2.25 days. The growth period was 63 days. The plants were illuminated around the clock, and the PAR intensity was 90–115 W/m^2. The plants were grown by the air subirrigation method without substrate (see Chapter 7, Section 4).

The planting area of the phytotron (4.5 m^2) was divided into 12 sections, and each of them in turn received an equal volume (65 l) medium. The nutrient solution was based on the Knop medium. The medium composition was corrected four times a day. The daily average inputs were: 14.3 g KNO_3, 3.0 g $Ca(NO_3)_2 \cdot 4H_2O$, 2.6 g $MgSO_4 \cdot 7H_2O$, 13.5 g HNO_3 (57%), 8.6 g H_3PO_4 (87.5%0, 0.42 g $FeC_6H_5O_7 \cdot H_2O$, 0.037 g trilon B.

In the experiment the phytotron produced an average of 118 l of gas a day, consuming about 110 l of CO_2 a day.

During the experiment the condition and productivity of the plants gradually worsened. A particularly sharp decrease was recorded in the productivity of food biomass—grain.

There could be a number of possible reasons behind this decrease in wheat productivity. The analysis of these reasons, however, suggested that there was only one factor responsible for the decrease—the accumulation of metabolites of associated algal flora and microflora in the nutrient solution.

The conditions were favorable for such accumulation: 1) there was no substrate serving as sorbent for water-dissolved metabolites, and 2) light penetrated into the root zone of plants through acrylic–plastic sowing plates and gaps between them, causing heavy growth of algae on roots. Later experiments with wheat outside of the closed system demonstrated that if algal growth on roots is precluded by using light-tight materials to make sowing plates and by shading the space between rows, plants can be grown on unchanged medium, showing no sign of depression.

Part of the grain grown in the phytotron was used as seeds, and the net yield, 20.3 g/day on the average, was used for baking bread, which was part of the test subject's ration.

The microbial cultivator included in the system not only utilized human solid wastes, but also oxidized metabolites of the algal link. Before solid wastes, previously comminuted in the tissue pulverizer, were fed into the microbial cultivator, they had been mixed with the centrifugate of the algal culture.

It was found in preliminary experiments that the optimum duration of the process in the microbial cultivator was 8 h at a temperature of 21–22°C, at an aeration rate of 4.0–4.5 l/l · min.

In the described experiment the daily average loading of the microbial cultivator consisted of the following components: 87.5 g (20.82 g on a dry mass basis) of solid wastes, 2928 g (including 10.7 g of dry matter, mostly NaCl) of the algal link centrifugate, 911 g of condensate from the hermetically sealed cabin, 208 g (26.23 g on the dry mass basis) of microbial mass of activated sludge. The sulfur correction of the medium was performed in every operation period by introducing 0.5 g of sulfuric acid into the microbial cultivator.

When the eighth processing period was over, the contents of the microbial cultivator were centrifuged. Then the centrifugate entered the nutrient medium of the algal link, and the separated activated sludge, together with the next portion of the condensate from the hermetically sealed cabin, was again placed into the microbial cultivator and was subjected to intensive aeration (to 20 l/l · min) for an hour. Due to such processing—recycling—the activated sludge retained its efficiency. After being recycled, the activated sludge suspension was kept at +5°C until the next operation period began.

The average increment in activated sludge biomass in one operation period was 6.3 g on a dry mass basis. This biomass was removed from the system. The microbial cultivator centrifugate, which entered the algal link after each operation period, contained, on average, 12.3 g of dissolved substances, mostly NaCl.

During the experiment the balance between the basic biogenic elements was estimated in each link of the system, including the microbial cultivator. The results obtained suggested a conclusion that during the operation of the microbial cultivator no appreciable decomposition of nitrogen to the molecular form occurred.

An estimation of the balance between the main biogenic elements throughout the system proved that mass exchange in the experiment was basically balanced. Some recorded accumulation of sodium chloride in the algal and microbial cultivators was asymptotic. The total daily average input of substances was 4657 g, including 4140 g of water. The total daily average mass exchange imbalance was 29 g for all substances, including 17 g for water.

This mass exchange imbalance in the system is the difference between the recorded inputs and outputs of substances. The experiment, however, did not reveal any deviations in the operating conditions of the links, which could be accounted for by a mass exchange imbalance.

It is worth detailing the mass exchange functions of the microbial cultivator in the system. In the experiment described above the microbial cultivator on average produced 101 liters of carbon dioxide daily. To fix this amount of CO_2 it was necessary to grow 10.3 g of algae more than was needed to assimilate carbon dioxide produced by the human every day, i.e. to increase the capacity of the algal cultivator in this experiment by about 2.7%. Besides, to grow additional quantities of algae an additional input of mineral substances was needed as mineralization of human solid wastes did not yield enough mineral substances to synthesize an additional amount of algae.

The increase in activated sludge microbial biomass from the microbial cultivator is a by-product produced with the expenditure of electrical power and some substances, such as H_2SO_4. Additional algal biomass generated in the system upon the introduction of the microbial cultivator is not a useful product either, as it finds no application in the system.

It is hoped that under the improved conditions of microbial cultivator operation human solid wastes can be mineralized in such a way that sludge biomass will not increase. If, however, human solid wastes are mineralized completely, the resulting mineral elements will be hardly enough to grow an additional amount of algae that could assimilate the CO_2 emitted during mineralization. The mass of additionally grown biomass will be approximately the same as that of mineralized releases.

Thus, the microbial link fits well into the system where the regenerative link is represented by algae and, in part, higher plants. It does not affect adversely the gas and water exchange of the system, but ultimately it merely transforms unneeded fecal biomass to equally unneeded algal biomass. Evidently, the link of human solid waste mineralization is quite a costly luxury for this system.

There is another, more general reason why it is pointless now to introduce into the system the link of solid waste mineralization: mass exchange between the system and surrounding environs must be in balance.

As long as the system receives part of its food (e.g. animal foodstuffs) from the outside (or from storage), equal amounts of organic matter have to be removed from the system. The human mean daily requirement for animal food is about 30–40 g of fat and 40–50 g of protein. Thus even for the system "human–higher plants" with complete regeneration of vegetable food the mass of organic matter removed from the system (70–90 g/day) is significantly larger than the dry mass of human solid wastes (25–30 g/day). Hence it follows that until the system with full regeneration of vegetable and animal food is created, there is no point in adding a specialized link for human solid wastes. A simpler way is to remove dried wastes, together with part of the inedible plant biomass from the system to compensate for animal or other foodstuffs brought into the system.

This system has a more urgent requirement—for biological or physicochemical mineralization of inedible plant biomass. The edible fraction can hardly exceed 50–55% of the total yield, vegetable food constituting 80–90% of human diet. Therefore, in-the-system production of all vegetable food required by humans would result in upsetting the balance, production of excess inedible biomass, excess oxygen, and lack of carbon dioxide. To keep the material cycle of this system in balance, it will be essential to oxidize a significant part of the inedible plant biomass within the system, to consume the excess oxygen and regenerate the carbon dioxide and mineral components required by plants. Consequently, as the system gets more closed, greater and greater importance is attached to the development of the link of biological or physicochemical mineralization of inedible plant biomass rather than human solid wastes.

CHAPTER 9

Experiments Testing Prolonged Human Life in the Bios-3 Life Support System

9.1 EXPERIMENTAL COMPLEX BIOS-3

Regenerating Atmosphere, Water, and Nutrients to Sustain Three Human Lives in a Closed Ecosystem

Creating a continuous system that can remain active over the course of many days and that can be controlled by one person is difficult enough. But in reality, the future of space travel belongs to cooperative teams, not to those who fly solo. A striking example of team effort in space travel is the way the Shuttle team worked together to repair the Hubble telescope. It is for this reason that our life support research veered towards developing a closed system that could sustain the minimum necessary and sufficient number of human inhabitants—2 to 3 persons. This was and is the goal of Bios-3.

In order to conduct experiments with a crew of several people living in the system, Experimental Complex Bios-3 was assembled during the 1960s at the Biophysics Division of the Institute of Physics of the Siberian Branch of the Soviet Academy of Sciences. (This division was later renamed as the Institute of Biophysics.) Autonomy of control is a special feature of this system, and it is a feature that has not since been repeated in any other experimental complex. From the very beginning Bios-3 was designed to be governed by those living in it, thus simulating conditions that would adhere in life support systems created for use in space.

Bios-3 not only makes it possible to study how several people can work together to control the closed system in which they are living, but it also allows the research of the problem of the reciprocal influence that microbes and humans have upon one another, as well as other problems that arise when more than one person inhabits a closed ecosystem.

Three things allowed researchers to proceed to the Bios-3 stage in the development of closed systems—that is, to the creation of systems with autonomous internal control: 1) the favorable results of medical tests conducted on the human test subjects in Bios-1 and Bios-2, 2) biosynthetic processes of sufficient intensity where this intensity is enabled by a well-developed cultivation technology, and 3) the high degree of self-sustainability and reliability found in systemic biological links which manifested themselves over the many years spent studying the controlled cultivation of microalgae and higher plants.

In all experimental ecosystems implemented before Bios-3, the human test subject participated in the system as nothing more than a metabolic link. No system control was accorded to this tester. Instead, the sometimes numerous staff that serviced the system from the outside were those who controlled the system. The design of the metabolic control for such an outwardly-controlled system was based on the results of analyses conducted with respect to a multitude of systemic processes. Analytical chemists, biochemists, bacteriologists, algologists, agronomists, and doctors have performed these analyses according to a complex, specially developed regimen. According to their data, control decisions were made and implemented by those operators outside the system who directed the system's technological processes. Since more people control the ecosystem than live in it, the ecosystem is clearly not closed with respect to its informational–control linkages and therefore cannot exist as an independent entity.

Not only is a system controlled from outside not an independent entity, but in practical terms, a closed ecosystem controlled from outside does not satisfy the main functional specification for which it was created in the first place—facilitating life support for human teams that are far away from their native planet Earth.

Interestingly, the material and informational closure of the system are interconnected. The necessity to remove a noticeable quantity of substances from the system for external analyses and technological operations impedes its material closure.

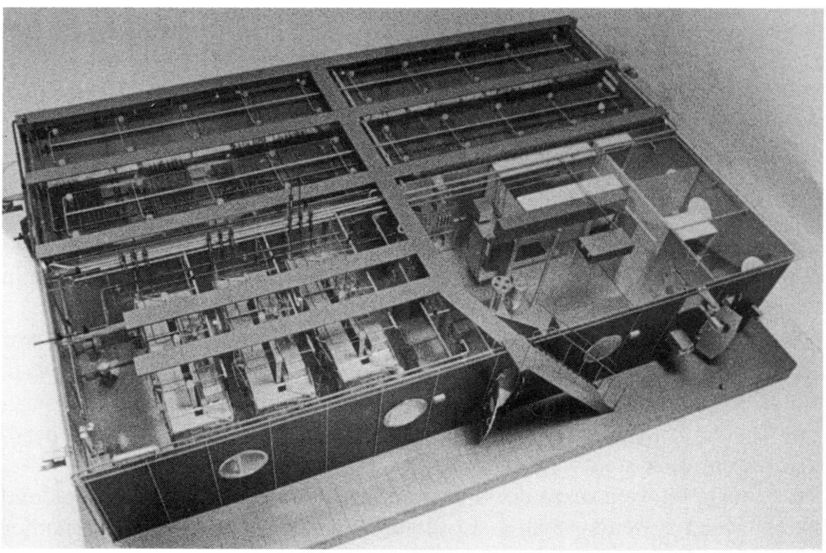

Figure 9.1 General view of the Bios-3 (model with transparent roof). Front left—algal compartment; right—crew compartment; back—two higher-plant compartments. Light sources are on the roof of installation. Ladders and gangways on the roof are for servicing light sources.
On the front wall, to the right—entrance of one of crew's cabins. To the right and left of it—locks for passing tools, chemical reagents and other things in and out.

Transferring control to within the system required the development of new technological variants and bioregenerative techniques: it was necessary to 1) sharply reduce the hours spent on technological operations, 2) minimize the types of analyses essential to system governance, and 3) prepare the system inhabitants to work independently in maintaining the life support system.

These problems were solved in parallel and their solution has led to the creation of the Experimental Complex Bios-3. Bios-3 is perfectly suited to implementing and researching experimental ecosystems with internal control.

Figure 9.2 Bios-3 (model, without roof). Top view. On the left—two higher-plant compartments. Under wheat trays can be seen pumps and fans of internal cooling systems and plant mineral nutrition. On the right in the foreground—algal compartment with three cultivators; at the left wall—fans of the atmosphere cooling system; at the nearest wall—separator and gas blower; at the far wall—power panel and cultivator control desk. In the background—living quarters.

Experimental Complex Bios-3: an Overview

Experimental complex Bios-3 is illustrated in Figures 9.1 and 9.2. Bios-3 is enclosed in a stainless steel, hermetic, rectangular welded housing of dimensions $14.9 \times 9 \times 2.5$ m. It is partitioned into four equal and hermetic compartments. The higher plants (phytotrons) are accommodated in two of these compartments, one is occupied by the unicellular algae growing in cultivators, and the last is inhabited by a crew of three people (see Figure 9.3). In the crew's compartment, there are three individual cabins, a kitchen–dining room, and a room equipped with a shower and a toilet. One of the cabins can be hermetically sealed to study the metabolism of the person located within it. The toilet serves as the entrance to the whole complex through an entryway compression lock. The crew's compartment also includes a common area that functions as a laboratory, studio, and recreational room.

Figure 9.3 Living quarters of the Bios-3 (model). On the left there are control boards for processes in higher plant compartments, tanks and ion exchange columns of the system for additional purification of drinking water; in the center there are a drying oven and a thrasher; on the top right there are a kitchen, and a dining-room, in which can be see a cooker, a table, a chair, a refrigerator, on the right there are doors of crew's cabins.

The overall dimensions of the complex are 315 m^3. The volume of each compartment is about 79 m^3. The volume of the assembled equipment within the complex makes up less than 1% of the complex volume, and for this reason the gas phase volume in calculations is taken as being equal to the overall volume.

All the compartments of the complex are joined by hermetic doors and there are hermetic doors that open to the outside of the complex from each compartment. Each door is designed to be opened by one person from the inside and the outside, when necessary, in no more than 20 s. This was extremely important for maintaining the safety of the test subjects: they were able to leave Bios-3 without delay and without external help in cases of danger from fire, etc.

Algal Cultivator Compartment

There are three algal cultivators in the compartment where unicellular algae are designated to grow (see Figure 9.4). Each of these cultivators has a light-receptive surface measuring 10 m^2, and the productivity measured in dry algal biomass reaches up to 800 g/day (see Figure 9.4). The algae-growing technology and cultivator design do not vary in principle from those described in Chapter 6, but these cultivators are to be serviced all at one time by one person.

Figure 9.4 Algal compartment of the Bios-3 (model). On the cultivators can be seen nutrient solution tanks, foam separators, outgoing gas coolers, thermoregulation system service lines.

Phytotron Compartments

Two identical phytotrons are in the Bios-3 complex in two different compartments. In each of these compartments, there are two systems for cultivating the plants, as well as an illumination system and a heat removal system.

The Main Cultivation System

The main cultivation system consists of 12 identical trays measuring 140 × 100 × 12 cm and a device that distributes the nutritive medium (see layout in Figure 9.5 and crew members at work in Figures 9.6, 9.7, and 9.8). All trays and service lines were manufactured from stainless steel. Two nutritive media can be used simultaneously in two groups of plant cultivation trays (see Chapter 7, Section 4).

Phytotron Compartment Heat Removal System (Figure 9.9)

The phytotron heat removal system is composed of two subsystems that use water as the heat-transfer agent. This enables the removal of heat through the metal phytotron compartment wall in two ways. From the inside, a pump (1) and a collector (2) gather the internal water that has condensed on the wall into a chute. Water condensation is suctioned by the pump (4) and through the collector (5) which is positioned under the ceiling. This water travels through the collector to the phytotron perimeter into the space between the wall and the metallic screen. In this same space, only from below, fans (10) cool the air. Warm water accumulates in another chute (6) from which it is again sent to be cooled, but cool and dry air enters the phytotron upper section. From the outside, water in the pipes cools the outer phytotron wall, and is then used to cool the light source—xenon lamps DKcTV6000. Phytotron atmospheric temperature is regulated by changing the flow rate of the primary cooling agent—the water in the pipes.

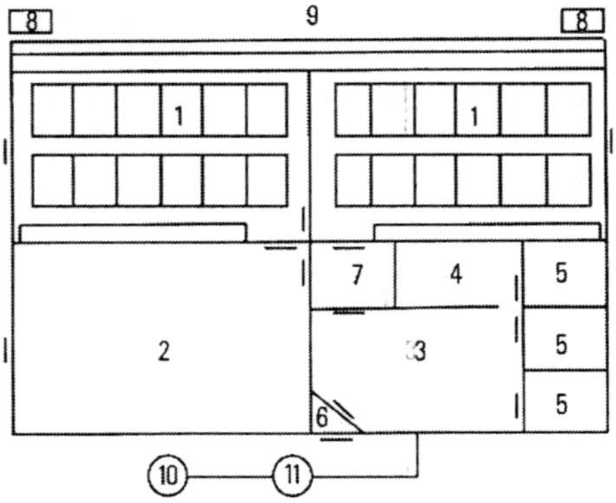

Figure 9.5 Design of the Bios-3 experimental system. 1—phytotrons; 2—algal cultivator compartment; 3—living quarters; 4—kitchen–dining-room; 5—cabins; 6—toilet lock; 7—vestibule; 8—pumps of the cooling system for light sources; 9—watering collector of the heat exchange wall of phytotrons; 10—pressurization compressor; 11—bacterial filter.

Figure 9.6 Harvesting in the phytotron vegetable section.

Figure 9.7 In the microalgal cultivator compartment.

Figure 9.8 Repairing of control system by a crew member.

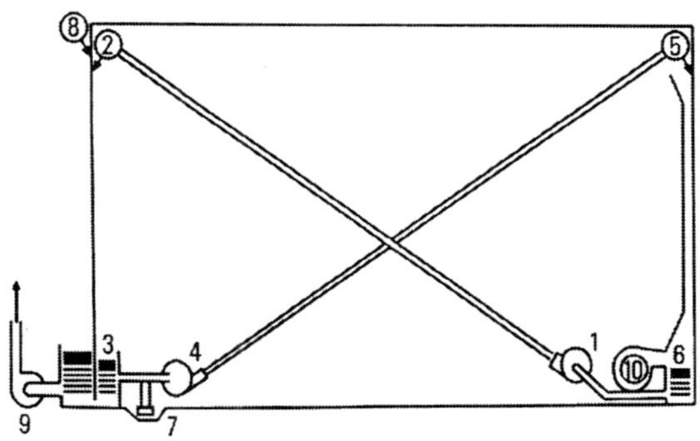

Figure 9.9 Phytotron heat removal schematic diagram (vertical section). 1—pump of heated heat-transfer agent; 2—collector for heated heat-transfer agent; 3—chute; 4—pump; 5—collector for cold heat-transfer agent; 6—receiver of heated heat-transfer agent; 7—float valve of condensate suction; 8—tap water collector; 9—pump for cooling 6-kW xenon lamps; 10—fan.

When cultivating plants in a phytotron compartment, transpirational water condenses on the cooling system, and for this reason there is a constant supply of water to make up for the lack in nutritive medium volume. The nutritive medium volume is adjusted by automatic relays and contact meters that measure the medium level.

Condensate that forms on the cooling system pipes inevitably gathers on the phytotron floor during its operation. The water that leaks through cooling pipe seals and glands, etc. is almost completely extracted from the system to retain its closure by a floating valve device (7) installed on the phytotron floor specially designed to gather this water. A pump draws the water through this device and into the cooling system. To increase the cooling system's reliability, the pumps (1, 4) are mirrored. Phytotron systems are governed by a control panel that is situated in the habitable compartment.

Both the taking of nutritive medium samples for analysis and the adding of corrective elements to the medium are accomplished from the habitable compartment by means of specially devised pipes.

The presence of the crew in the phytotrons is required only during the scheduled technical work and agrotechnic operations (harvest and sowing of plants, thinning them, etc.). The operating condition of the systems that supply the nutritive medium and that cool the phytotrons is automatically checked and requires no human monitoring. When one of the process parameters deviates from its targeted norm, then a light and sound indicator are activated on the control panel in the habitable compartment and on the outside of the phytotron. Conditions that might activate these indicators include: untimely supply of nutritive medium, a pressure decrease in the head pipes carrying the cooling agent or the nutritive medium, the switching off of energy-consumptive devices, etc.

Phytotron Illumination System

Research into the dependency of plant productivity and photosynthetic efficiency on irradiation intensity has shown that the maximum photosynthetic efficiency coefficient in a cenosis is reached when the irradiation intensity is 100–120 W/m^2 of photosynthetically active radiation (PAR). A further increase in irradiation intensity is accompanied by a growth in productivity, but the photosynthetic efficiency coefficient decreases.

Taking this dependency into consideration and assuming the projected oxygen output of the plants cultivated in phytotron compartments to be 1800–2000 l a day, the irradiation level of the plants was taken to be 140–180 W/m^2 PAR when designing and building the experimental complex Bios-3. Such an irradiation level ensures sufficiently high harvest yields for different cultures under the 7–8% average photosynthetic efficiency coefficient for plants. Under these conditions, the cultivation area has acceptable dimensions and is not too large.

Six-kilowatt, miniature, water-cooled xenon lamps DKsTV6000 were used to produce illumination (Marshak *et al.*, 1963). The xenon lamp DKsTV6000 is a powerful source of radiant energy with a basic stream of PAR of up to 800 W. The radiation spectrum in the visible region is close to that of sunlight. The lamp is water-cooled, and this simplifies the removal of excess heat.

Twenty lamps were placed in each of the two phytotron compartments of Bios-3. The maximum electrical power required by the lamps reached 240 kW, which made up 96% of the energy that was required to run the whole system. The lamps and other electrical power consumers were connected to city electric lines, and these had an alternating current and a voltage of 220/380. The energy associated with the photosynthetic process makes up an insignificant portion of the entire electrical power consumed by the system—about 1%. Almost all the energy introduced into the system is converted into heat energy that must be removed from the system.

Heat removal is accomplished by water at about 10°C that flows in pipes with a flow rate that reaches up to 8 m^3/h and that then travels between the two jackets that enclose the xenon light source. The cooling water removes about 75% of the heat energy produced by the lamp. A part of the heat energy radiates out through the phytotron ceilings. The ceiling temperature can reach up to 50°C when the lamps are in operation. The heat that radiates through the phytotron ceiling adds up to about 2500 kcal/h. The heat that remains in the phytotron is 29200 kcal/h. The total quantity of heat removed is 77400 kcal/h.

The ratio of PAR to infrared radiation in xenon lamps was initially 1:3. A negative effect of infrared radiation can be detected when the plants are radiated at intensities greater than 120–130 W/m^2 PAR. To decrease the portion of infrared radiation in lamp radiation, the design of the light fixture jackets was altered. The light source was installed into a quartz jacket. The quartz jackets were vertically affixed to the phytotron ceiling. The quartz jacket, in turn, was enclosed in a glass jacket. The ratio of PAR to infrared light was reduced to 1:1.25. The excess infrared and UV that radiates from the xenon lamps is absorbed by the glass jackets that help form the lamp cooling system.

Four control panels were built to govern 10 lamps each. The lamps were turned off and on individually. Each control panel was equipped with a starting device—an ignition device. An electronic blocking system to regulate the cooling water pressure and temperature was included in the control panel scheme.

The DKsTV6000 lamps were intended for use with distilled water as a cooling agent. However, the experience of many years has shown that it is possible to use water right out of the Yenisei River[1], which has a low temperature (3–10°C) and is sufficiently pure. Special pumps propelled water in pipes from interim vessels past the lamps, and this helped to reduce the number of lamps that switch off due to water pressure fluctuations in the water pipes. For this reason, the lamps in Bios-3 were cooled by pipe water from the interim vessels. The cooling water entered these vessels after leaving cooling towers and then cooling the phytotron wall outer surfaces.

The installation of lamps in each phytotron makes it possible to irradiate the plants from all sides. On the horizontal surface of the sown area in the phytotron compartments, about 70% of the radiation falls under an optical angle of 15–60°. This enables a more evenly distributed radiation for all plant phytoelements, and normal plant growth and development.

Habitable Compartment

The Bios-3 habitable compartment has an area of 4.5 × 7 m. There are three cabins for the crew located in this compartment (see Figure 9.10). In one experiment, a device for burning atmospheric volatiles and inedible biomass from plants was placed in one of the cabins, and the crew lived in the two remaining cabins. There was also a kitchen–dining room in this compartment, which was equipped with an electric stove and a refrigerator. The equipment needed to prepare drinking water was installed in the kitchen.

[1] The Yenisei River is a great Siberian river, one of the largest rivers in the world. The Krasnoyarsk Institute of Biophysics, which housed the experimental complex Bios-3, is situated on the banks of the Yenisei. After building a high dam and hydroelectric station on the Yenisei, the temperature in the part of the river that was lower than the dam did not climb higher than 12°C throughout the year and in winter time it can fall to 3–4°C.

The sanitary area is used for sanitary–hygienic procedures. At the same time, it served as the entrance to Bios-3. There was a sink, shower, electric water heater, and a sewage-disposal unit in the sanitary area. In the vestibule near the phytotron doors was a dustless thresher–winnower to grind the wheat and separate the grain from the chaff, and there was a grinder to make flour.

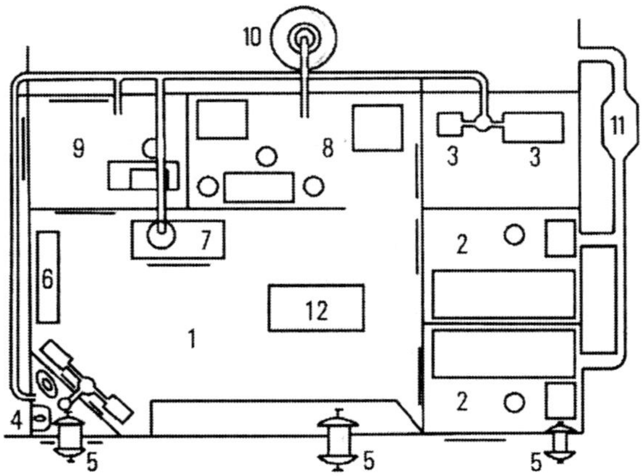

Figure 9.10 Design of Bios-3 living quarters. 1—working room; 2—cabins; 3—catalytic converter for incineration of atmospheric contaminants and inedible biomass; 4—toilet lock; 5—lock for exchanging items; 6—phytotron monitoring and control desk; 7—drying oven; 8—kitchen–dining-room; 9—vestibule; 10—circulation fan; 11—heat exchanger; 12—conditioning device (hanging at the ceiling).

There was also a work table in the habitable compartment, a control panel for the electricity consumers in the phytotron compartments, a drying cabinet, an air conditioner, and other auxiliary equipment.

To transfer objects back and forth during the experiment, the habitable compartment came with three compression chambers with external and internal air locks: one in the sanitary area to transfer sanitary vessels, one in the work area to transfer equipment, instruments, and so on, and one in the crew's cabin for medicinal purposes.

The atmosphere was circulated between the compartments during the Bios-3 experiments. The ventilation fan suctioned air from three points of the habitable compartment: the kitchen, the sanitary area, and the phytotron vestibule. Besides this, air was continually suctioned out of the drying cabinet and the thermocatalytic furnaces. The total air flow rate of air suctioned out of the habitable area equaled 0.5–1.5 m^3/min.

The gas exchange between the phytotron compartments was accomplished by means of an open door between them and the cooling system fans. When the ventilation fan was working, the phytotron pressure held at 10–15 mm of water higher than in the habitable compartment. The phytotron compartment atmosphere entered the crew cabins through the heat exchanger and from there flowed on to the work area.

The condensate that accumulated in the heat exchanger was used by the crew for preparing drinking and hygienic water.

When the crew left the habitable compartment to work in the phytotron compartments, the ventilation fan turned off, and after the pressure in the two compartments equalized, the passage door underwent decompression. This regimen was necessary to ward off a mass penetration of the phytotron atmosphere, which is rich in microflora and fungal spores, through the door into the habitable compartment.

The Device for Burning Atmospheric Volatiles and Inedible Plant Biomass

In the first experiment conducted using Bios-3, it was found that the Bios-3 atmosphere could be toxic for plants. The source of the toxicity was not identified, but it was assumed that these toxins were reduced or not fully oxidized volatile compounds in the atmosphere.

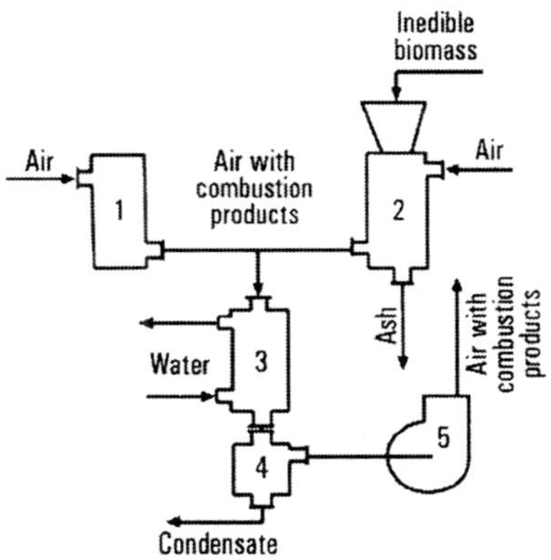

Figure 9.11 Block diagram of the installation for incineration of inedible biomass and atmospheric contaminants. 1—catalytic converter; 2—thermocatalytic converter; 3—heat exchanger; 4—condensate collector; 5—exhaust fan.

In order to oxidize these compounds up to higher oxides, and to increase the closure of the mass exchange in the life support system with respect to carbon, oxygen, and hydrogen, a device was introduced into the system that would burn up atmospheric volatiles and the inedible plant biomass (see Figure 9.11). The device consists of the following main assemblies: a catalytic furnace for the combustion of atmospheric volatiles (1); a thermocatalytic furnace for the combustion of inedible biomass (2); a heat exchanger (3); a condensate collector (4); and an exhaust fan (5).

The catalytic furnace for the combustion of atmospheric volatiles (see Figure 9.12) starts working when atmospheric air containing organic substances is suctioned through the inlet pipe (1) into the catalytic furnace. Having passed through the recuperative heat exchanger (2), the air enters the chamber containing the heating elements (3), where its temperature is raised to 600–650°C. After the heating elements, the hot air enters the chamber containing the catalyst (4), where the volatiles are oxidized up to CO_2, H_2O, SO_3, and NO_2. The catalyst is arranged in layers. The first layer is composed of ferrochromic compounds (TU6-03-317-72) and measures 250 × 250 × 100 mm; the second layer is composed of zinc–chromic compounds (TU6-03-345-73) and measures 250 × 250 × 200 mm; the third layer is composed of pentoxide, vanadium, and titanium dioxide compounds (TU-38-10-278-75) and measures 250 × 250 × 100 mm; and the fourth layer is composed of aluminum platinum compounds (TU-38-101-486-74) and measures 250 × 250 × 200 mm.

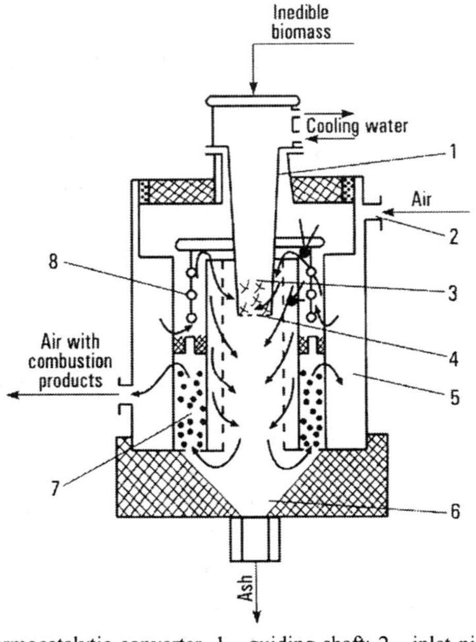

Figure 9.12 Scheme of thermocatalytic converter. 1—guiding shaft; 2—inlet pipe; 3—shaft (combustion zone); 4—grate; 5—recuperator; 6—ash collector; 8—heating element.

After the catalyst chamber, the air with the oxidation products passes through the recuperative heat exchanger and gives off part of its heat, then it enters the main heat exchanger, after which a fan sends it into the phytotron atmosphere. The air flow rate through the catalytic furnace is 300–400 l/min. It should be noted that with this rate of air flow and with the total volume of the atmosphere in the system amounting to 300 m^3, the process of complete oxidation of a one-time input of toxic volatiles into the system lasts 36–48 h. If these volatiles are released gradually, the system's atmosphere will constantly contain some background amounts of pollutants. This allows a suggestion that in spite of the continuous operation of the catalytic furnace, the system's atmosphere could contain small amounts of volatile organic pollutants, e.g. ethylene. Unfortunately, during the experiment, there was no chemical control of these microimpurities.

During thermocatalytic furnace operation, ground biomass (the largest particle size is 10–20 mm) from the loading bin travels along the input shaft and pours into the shaft (3) where it is held on a strainer (4). Air passes through the recuperative heat exchanger (5) and the heating element chamber (8) to be suctioned through the input pipe (2) and continue through the shaft (3). This air has been heated to 680–700°C, and when it comes into contact with the biomass in the shaft, the biomass burns. The ashes that form during combustion fall out into the ash collector (6). The gaseous combustion products enter the catalyst chamber (7) with the air. The composition and volume of each catalyst blend is the same as for the catalytic furnace. Not fully oxidized combustion by-products are oxidized here up to CO_2, H_2O, SO_3, and NO_2. Having passed through the recuperative heat exchanger, which is a helical heat exchanger, and having given off part of its heat, the air with gaseous combustion by-products enters the main heat exchanger and then flows into the phytotron atmosphere. The air flow rate through the thermocatalytic furnace is 200–300 l/min.

Thermocatalytic furnace operation for the combustion of inedible biomass requires the participation of one crew member. Once engaged, the catalytic furnace for the combustion of atmospheric volatiles works according to a regular regime without the participation of a crew member.

Bios-3 External Systems

Bios-3 external systems include the energy supply—for the complex as a whole and for the light sources separately—and an external cooling system that removes heat from phytotron outer walls and from the light sources. Fluctuations in outer atmospheric pressure could possibly destroy Bios-3 walls if a special system to equalize the inner and outer air pressures were not installed. This system equalizes the air pressure without allowing any air to pass between Bios-3 and the outside world. A convertible pressurization–depressurization system is used to accomplish this. It consists of a membrane compressor, gas cylinders, gas meter, and microbiological filter. This pressurization system is regulated by membrane pressure meters, which are installed in the habitable compartment of the complex. As a result of pressurization, the pressure in the habitable compartment is maintained slightly higher than atmospheric pressure by 15–30 mm of water. This prevents external microbes from entering the complex during the running of the experiment.

The external systems, as opposed to the internal systems that were described above, were governed and serviced by the external systems operator on duty. All Bios-3 compartments and crew cabins were equipped with a selector line to the control panel of the external systems operator on duty and to the doctor on duty. Radio and TV communications were a reserve channel. Besides this, the crew cabins were equipped with a telephone line to the doctor on duty.

The experimental Bios-3 complex was equipped with industrial television. The cameras in this system were installed on the outside of the complex at the windows of each compartment and were controlled by the external systems operator on duty.

As is clear from the description, the technological equipment that enabled the functioning of the higher plants and the crew was situated inside the Bios-3 housing, and during the experiment it was serviced only by members of the crew. The only thing available from outside the complex was technical consultations with specialists.

The design of the experimental Bios-3 complex was developed under the direction of B.G. Kovrov by an engineering group. A.N. Biriukov, L.A. Veber, G.B. Denisov, Yu.P. Zub, V.P. Lashinsky, V.M. Losev, E.S. Melnikov, G.S. Petrov, A.A. Shtol, and V.V. Yakubovsky took part in this engineering effort.

9.2 EXPERIMENTS TESTING PROLONGED HUMAN LIFE SUPPORT IN THE BIOS-3 SYSTEM[2]

Bios-3 has been the first and until now the only experimental system in which human requirements for gas, water, and vegetation were fully provided by the material cycle in the system by means of biological processes, especially photosynthesis. That is why we have considered it worthwhile to give a rather detailed description of the life of the humans in this system during each of the four-, five-, and six-month experiments.

The main differences in the Bios-3 experiments as opposed to the previous Bios-1 and Bios-2 experiments are:

1. Not just one person lived in the system, but a team of 2–3 people. This fact alone substantially alters the dynamic character of the system's metabolism and introduces a new ecological factor in the exchange of microflora between humans, and also alters the psychological situation by diminishing the problem of loneliness. However, new problems arose with Bios-3: crew member compatibility and role distribution.
2. For the first time, the system became autonomous, that is, governed and serviced by the humans living in it.
3. The closure of the system was significantly increased with respect to food. During some experimental periods, practically all the vegetative portion of the food ration was produced within the system.
4. The high closure of the system allowed the conducting of the first Bios-3 experiments with an eye to observing the autonomous dynamics climax of the system microflora without external contamination.

The novelty of this living habitat sets an overriding problem before the researcher of the experimental ecosystem: to ensure the complete safety of the participants living in the system. This problem demands conducting multifaceted tests on the human participants, including an analysis of the functioning of vital body organs and physiological systems and meticulously maintaining the hygiene of the habitat.

[2] This section was written by Yu.N. Okladnikov and I.I. Gitelson.

Humans in an experimental ecological system perform a double role. They are the reference-element link that determines the system's structure, and at the same time, they are the most sensitive indicator of the completeness of its intermediate functions. No matter to what level of detail ecosystem links are investigated, at the last stages of system creation humans remain irreplaceable as a metabolic link that cannot be imitated by any other creature due to human complexity. Humans are also irreplaceable as the determinant of biological compatibility for the system, since they are an absolutely unique species.

After the system had been subjected to prolonged preliminary tests and improvements, and we could be sure of its reliability and safety for its inhabitants, on December 24, 1972 three researchers of the Institute, preliminarily selected by a competition and subjected to thorough medical examination, N. Petrov, M. Shilenko, and N. Bugreev, entered the Bios-3, hermetically sealed the hatch, and spent 180 days inside.

Prior to the introduction of humans into the system, plant conveyer in the phytotron was brought to the steady state.

The first experiment was followed by two more, in 1977 (with M. Shilenko, N. Bugreev, and G. Asinyarov as test subjects) and 1983/84 (with N. Alekseev, L. Mozgovoy, and N. Bugreev). The results made us confident of the feasibility of creating a reliable and safe biological human life support system.

When a crew of several people is introduced into the system, a series of medical, psychological, and even ethical problems arise. While solving them, the following principles were assumed:

1. The experiment participants are laboratory researchers who took part in Bios-3 system development. This ensures trust on the part of the test subject toward a very familiar system, knowledge of its conditions, a considerable interest in the course of the experiment, an understanding of the importance of all operations, and the ability to make valuable observations that can be used to further perfect the system.

2. The experiment directors make the final choice of test subjects from among the volunteers on the basis of the conclusions of a medical committee and taking into consideration the recommendation of the operator group that will work together with the crew from the outside during the experiment.

3. The test subject can stop the experiment and abandon the system at any time. The test subject knows this option is ensured by a system of hatches that permit opening any hatch for a few seconds to enable exiting from the system without outside help.

4. Since the problem of mental isolation was not investigated in these experiments, contact with the outside world is predominantly available to the test subjects.

5. The test subjects are informed about all tasks and details of the experiment in progress, and they actively participate in tests and analyses. Decisions concerning program changes and the course of the experiment are made with crew participation.

6. During the whole experiment, a doctor must be on call at all times. The test subject must report on all changes in their state to the doctor on duty or when called upon to do so by the director of the experiment's medical program.

These principles turned out to be solid enough. During the experiments, the total length of which added up to more than two years, there were no serious difficulties with respect to psychological issues. Not one of the experiments was broken off due to a test subject not wishing to continue. Probably the feeling of being immured disappeared in the face of the high degree of responsibility and interest in the course of the experiment, and this ensured the good psychological climate that reigned during the experiments (Gitelson and Okladnikov, 1997).

The length of the experiments grew gradually from a few hours to a half-year of continuous residence in the system. The basis on which the duration of the experiment was lengthened and new links were introduced into the system was the absence of any change in the basic indicators that characterized the state of the crew at the end of each stay in the experimental ecosystem.

Bios-3 Half-Year Life Support Experiment for a Three-Member Crew

The Bios-3 experiment that began on December 24, 1972 ended according to schedule on the 180th day—June 22, 1973. It consisted of 3 phases, each two months long, which differed in their mass transfer characteristics. The gas and water exchange variables for the three experimental phases are depicted in Figures 9.13–9.15.

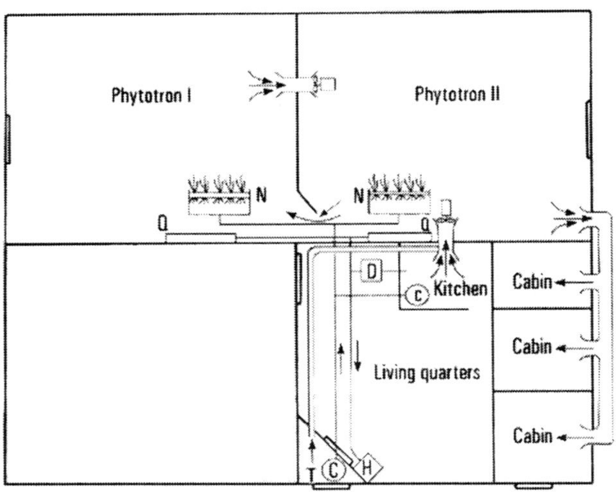

Figure 9.13 Schematic diagram of gas and water exchange in the Bios-3 installation at Phase I of the experiment. Q—collector for transpirational moisture condensate in the phytotron; N—wheat nutrient medium; D—unit for additional treatment of drinking water by sorption; C—sewage and gray water collectors in the toilet and kitchen; T—toilet and shower; H—vessel for boiling and storing hygiene water. Arrows show the direction of liquid and gas flows.

During the first phase of the experiment, the system consisted of 2 phytotrons containing both a wheat culture and a selection of vegetables and a living compartment for the crew. All crew requirements for gas and water were satisfied by the higher plants. Gray water from the living compartment was pumped into the wheat's nutrient medium. Solid and liquid wastes from the crew were removed from the Bios-3 complex. Crew food requirements were satisfied in two ways. The grain (bread) and vegetables produced by the higher plants provided fresh food, and the freeze-dried (lyophilized) food stores stocked in the Bios-3 complex at the beginning of the experiment were sources of animal protein and other processed food.

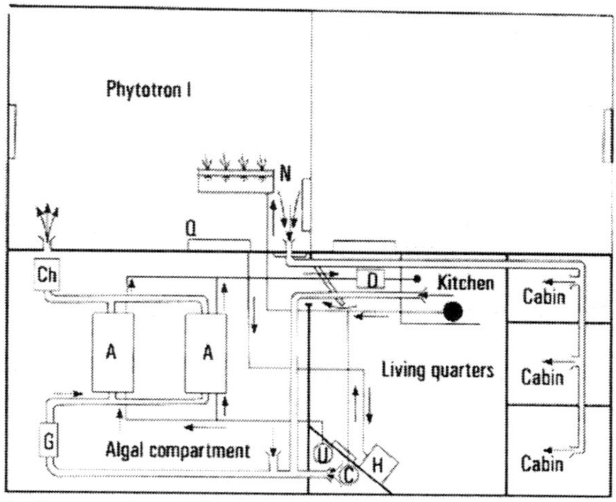

Figure 9.14 Schematic diagram of gas and water exchange in the Bios-3 installation at Phase II of the experiment. A—microalgal cultivator; G—gas blower; Ch—charcoal filter; C—sewage and gray water collection in the toilet and kitchen; N—wheat nutrient medium; Q—collector for transpirational moisture condensate in the phytotron; H—vessel for boiling and storing hygiene water; U—urine collector; D—unit for additional treatment of drinking water by sorption. Arrows show direction of liquid and gas flows.

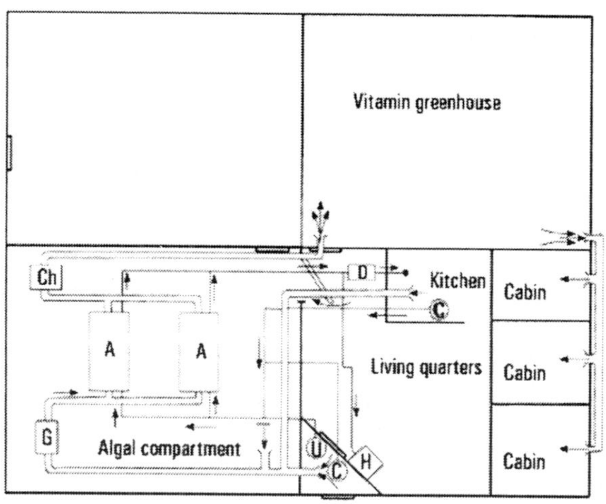

Figure 9.15 Schematic diagram of gas and water exchanges in the Bios-3 installation at Phase III of the experiment. The same notations are used as in Figure 9.14. (Phase II).

During Phase II of the experiment, one phytotron was removed from the system and a compartment containing *Chlorella* cultivators was introduced in its place. The cultivator photoreactor enabled the system to attain a higher degree of closure with respect to gas and water exchange. Now human gas and water exchange requirements were satisfied by the combined photosynthetic activity of the photoreactors and the phytotron.

Crew liquid wastes were consumed by the photoreactors, the solid wastes were dried so that water could be returned to the system. The nutrition system supplying the plants remained exactly as it was in Phase I.

In Phase III, the phytotron containing wheat and vegetables was replaced with a phytotron containing only a selection of vegetable crops.

Phase I: Higher Plants and Humans

Phase I was conducted to study gas and water exchange between higher plants and humans. The higher plant link consisted of wheat and vegetables. The short-stemmed wheat cultivar 232 produced in the selective culture of G.M. Lisovsky and a selection of 9 vegetable crops were used. The 9 vegetable crops were: Bordeaux table beet, Chantene carrot, dill, Petrov turnip, Peking kale, white Virov radish, spring onion, Din-zo-sn cucumber, and sorrel.

A continuous wheat culture was grown in a 14-stage "conveyer" (14 age-groups of continuous wheat culture were grown simultaneously: see Chapter 7, Section 4). The vegetables were grown in 6 different stages—except for onion, sorrel, and cucumber, which were grown without a conveyer.

During plant cultivation, the illumination was continuous and the radiant flux reached 145 W/m^2 in the baseline period before the experiment. During the experiment, this continuous flux reached 180 W/m^2 and was altered according to a specific schedule. The CO_2 concentration in the air was about 0.2% in the baseline period, and it ranged from 0.2 to 1.5% in agreement with changes in the experimental schedule.

The nutrient solution in the experiment did not change. The loss in water in the nutrient medium caused by evapotranspiration was compensated by returning condensed transpirational water to the nutrient medium. The crew corrected the nutrient medium twice a day for wheat and once a day for vegetables to ensure that the essential concentration of all elements in the nutrient medium was maintained constant and that its pH level was held stable within the range 6.2–6.8.

The productivity of each crop was evaluated after establishing a steady-state condition for each crop conveyer. Crop yield and productivity achieved during steady state are listed in Tables 9.1 and 9.2. In terms of total biomass, the most productive crops are wheat, carrot, and turnip. Their harvests exceeded the yield levels expected for field conditions. The ratio of edible biomass in the total wheat harvest was less than that for vegetables, but with respect to absolute productivity, which was calculated in terms of dry edible biomass, wheat deferred only slightly to carrot and turnip and exceeded the yield of other crops. The productivity achieved was close to the expected level. These values of crop productivity have been confirmed in the papers of Bugbee, Salisbury, Tibbitts, Wheeler, Knott, Sager, Mitchell, and others (see Chapter 3).

Table 9.1 Crop yield and productivity in the control period.

Crop	Area, m^2	Growth period, days	Vegetable yield, g/m^2		Daily productivity, g/m^2, of dry biomass		Edible biomass harvest, %
			Edible dry biomass	Total dry biomass	Edible	Total	
Wheat	33.6	63	1118	3182	17.7	50.5	35.1
Beet	1.8	80	927	1813	11.6	22.9	50.6
Carrot	2.4	80	1709	3491	21.3	43.6	49 0
Dill[a]		15–40	331	331	4.1	4.1	100.0
Turnip	1.5	53	989	1980	18.6	37.4	49.9
Kale[a]		15–30	297	312	5.6	5.9	95.2
Radish	0.75	27	341	624	12.6	23.1	54.7
Onion	0.35	80	830	964	10.4	12.0	86.3

[a]Dill and kale were not planted in their own isolated areas, but were sown as companion crops between rows of carrots and turnips.

The CO_2/O_2 assimilatory quotient for the higher plant link was around 0.94 in the baseline period. For wheat the photosynthetic efficiency for biomass synthesis was 7.4–9% of incident PAR, and it was thus not less than that for the microalga *Chlorella*.

The productivity reached by the higher plant link for the first phase of the experiment is detailed in Table 9.2. Each day the plants provided 30% of crew food requirements. On average, each of the three crew members consumed 200 g of air-dried grain and 388 g of fresh vegetables. With the exception of seeds retained for subsequent plantings and for analysis, the major portion of the grain yielded by the system was eaten by the crew members after being ground and baked into bread. The vegetables were eaten fresh and cooked in various dishes, with the exception of small quantities that were removed from the system for analysis and in the form of kitchen waste.

The total productivity of the higher plant link in Bios-3 was high enough to supply the required amount of oxygen for 3 people—about 1500 l/day. It was found that oxygen productivity could be increased to 2300 l/day if the illumination were increased. This afforded an opportunity to control the system's atmospheric composition during the course of the experiment by changing the intensity with which the plants in the phytotrons were illuminated. Governing phototrophic link productivity and system atmospheric gas composition through changes in illumination confirmed the accuracy of initial calculations made with respect to these parameters and which were made on the basis of physiological research performed earlier on higher plants. (See Chapter 7.)

Table 9.2 Crop yield and productivity in the two-month experiment "humans–higher plants."

Crop	Area, m^2	Growth period, days	Vegetable yield, g/m^2		Daily productivity of dry biomass, g/m^2		Edible biomass harvest, %
			Edible dry biomass	Total dry biomass	Edible	Total	
Wheat	33.60	63	967	3228	15.4	51.2	30.0
Beet	1.35	78	1298	2170	16.6	27.8	59.8
Carrot	2.07	78	1770	3669	22.7	47.0	48.3
Dill[a]		15–30	165	178	5.5	6.0	91.7
Turnip	0.36	52	1097	2917	21.1	56.1	37.6
Kale[a]		15–30	184	219	6.1	7.3	84.3
Radish	1.26	26	201	443	7.7	17.1	45.3
Cucumber	1.20	90	397	2259	4.4	25.1	17.6
Onion	0.69	90	532	532	4.1	4.1	100.0
Sorrel	0.27	90	2128	2128	16.7	16.7	100.0

[a]Dill and kale did not occupy their own area, but were sown as companion crops between rows of carrots and turnips.

Several higher plant characteristics that could not be foreseen and that were not apparent during their separate cultivation were unearthed when the higher plants were cultivated in the closed experimental system. Data on the limits of higher plant compatibility with respect to the remaining systemic links and with respect to the technological processes in the system proved the most interesting.

Phase II: Microalgae, Higher Plants, and Humans

The second phototrophic link, *Chlorella*, was included in the system during the second phase of the experiment. The microalgal link consisted of 3 cultivators, each cultivator working independently of the other and containing one photoreactor.

The continuous culture of a thermophilic strain of *Chlorella* served as the basic cultivation process. The technique and technology of this culture is described in Chapter 6. All technological cultivation operations were conducted within the system: centrifuging, biomass dehydration with a return of medium water to the system, and ongoing correction of the nutrient solution. Each cultivator was irradiated by a PAR flux having an intensity of 280–300 W/m^2. The photosynthetic efficiency was 6–8%. Each cultivator could satisfy the oxygen requirements and absorb the carbon dioxide produced by one person.

The compartment housing the algal cultivators was added to the system while one of the phytotron compartments was removed (see Figure 9.2). Sewage–sanitation water continued to enter the nutrient solution of the remaining phytotron in full volume, that is, in twice as strong a dosage as in the first phase.

With the introduction of *Chlorella* into the system to regenerate the atmosphere and with the doubling of the sewage–sanitary water dosage fed to the wheat, the state of the wheat worsened. The manifestations of this worsened state were: 1) the plants stopped gaining height, 2) their spikes became progressively sterile, 3) the leaves exhibited signs of general chlorosis—even among the young plants—and 4) root growth was curbed. In the end, root growth retardation brought about the destruction of the plants. The design of the experiment did not allow the distinguishing of the degree to which the plants were affected by the change in the nutrient medium that occurred due to the heightened dose of sewage–sanitation water as opposed to other factors.

Phase III: Microalgae, Vegetables, and Humans

During the third phase (Figure 9.15), instead of using a phytotron containing both wheat and vegetables, a phytotron containing only vegetables was used in the system, and the sewage–sanitation water was not used in the nutrient medium. Before this all-vegetable phytotron was joined to the system, the air in the experimental complex was regenerated for 65 days using *Chlorella* alone. When the phytotron was attached, the atmosphere proved toxic for the vegetables. For 2–3 days, the young potato and tomato plants stopped growing, and their leaves curled, although they did not dry out. The beet experienced a strong accumulation of anthocyan in its leaves as well as local necrosis. The leaves on the cucumbers yellowed and began to die, and the cucumbers did not flower. After the phytotron was closed off from the system and it was ventilated with the usual atmosphere, the young tomato and potato plants lost all signs of toxicity after 2–4 days. After 5–8 days, the cucumbers began to flower again. Repeatedly joining the all-vegetable phytotron to the system resulted in the same state of toxicity as would be induced by some strongly-acting agent. It was at this stage of the experiment that the *Chlorella*-regenerated atmosphere was positively identified as toxic to the vegetables. The toxic substance was not isolated at this time, and its source was not determined. But further observation of the effect of some volatile pollutants on plants showed that the effect closest to the above was produced by low ethylene concentrations.

System Control

Analyses that were necessary to control the system were the responsibility of the crew living within the Bios-3 complex. These tasks were kept to a minimum to ensure that the least amount of work possible was spent by test subjects to control the life support system by those living in the system. Work hours spent by test subjects to control the life support system made up not more than 20% of their time.

This provided the test subjects with more time for other tasks. In the future, time spent maintaining the system will be substantially reduced by introducing a higher degree of automation into the system. Time economy will prove very important in upcoming space flights when the astronauts will need the time saved on life support to conduct experiments directly related to the space flight missions.

Those researchers conducting the experiment from the outside were naturally interested in a wider circle of investigations than the information about internal system processes achieved by tests within the system. Therefore, laboratory analyses external to the system were conducted in tandem with those analyses occurring on the inside, and in some cases analysis grids were actually intensified for the external analyses. The main harvest for researchers was not a material product such as grain or oxygen, but it was the information gathered about system functionality. For this reason, the number of analyses performed significantly exceeded the number actually needed to govern and sustain the system.

Results of the Experiment

Mass Transfer

System mass transfer with the surrounding environs and between links was studied according to a complicated analysis grid. According to the results of these analyses, mass transfer was described as the daily movement of mass throughout the system.

The human role in the mass transfer of an autonomous system does not differ from this same human role in a nonautonomous ecological system. This role has been described in previous chapters and is therefore not reviewed here. We would only like to note that since crew exometabolites mingle with one another, the crew acts like one single ecosystem link. A link composed of three people rather than one person is advantageous in the sense that temporary fluctuations in an individual's metabolism are smoothed over by the total metabolism of the three people taken together.

Trace Element Dynamics

Trace element dynamics in the system were a special research topic (based on the data of I. Gribovskaya). The trace elements were: iron, lead, nickel, chromium, aluminum, titanium, molybdenum, boron, copper, zinc, and manganese. It was found that macroelement dynamics—for instance, the dynamics of nitrogen and phosphorous—are controllable and balanced, as opposed to trace element dynamics that cannot be governed. Trace elements are emitted from certain components in the system, and mainly from building materials, after which they can accumulate in other places in the system. In part, trace elements accumulate in the edible parts of plants. During the length of the experiment, no toxic effects were discovered due to uncontrolled trace element dynamics, but the concentration of some of them in the solutions circulated within the system rose or fell from 2 to 10 times. The imbalance in the movement of these trace elements poses a potential threat to the existence of the system. In the future, it would behoove researchers to pay special attention to the trace element dynamics of a closed system and the ways to control them.

Table 9.3 Requirements of the crew and 2-link life support system, which includes the crew in its supply of substances.

Crew requirements	g/day	Ecosystem requirements	g/day
Lyophilized food	1890	Lyophilized food	1531
Oxygen	2441.8	Chemical elements	226
Drinking water	9629	External water	4420
Sanitation water	29760	Activated charcoal for purifying gases	6.6
Soap, toothpaste, toothpowder	34	Soap, toothpaste, toothpowder	34
Total	34395.8	Total	6128.2

Note: The value of the closure of the material cycle (R) was calculated as follows:
$R = (1 - 6128.2/34754.9) \cdot 100\% = 82.4\%$.

Material Exchange

The closure of the material exchange cycle in the system is evaluated by the cumulative index $R = (1 - m/M) \cdot 100\%$.

As is readily apparent from Table 9.3, system closure during Phase I for the "humans–higher plants" variant was evaluated at 82%. When a third link, *Chlorella*, that consumes human liquid excretions, was introduced into the system, then closure grew to 91%.

Gas Exchange

The composition of the atmosphere regenerated by the system is distinguished by the presence of a series of organic volatiles and a heightened carbon dioxide content—from 0.4 to 1% for short-lived spikes to 2.3% at various phases of the experiment. As Figure 9.16 shows, volatile concentrations quickly reach equilibrium values, and they fluctuate around these values during the course of the remainder of the experiment.

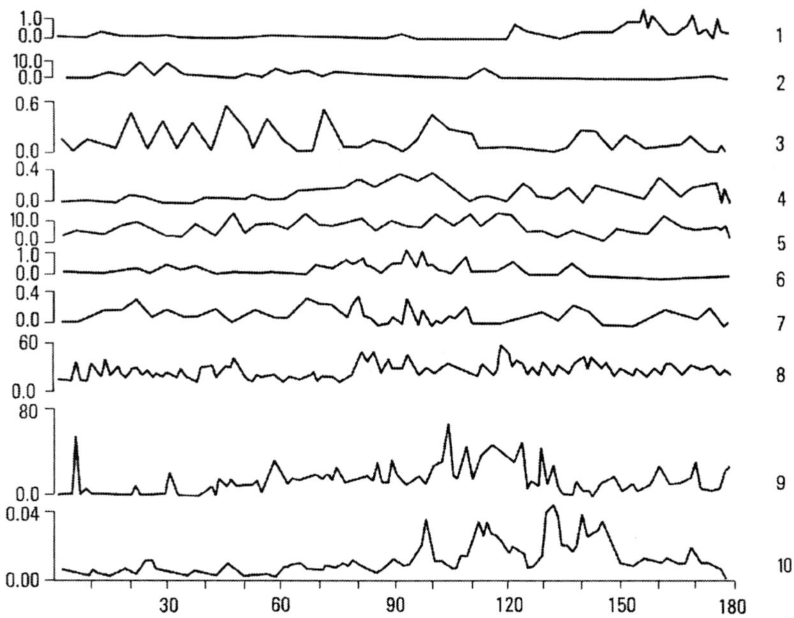

Figure 9.16 Changes in potentially toxic gases in atmosphere of a biological system: 1—ammonia, mg/m^3; 2—acetic acid, mg/m^3; 3—acrolein, mg/m^3; 4—aldegydes, mg/m^3; 5—alcohols, mg/m^3; 6—hydrogen sulfide, mg/m^3; 7—mercaptans, mg/m^3; 8—organic substances (water soluble), mg O_2/m^3; 9—organic substances (water insoluble), mg O_2/m^3; 10—carbon monoxide, mg/l.

An exception to this are the short-lived spikes in carbon monoxide concentrations or other carbon compounds, which occur during emergency situations and which are subsequently corrected by the crew. After the system was returned to its normal state, the CO concentration was maintained on a low equilibrium level by internal system procedures.

These very significant observations demonstrate that there are not only producers but consumers of CO in the system. The CO level could be maintained steady by processes internal to the system only if this were the case. This conclusion is applicable to other toxic gases measured in the system atmosphere.

Water Exchange

At various stages of the experiment the condensate from the phytotrons and algal cultivators served as a source of drinking and hygienic water. As is clear from Tables 9.4 and 9.5, water is procured from these sources and then conditioned over ion-exchanging resins and charcoal. After conditioning, this water attains the hygienic standards required for it to be classed as drinking water.

Table 9.4 Quality of drinking water and phytotron condensate.

Sample	Permanganate oxidation, mg/l O_2	Bichromic oxidation, mg/l O_2	Odor, count	pH
Government standard	2–10	20	2	6.5–9.5
Condensate, day 63	4.65	8.0	4	5.8
Drinking water, day 63	2.08	6.4	2	–

Table 9.5 Element content in water regenerated in the Bios-3 system.

Element	Content, mg/l		
$Na^{\cdot}$	100	1	Traces
$K^{\cdot}$	200	4	Traces
$Ca^{\cdot\cdot}$	100	18	2
$Mg^{\cdot\cdot}$	100	1	Traces
$SO_4^{\cdot\cdot}$	100	0	0
$PO_4^{\cdot\cdot\cdot}$	Traces –0.01	0	0
$NO_3^{\cdot}$	15–40	0	0
$NO_2^{\cdot}$	Traces –0.01	0	0
$NH_4^{\cdot}$	0.2–1.0	0	0
N_{tot}	6–8	0	0

Human Food Rations

The food consumed (rations and vitamins included) in a system containing 2 phytotrons provided 3 people with 26% of their carbohydrates, 14% of their proteins, and 2.3% of their fats.

The foods yielded from plants grown in the phytotrons— the bread baked by the test subjects and the vegetables—did not differ in their biochemical composition and taste characteristics from high quality food produced by ordinary agricultural methods. Inclusion of these products in the food ration gave rise to positive emotions from the test subjects.

Human Life in the Hermetically Sealed Bios-3 Complex

The crew entered the system after 3–5 days' quarantine. For the entire time of the experiment the crew was under around-the-clock medical surveillance with results recorded at regular intervals. Additional information about the health condition of the crew was obtained by medical–physiological and laboratory studies made at regular intervals. Each crewman was obliged to report to the doctor on duty or, at his discretion, doctor in charge of the medical–hygienic program, all changes in his (her) physical well-being. The list of events to ensure safe performance of the experiment involves permanent monitoring of the habitat quality.

For sleep the crewmen wore biotelemetry sensors to record their pulse and respiration and to do electrocardiograms of the first standard leads. The quality of sleep was evaluated by actograms.

An efficient work and rest routine is an essential condition for ensuring high working ability, normalizing subjective and objective condition of the organism. The presence of a human in a room of small volume with hypodynamia, isolation, and altered chemical composition of the atmosphere is a factor capable itself of rendering adverse impact on the organism. This circumstance can considerably impair the correct estimation of human reactions, which may be due to the effect of the specific environment created by the biological regeneration system. Therefore, to choose the routine for the crew, special attention was paid to preventing or reducing to the minimum the negative effect of the above mentioned factors. The benign effect of physical exercise on the human physiological condition and well-being is well known. Muscular work induces reactions nonspecifically enhancing the organism's tolerance to adverse effect of the environment. So we paid particular attention to the measured physical load of the crew, at the phase of system investigation with outside servicing of the regenerator, when the labor of the crewmen was limited. The routine included physical exercises engaging major muscle groups and differing in effect and intensity. Among the sports gear used were springs, weights, and skipping ropes. Exercises with physical load were excluded from the daily routine of the autonomous Biological Closed Life Support System (BCLSS) crew, as servicing the system required strenuous muscular effort. Autonomous operation of the BCLSS presupposed around-the-clock control of regeneration components performance by the crew, largely determining the routine pattern of the test subjects. Figure 9.17 shows basic elements of the daily routine for a 3-member crew ensuring around-the-clock duty, 8 hours' sleep in 24 h, a sufficient period when the entire crew is awake to perform work requiring the efforts of all crew members, meals, etc. The daily routine pattern for a 2-member crew was even more complicated. To perform the same duties the sleep and wakefulness periods were in 8 h shifts (Figure 9.18). Owing to this change of circadian rhythm during the first week this test subject was found to have reduced working ability and impaired well-being. With this routine the cooking time increased by 1.5 times and put the crew members in different psychological situations; certain difficulties arose in conducting medical–physiological examinations.

However, this routine pattern created no complications in the relations between the test subjects and between the crew and researchers on the outside. The crew followed the new rhythm punctually, so the adaptation was fast: physical working ability completely restored within 7–8 days; new pulse rate and good sleep stabilized by the end of the same period. At the end of the second week the body temperature rhythm, the most inert parameter, began to shift.

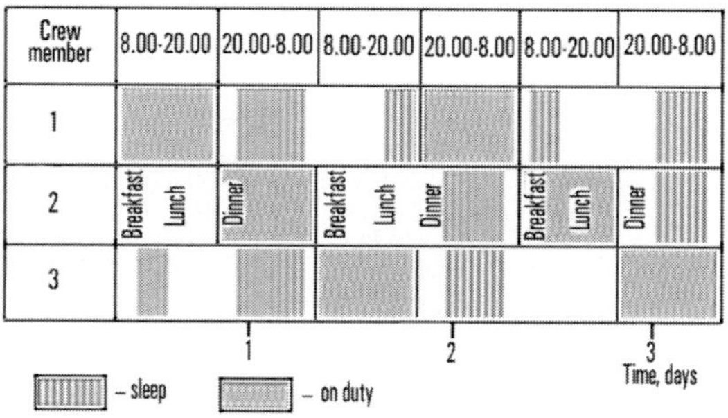

Figure 9.17 Test routine for 3 crew members.

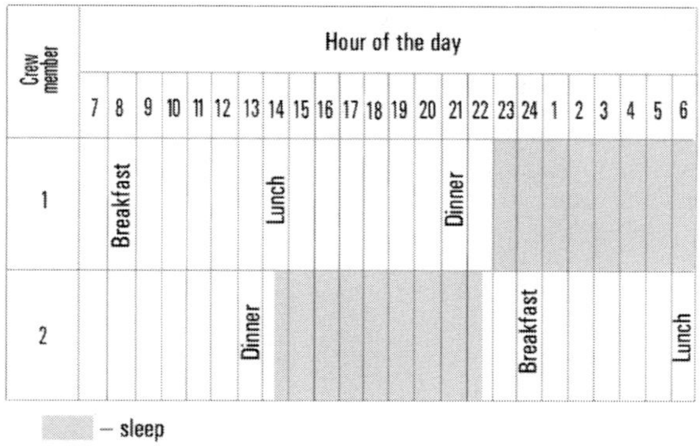

Figure 9.18 Daily routine for 2 crew members.

Pictures in Figure 9.19 show some test subjects in the Bios-3 at work.

The sanitary–hygiene routine implied personal hygiene procedures conducted to a habitual extent and with a habitual periodicity.

Table and kitchen appliances were washed in hot soapy water followed by rinsing with clear water, drying and sterilization in the oven for an hour at 160–180°C.

Every day all surfaces in the living compartment were vacuum-cleaned, and the floor was cleaned with a wet cloth. All surfaces of the shower-and-lavatory unit were disinfected with a 3%-solution of hydrogen peroxide. Every 7–10 days the living compartment underwent general cleaning, when the floor and walls as well as the superficial area of furniture and equipment were wet cleaned. All sanitary needs were satisfied with intrasystem water and table and kitchen appliances.

Diet

To determine the optimal human diet we used the standards worked out in the All-Union Institute of Nutrition. Note that investigations conducted for almost half a century since have resulted in nearly the same requirements on the astronauts' diet, as is reported in the Life Support and Habitability Manual by the European Space Agency (1991).

Figure 9.19 Wheat harvesting (A) and thrashing (B); baking (C) and tasting (D) of fresh bread.

At the first phase of research the crew's diet consisted of prestored lyophilized or canned products. However, after introducing higher plants into the system the fraction of the prestored food decreased, and, finally, only products of animal origin were stocked. The vegetable part of the diet was fully synthesized inside the system. It comprised of wheat, chufa (a source of polyunsaturated fatty acids), and a large number of vegetable crops (Table 9.6). The diet was drawn according to the recommendations for balanced nutrition with account of the individual requirements of the test subjects, with respect to sex, age, body mass, and intensity of physical load. Proteins, fats, and carbohydrates in the diet were related as 1:1:3.5–4 in terms of mass. The fraction of proteins of animal origin amounted to 40–50% of the total amount of protein. Forty-four products were used in cooking, affording 4 diversified meals a day, the menu changing in 5-day shifts. Lyophilized and canned products in ready and prefabricated forms were used together with vegetables;

wheat was baked into bread using a waste-free technology. To supply vegetable lipids, chufa nodules were used in the diet as fine powder, enough so as to meet the human requirement for linolic acid. In this form chufa can be a component of nearly all food and products. Properly dosed, the chufa powder does not change the organoleptic properties of food, imparting to it a somewhat peculiar taste positively assessed by all crew members. The largest fraction of chufa powder can be contained in bread and dairy products. Experiments indicate that daily consumption of about 1.9 g of chufa is not associated with any adverse effect of a clinical, physiological, or psychological nature.

Table 9.6 Consumption of edible biomass in the 5-month experiment, g/2 persons per day.

Crop	g/2 persons per day
Wheat, dry mass	495
Chufa, dry mass	120
Pea, dry mass	26
Radish, fresh mass	266
Beet, fresh mass	236
Carrot, fresh mass	132
Potato, fresh mass	21
Tomato, fresh mass	87
Cucumber, fresh mass	277
Cabbage, fresh mass	164
Onion, fresh mass	110

Rationing concerned only the products of animal origin, as they were stocked in advance in accordance with the average daily individual requirements of each test subject. Use of the phytotron produce was at the crew's discretion, depending on the amount and assortment available. This was not the same all the time because the "green conveyer" supplied them at regular intervals. Consumption of products grown inside the system was, of course, punctually recorded.

Objective evidence of the experiments and the opinion of the crew suggest that the diet was sufficient, of full value, and quite diverse.

Labor Expenditures

Biological Closed Life Support System researchers are concerned with the issue of the evaluation of the labor expenditures of crew members in an autonomous operating system. Table 9.7 shows chronometrical measurements of the principal jobs and procedures performed inside the system. The values presented specify this system only and are of tentative significance. Labor expenditures depend on the

Table 9.7 Labor expenditures of the crew in the LSS.

Procedure	3-member crew		2-member crew	
	Man/h · day	Fraction of total time reserve, %	Man/h · day	Fraction of total time reserve, %
Maintaining operation of the higher plant link				
Sowing and harvesting plants, wheat thrashing	2.42		3.63	
Monitoring of plants' conditions, equipment maintenance	2.58		3.87	
Preparation of correcting solutions	1.23		1.35	
Sanitary scrubbing of phytotrons	0.24		0.36	
Total	6.47	8.99	9.21	19.19
Maintaining algal cultivator operation				
Cultivator operation control, equipment maintenance	2.65		–	
Preparation of correcting solutions	0.66		–	
Centrifugation, dehydration of harvest biomass	2.50		–	
Collection of samples for analyses and conducting analyses	2.28		–	
Sanitary cleaning of the algal compartment	0.68		–	
Total	8.77	12.18	–	–
Workaday procedures				
Cooking and having meals, bread baking, washing the dishes, galley tidying up	5.10		7.65	
Preparation of quality standardized water	0.42		0.42	
Personal hygiene	1.17		0.78	
Hygiene procedures in the living compartment	0.81		0.82	
Total	7.50	10.42	9.67	20.15
Servicing inedible biomass incinerator	–		2.10	4.37
Sleep	24.0	33.33	16.0	33.33

Notes: 1. In the described experiment with a 3-member crew, 2 phytotrons were in operation.
2. Data of another experiment (1985) are given for comparison (2-member crew maintained operation of 3 phytotrons).
3. Total reserve of time for 3 persons is 72 h a day; for 2 persons—48 h.

system structure, the automation of the processes, professionalism of the crew, and refinement of their skills in servicing the system operation. It should be noted that while technological and housekeeping jobs were fulfilled by the crew out of necessity, all work in the phytotrons, in direct contact with plants, was a pleasure for them.

Human Physiology and Biochemistry Tests

Numerous studies done by many authors indicate that the presence of humans in a room of limited volume for a long time brings forth the so-called asthenic syndrome—a set of functional disorders specified by cachexia, impaired functioning of some organs or their systems, the cardiovascular included. It is generally agreed that the main cause of these changes is hypodynamia. In view of this potential risk we thoroughly examined the functional condition of the circulation organs of the Bios-3 crew.

The dynamics of the pulse rate and arterial pressure in the course of the 3-month experiment is given in Figure 9.20. The parameters presented reveal no direct changes. The pulse rate measured at regular intervals exhibits the persistence of its diurnal dynamics (Figure 9.21). The measured physical load test demonstrates (Table 9.8) persistence of the cardiovascular system function.

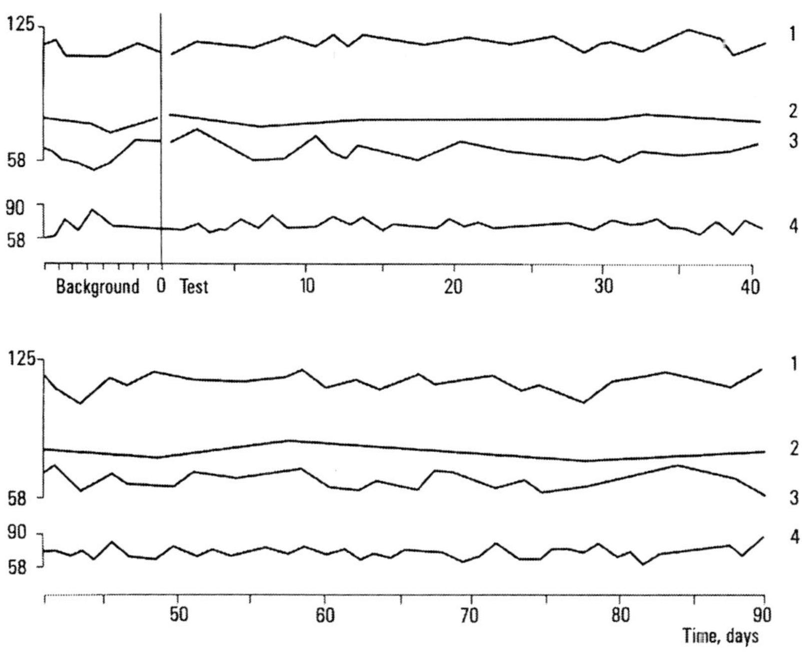

Figure 9.20 Crewman's hemadynamics in experiment. 1—maximum arterial pressure, mm Hg; 2—average arterial pressure, mm Hg; 3—minimum arterial pressure, mm Hg; 4—pulse rate, per min.

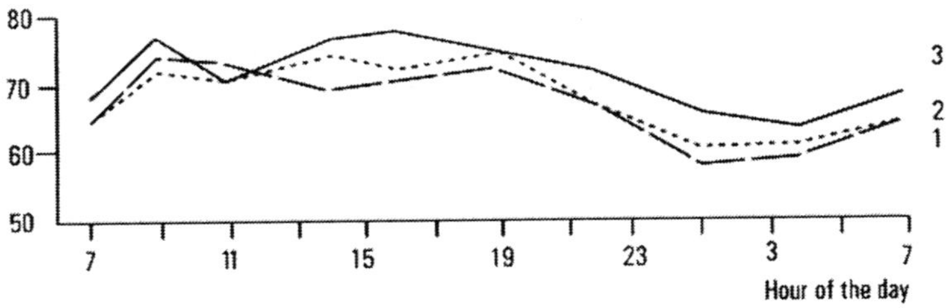

Figure 9.21 Crewman's average pulse rate per min in the course of a day 1–3 months of experiment.

Table 9.8 Test subject M's arterial pressure and pulse rate examined under physical load.

	Parameter		Background values	Day of experiment	
				1–45	46–90
Arterial pressure, mmHg	at rest	minimum	60±9	61±2	63±3
		maximum	112±3	117±3	116±5
	after 15 squats	minimum	67±7	71±5	62±4
		maximum	124±7	130±8	128±6
	3 min after exertion	minimum	63±4	64±4	60±4
		maximum	113±4	119±4	113±5
Pulse rate	at rest		69±9	76±6	74±8
	after 15 squats		89±12	113±7	97±10
	3 min after exertion		68±5	75±6	73±6

The results of investigating the function of the cardiovascular system indicate that neither hemodynamic disorders nor the impairment of the functioning of the circulation organs could be recorded in the humans who stayed for many months in the specific conditions of the closed ecological system. It may mean that we have elaborated proper daily routines for the crew.

Table 9.9 presents data on the morphological composition of the peripheral blood in the crew staying inside the closed system for several months. As opposed to the background, no parameter under study manifested undirectional changes in the course of the experiments. The error on the averaged values is statistically uncertain.

Since the diet for a prolonged period included considerable amounts of chufa nodules (up to 130 g a day), as a source of polyunsaturated fatty acids, and lyophilized products (unconventional foodstuffs for everyday life), alimentary duct functions had to be studied. Investigations concerned several parameters specifying the alimentary properties of the stomach content, enzymatic activity of the gastric juice, and, finally, the assimilation of the nutrients as an integral index of the interaction between the food and alimentary duct function. The "test breakfast" was lyophilized cabbage juice, its stimulating effect on the principal and oxyntic cells of the stomach mucus being identical to submaximum histamine stimulants.

Table 9.9 Morphological composition of peripheral blood.

Index	Test subject Tr.		Test subject Br.	
	Experiment duration			
	6 months		5 months	
	Background values	Experiment	Background values	Experiment
Erythrocytes, thousand per mm^3	4683±101	4830±131	4381±270	4236±165
Hemoglobin, g/l	141±12.7	148±6.5	150±7.2	157±6.0
Leukocytes per mm^3	6130±1120	7910±1719	6870±425	7240±715
Bacilinuclear, %	1.3±0.63	2.6±2.24	4.0±1.60	2.8±1.31
Segmento-nuclear, %	45.2±5.64	51.4±4.48	57.0±6.70	56.0±6.70
Lymphocytes, %	37.8±5.55	32.3±3.84	30.9±9.50	35.0±4.90
Monocytes, %	8.2±1.12	6.9±2.86	6.5±1.20	4.2±2.86
Eosinophiles, %	7.5±2.32	6.8±2.48	1.6±0.70	2.0±1.41

In the course of the experiment all of the test subjects exhibited a certain increase in their stomach secretory activity. Table 9.10 presents the average data on the total acidity of the gastric juice and free hydrochloric acid content at the secretion peak in the neuro-humoral phase as the most informative indicators of stomach mucus condition. No changes in proteolytic activity of the stomach content were recorded either by the evaluation of pepsin activity or by indirect data—uropepsin content in urine.

Assimilation of proteins, fats, and carbohydrates was evaluated on the so-called "exchange," or "metabolic," days uniformly distributed over the experiment—the last ten days of each month. On these days, all products consumed and excreted by a human were thoroughly measured. Results of these studies are presented in Table 9.11. The experiments all yielded identical results, the gastro-enteric duct function did not vary with the diet used.

Table 9.12 lists data on the renal urogenous function of the test subjects in the 6-month experiment. The table shows that diurnal diuresis varies within a small

Table 9.10 Total acidity and free hydrochloric acid in gastric juice.

Test subject	Total acidity, titration units	Free HCl, titration units
Test subject Tr.:		
Background	90	70
Day 108 of experiment	108	84
Day 180 of experiment	132	108
Test subject Pn.:		
Background	122	102
Day 50 of experiment	150	120
Day 101 of experiment	148	120
Test subject Br.:		
Background	96	76
Day 40 of experiment	128	98
Day 180 of experiment	116	100

Table 9.11 Assimilation of basic nutrients, averaged over the crew.

Month of experiment	Assimilation, %		
	Proteins	Fats	Carbohydrates
1–2	91.9	95.2	99.4
3–4	91.5	94.7	99.5
5–6	93.4	95.9	99.7

range (not more than 10% of the average measured for 60 days). Depending on the volume, the specific mass of urine varies in inverse proportion, suggesting a normal condition of the filtering and concentration renal functions. In the course of the experiment the test subjects produced an average of 20.6, 22.1, 22.2, and 21.8 ml of urine per kg of body mass. Table 9.13 presents results specifying the renal function of purifying the organism from some metabolites. Analysis of the data suggests that this principal ability of the kidneys underwent no changes in the course of the experiment. The parameters under study varied within the normally admissible range.

One of prerequisites for achieving maximum plant productivity in the BCLSS is optimal CO_2 concentration in the system's air. The results of preliminary experiments showed that unicellular algae (*Chlorella*) achieve maximum photosynthesis with atmospheric CO_2 content not less than 1% volume; for higher plants it must be about 0.3% volume. Carbon dioxide concentrations below these levels reduce the efficiency of the regenerators, and to attain balanced human gas exchange it may be required to increase the performance of algal cultivators or to enlarge the area under plants. Therefore, we maintained in the system a compromise CO_2 concentration—fairly high to sustain photosynthesis, but not exceeding the levels admissible for humans staying in the system for prolonged periods.

Table 9.12 Renal urogenous function, average daily data for exchange periods.

Test subject, parameter	Days of experiment					
	19–28	50–59	80–89	110–119	143–152	171–180
Test subject Tr.:						
Urine volume, ml	1560	1406	1390	1440	1440	1550
Specific mass, kg/l	1.021	1.021	1.021	1.019	1.019	1.017
Test subject Pn.:						
Urine volume, ml	1826	2180	1880	1980	–	–
Specific mass, kg/l	1.020	1.018	1.019	1.018	–	–
Test subject Sh.:						
Urine volume, ml	1300	1518	–	–	1225	1200
Specific mass, kg/l	1.022	1.020	–	–	1.019	1.019
Test subject Br.:						
Urine volume, ml	–	–	1480	1480	1540	1510
Specific mass, kg/l	–	–	1.023	1.022	1.021	1.019

Thus, BCLSS studies demonstrate that in the course of long-duration experiments the CO_2 content in the regenerated atmosphere of the system of the tested design vary within uncertain limits. That is why we had to study the human response to varying CO_2 concentrations in the system's atmosphere. Two groups of parameters were monitored: external respiration and indices specifying the acid–alkali equilibrium of the organism's internal medium.

Figure 9.22 presents the dynamics of the basic parameter of external respiration—pulmonary ventilation (minute respiratory volume—MRV) in the test subjects, depending on the varying CO_2 concentration in the inhaled air. This human

Table 9.13 Organism renal purification of metabolites.

Metabolite	Test subject Tr.		Test subject Pr.	
	excreted with urine, g/day			
	background	experiment	background	experiment
Creatinine	1.30±0.29	1.40±0.73	1.20±0.26	1.40±0.29
Uric acid	517±141.1	662±145.4	690±182.1	752±208.1
Urea	26.9±7.55	30.02±6.07	24.7±6.60	33.1±7.33
Metabolite	content in plasma, mg%			
	background	experiment	background	experiment
Creatinine	–	–	–	–
Uric acid	3.0±0.21	3.02±0.47	–	–
Urea	32.0±2.00	32.3±3.49	29.5±1.50	33.4±3.57

Table 9.14 Acid–alkali equilibrium in test subject's organism internal medium.

Parameter	Test subject Tr.		Test subject Pn.	
	Background	180 days in LSS	Background	120 days in LSS
CO_2 in inhaled air, vol.%	up to 0.1	1.15±0.4	up to 0.1	1.15±0.4
Blood pH	7.36±0.01	7.35±0.01	7.36±0.01	7.36±0.01
pCO in alveolar air, mmHG	36.0±0.82	36.9±1.14	29.4±0.58	40.0±1.60
Alkali reserve of blood, %CO_2	59.6±1.16	61.4±1.71	61.8±3.74	61.8±1.53
Urine pH	6.60±0.15	6.20±0.16	6.00±0.12	6.20±0.50
Bicarbonates excreted with urine, g/day	0.39±0.05	0.47±0.03	0.48±0.05	0.48±0.03
Ammonia excreted with urine, g/day	1.50±0.43	1.30±0.37	1.10±0	1.50±1.08

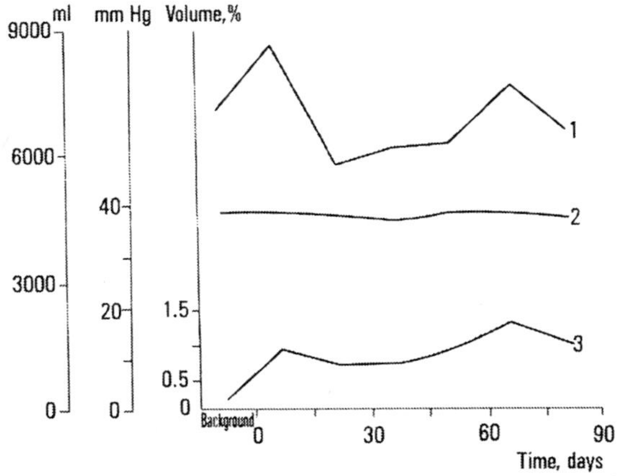

Figure 9.22 Dynamics of pulmonary ventilation and CO_2 partial pressure in the alveolar of test subjects. 1—pulmonary ventilation (MRV), ml; 2—CO_2 partial pressure in alveolar air, mm Hg; 3—CO_2 concentration in the inspired air, volume %.

response was sufficient to maintain essentially the same level of CO_2, which in turn determined stable active blood pressure and acid–alkali equilibrium (Table 9.14). A similar pattern was recorded in all experiments. We should also note that such a response of external respiration persists for many months.

So it has been proved that the external respiration function of a human staying for a long time in an atmosphere with elevated CO_2 concentration of about 1.05–1.15±0.3–0.4 volume% is not adversely affected. The active response of the internal medium and acid–alkali equilibrium remain unchanged. The adaptive changes of external respiration are flexible and do not demonstrate any signs of fatigue for a long time. The explanation can probably be that periods of respiration of the air with elevated CO_2 concentration capable of causing changes in the active response of the internal medium of the organism are short and alternate with periods when the CO_2 concentration in the respired air is neutral for humans. Such periodic variations in the CO_2 content of the regenerated atmosphere do not allow such serious consequences as changes in the organism's internal medium.

The composition of blood is a kind of metabolic mirror for the organism. Monitoring of this parameter can provide rather complete evidence about the exchange of substances and possible disorders.

Protein exchange was evaluated by the assimilation of proteins of the food supplied to the organism. Table 9.11 shows that the assimilation of the proteins of the food remained fairly high throughout the observation period. It varied little throughout the experiment, suggesting that the test subjects were fully provided with protein substances. As seen in Table 9.15, none of the parameters of protein exchange in the organism changed appreciably in the course of the experiment. Average values of these indices and their variations are within physiological limits. Calculations of nitrogenous equilibrium averaged over 5 exchange periods yielded a negative value equal to 2.4%, which is within the limits of the error of the method used for the calculation.

Table 9.15 Test subjects' protein metabolism (some indices).

Index	Test subject Tr.		Test subject Pn.	
	Background	Experiment	Background	Experiment
Total protein in blood serum, %	8.0±0.27	8.6±0.44	7.9±0.28	7.5±0.27
Albumins, % of protein	58.1±2.00	56.9±2.30	57.1±1.82	59.2±0.89
Globulins, % of protein:				
α_1	60.±0.36	6.3±0.43	6.6±0.27	6.1±0.53
α_2	9.0±0.81	8.7±0.85	9.1±0.91	8.7±0.49
β	12.7±0.76	13.5±2.13	12.9±0.43	12.5±0.34
γ	14.2±0.72	14.6±0.54	14.3±0.38	13.5±0.38
Residual nitrogen in blood, %	27.4±2.50	28.0±5.04	22.0±1.76	26.2±4.7
Urea in blood, mg%	32.0±2.00	32.3±3.49	29.5±1.50	33.4±3.57
Urea in urine, g/day	26.9±7.55	30.2±6.07	24.7±6.60	33.1±7.33
Creatinine in urine, g/day	1.3±0.29	1.4±0.73	1.2±0.26	1.4±0.29

An autonomous LSS operator should be emotionally hardy, enterprising, persistent, highly perceptive and attentive, capable of keeping in his/her memory a large body of information, quickly mastering and remastering skills, and making decisions within a short time. Table 9.19 lists some parameters specifying human psychic functions. No deterioration of any parameter was recorded during the experiment, which suggests the stable mental working ability of the test subjects.

Data on lipid exchange are presented in Table 9.16. No disorders were recorded in this respect. Fluctuations of average values are statistically uncertain and lie within the accepted standard limits.

Similar results have been obtained in all experiments.

To perform the functions of an operator of all processes running in the system, each member of the crew must have a high physical and mental working ability, which is especially important with the altered routine of rest and work. Some investigation results are given in Table 9.17. Analysis of the data demonstrates that nearly all the parameters under study remained as the background levels. The test subjects had slight individual deviations (both positive and negative) in reactions. The recorded deviations are of an individual nature. The test subjects Br. and Sh., for example, had an increased latent period of the acoustic–muscular reaction in the experiment, but this parameter remained at a steady state throughout the experiment; the test subject Pr. had no changes in this index; in the test subject Tr. this parameter even decreased by the end of the experiment. The test subjects Tr. and Pn. had a distinct (though also remaining at a steady state) reduction in the maximal rate of movements of the wrist joint against the background values, which may be a direct indication of an altered rate of excitation and inhibition in the central nervous system. The total volume of the test subjects' muscle mass and power did not change (Table 9.18). The results obtained indicate that the regime of physical activity and diversity of work requiring muscular strain can maintain fairly high physical working activity of the crew inside an autonomous Life Support System (LSS) for quite a long time.

Table 9.16 Lipid metabolism in test subjects' blood plasma in several months-long experiment, mg%.

Index	Test subject Tr.		Test subject Pn.	
	Background	Experiment	Background	Experiment
Total lipids	475.6±16.03	91.0±20.29	477.5±20.62	507.2±50.77
Tri-glycerides	187.3±19.44	174.0±29.34	120.0±22.82	154.4±60.37
Total cholesterol	175.0±14.59	191.6±22.22	200.8±27.21	220.9±38.23
Free cholesterol	61.1±12.27	73.1±17.81	72.5±16.58	83.73±18.88
Ether-bound cholesterol	106.0±6.21	107.2±26.89	142.2±6.9	135.6±26.86
Phospho-lipins	110.0±12.25	123.1±22.91	119.5±17.87	129.7±15.43
Ether-bound fatty acids	273.8±27.51	282.5±16.41	270.0±24.14	306.3±22.38
Lipase activity, cond. units	0.95±0.058	0.91±0.066	0.98±0.014	0.91±0.098

Throughout the experiment an expert medical board observed and examined the crew. The doctor who was a member of the crew could examine the other members using objective examination methods—palpation, percussion, and auscultation. Besides, to estimate the condition of the crew, clinical physiological and laboratory observation data were used during the experiment and one month after its completion. The recorded diseases (test subject Tr.'s finger paronychia that called for surgical operation without depressurizing the system, and Pn.'s acute pharyngitis cured in 8 days) could not be regarded as specific for the system. Headaches and dry coughs that the test subjects had for a short time, coincident with malfunction of the algal cultivator, were associated with the elevated hydrogen sulfide concentration in the system's atmosphere, or with intensive physical work under CO_2 concentration in the system elevated to 1.8–2.0%. During the 3–4 days after the experiment the test subjects all recorded pain in leg and foot muscles when going upstairs, which is probably due to the detraining of certain muscles.

Table 9.17 Some parameters of test subjects' working ability in the six-month experiment.

Parameter			Test subject Tr.		Test subject Pn.	
			Background	Experiment	Background	Experiment
Latent period of visual–muscular reaction, s			–	0.27±0.01	–	0.30±0.01
Latent period of acoustic–muscular reaction, s			0.20±0.05	0.17±0.01	0.18±0.02	0.19±0.02
Maximum rate of active movements of the wrist joint			388±16	349±19	348±7	317±22
Precision of movement reproduction, deflection from the given point, mm		right sector	21±2.8	28±12.4	17±2.6	16±2.5
		left sector	15±1.0	15±2.1	25±2.1	21±5.4
Coordination of small amplitude movements		test time, s	130±17.2	112±2.2	117±8.1	101±11.8
		errors	50±13.2	63±12.8	99±6.5	80±15.7
Dynamometry	wrist	left	76±2.2	75±2.8	87±3.1	84±2.3
		right	68±0	65±1.2	82±2.2	77±4.4
	torso		158±6.4	158±4.5	154±9.3	145±7.1

Table 9.18 Experiment erography (finger).

Parameter	Test subject Br.		Test subject Sh.	
	Background	Experiment	Background	Experiment
Total volume of work, kGm	3.4±11.02	40.0±2.55	26.6±6.83	29.1±9.52
Time, s	90.9±6.11	104.1±11.05	80.0±10.94	89.5±8.15
Power, kGm/s	0.37±0.092	0.37±0.110	0.33±0.025	0.32±0.046

Table 9.19 Some parameters of test subjects' working ability in six-month experiment.

Parameter		Observation period	Test subject Tr.	Test subject Pn.
1		Background	3.2±0.5	3.6±0.6
		Experiment	3.3±04	3.6±0.7
2		Background	141±4.2	207±15.5
		Experiment	143±5.9	194±11.4
3		Background	1553±452	1134±275
		Experiment	1665±241	994±663
4	4a	Background	652±28	851±26
		Experiment	588±71	756±60
	4b	Background	0.7±0.21	3.3±1.53
		Experiment	0.5±0.32	2.3±1.12
5	5a	Background	6.7±1.50	5.6±0.60
		Experiment	6.4±1.22	7.1±0.61
	5b	Background	5.7±1.75	4.3±1.51
		Experiment	5.4±1.70	5.9±1.08

Notes:
1. Volume of attention, characters captured after instantaneous presentation.
2. Distribution and change-over of attention, test time, seconds.
3. Stability, attention exhaustion, thinking productivity (by the correction table), conditional units.
4. Quality of thinking process (regular connections established).
   4a) test time, seconds;
   4b) errors.
5. Mechanistic memory, numbers reproduced correctly.
   5a) reproduction with deficient time immediately after memorizing;
   5b) reproduction 15 min after memorizing.

The physical well being of the crew during the experiment was at a rather good level. Re-adaptation to ordinary conditions after completion of the experiment took 4–7 days and caused no complications.

It is worth mentioning the so-called "autonomy feeling" recorded in the experiment. It manifests itself in higher confidence of the crew, increasing the independence of their actions, a wish to reduce consultations with experts and to emphasize the superiority of their expertise, in some cases—a condescending attitude to the knowledge and experience of the system consultants. This phenomenon has both good and bad sides. The good ones are obvious and need no special discussion. The bad sides are as follows. The crew, practically understanding the operation of the system and perfectly performing all of the servicing procedures, fails to realize the fact that in the course of the operation of any system, an experimental one in particular, it may have to be subjected to some alterations of its structure, construction or technology in order to verify assumptions, develop new elements, etc. These alterations change the course of the processes and should upset the stereotype the crew has formed.

The members of the crew have a firm conviction that their main objective is to maintain reliable functional conditions of the system to bring the experiment to its successful completion. They tend to underestimate the research goals. Staying inside the system, they gradually lose sense of the importance of their work for the future. We think that this may be the reason for the crew's reluctance to make changes to the operation of the system and experiment design.

Gas Exchange Balance Between Humans and Links That Regenerate the Closed System Habitat

To bring the human and plant gas exchange into qualitative agreement, we have to solve a complicated problem of equalizing CO_2 and O_2 flows in respiration and photosynthesis.

One of the advantages in regenerating the habitat by means of photosynthesis is that carbon dioxide consumption and oxygen evolution are taken care of by a single process. However, this process has its own disadvantages. Since there is a difference in the metabolisms of humans and algae, the flow of carbon dioxide and oxygen that they exchange may not be quantitatively commensurate. This occurs because their gas exchange coefficients—that is, the human respiratory quotient and the algal assimilation quotient—are not equal.

Therefore, one of the main tasks of the Bios-3 experiment was to research the possible ways to achieve the best quantitative complementarity between humans and those agents that biologically regenerate the atmosphere, specifically taking into account the continuous culture of the unicellular algae *Chlorella vulgaris*. Only an exact determination of human and algal gas exchange characteristics exhibited when they are confined to a closed hermetically sealed enclosure can serve as a basis for solving this problem. While solving this problem, data were procured that enabled the calculating of a cultivator productivity that would facilitate providing for human gas exchange requirements.

In the given instance, the total quantity of CO_2 absorbed by the algal culture, the total amount of oxygen released by the algae into the regenerated atmosphere, and the ratio of these two values are of great importance. These values represent a combined characteristic that describes 1) the photosynthesis and respiration of the algae and accompanying microflora, 2) the oxidation of organic substances contained in the culture medium, and so on. These processes determine the functional photosynthetic efficiency in the biological system.

Algal gas exchange could be satisfactorily described by the following parameters:

1. CO_2 consumption coefficient: $CC_{CO2} = CO_2^{AlQ}/C$;
2. O_2 evolution coefficient: $EC_{O2} = O_2^{AlQ}/C$;
3. Assimilation quotient: $AQ = CO_2^{AlQ}/ O_2^{AlQ}$;

where CO_2^{AlQ} is the quantity of carbon dioxide assimilated by the algal culture from the air; O_2^{AlQ} is the quantity of oxygen evolved into the air during the algal cultivation process; and C is the increase in the algal–bacterial biomass.

All of these values are expressed in the same unit of time.

The following average values were experimentally determined: AQ = 0.89; CC_{CO2} = 0.976 nl/g; EC_{O2} = 1.099 nl/g. No changes in these values were discovered when up to 90% of the nutrient medium was recycled.

Human gas exchange for the test subject in the biological system was determined by conducting all tests in a single room (Kreps *et al.*, 1956). In the end, test data were amassed that characterized human gas exchange intensity for various activity schedules within the closed system. The values obtained were used to calculate the respiratory quotient (RQ) and the calorie content of an energy-balanced food ration.

Having determined the RQ and ration calorie content, it was now possible to compare the cultivator output with human gas exchange requirements and predict the joint gas exchange balance for humans and an algae culture with respect to CO_2 and O_2.

Calculating cultivator output can be best achieved by measuring the quantity of carbon dioxide assimilated over a given time. This approach is justified by the fact that the permissible fluctuation range for carbon dioxide present in inspiratory air is much less than that of oxygen. In other words, if there is a quantitative disparity in the gas exchange between humans and the link that regenerates the atmosphere, then carbon dioxide reaches a concentration dangerous to human respiration much earlier than does oxygen.

The gas balance characteristic for the closed system can be conveniently represented as the system gas exchange coefficient (GEC) that expresses the relationship between human RQ and algal culture AQ (or the AQ of any link that regenerates the atmosphere).

Then

$$GEC = RQ/AQ = (CO_2^{HQ} \cdot O_2^{AlQ}) / (O_2^{HQ} \cdot CO_2^{AlQ}),$$

where CO_2^{HQ} is the quantity of carbon dioxide produced by humans, and O_2^{HQ} is the quantity of oxygen consumed by humans.

System balance with respect to oxygen and carbon dioxide in its gaseous phase can be reached only when the following equalities adhere: $CO_2^{HQ} = CO_2^{AlQ}$, $O_2^{HQ} = O_2^{AlQ}$, that is, when AQ = RQ; that is, when GEC = 1.

When the GEC deviates from unity, then there is a disparity in the gas exchange between the humans in the system and the regenerative link.

As was stated earlier, cultivator average daily output was calculated based on the quantity of CO_2 produced by humans inhabiting the system for that time period. It follows that $CO_2^{HQ} \geq CO_2^{AlQ}$, and therefore the GEC can be less than unity only when $O_2^{HQ} > O_2^{AlQ}$. Under these conditions, the O_2 content in the life support system atmosphere will decrease. The inverse is true when GEC > 1.

System balance with respect to CO_2 can be calculated for any time period based on cultivator productivity in relation to the level of human activity, that is in relation to the amount of carbon dioxide produced by the test subjects. When cultivator productivity is uniform over the course of the day, CO_2 accumulates in the air of the hermetically sealed cabin since heightened human physical activity during waking hours leads to the CO_2^{HQ} value exceeding the CO_2^{AIQ} value. During the human sleep period, the ratio inverts and produces a decrease in the level of carbon dioxide. These CO_2 dynamics were well-marked in each experiment (see Figure 9.23).

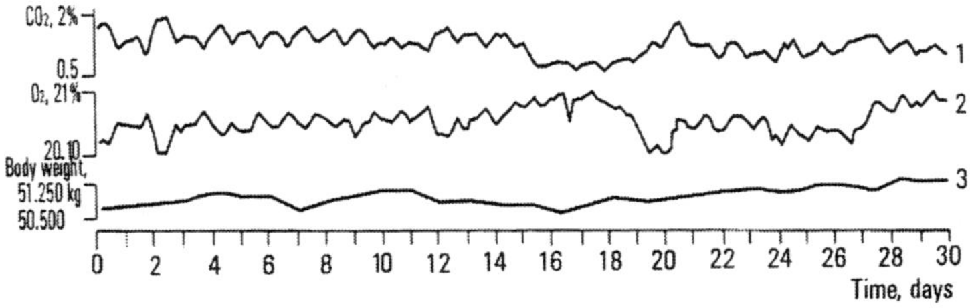

Figure 9.23 Dynamics of the carbon dioxide and oxygen concentration and the weight of a man under observation in a 30-day experiment. 1—CO_2 concentration; 2—absolute O_2 concentration; 3—body mass.

In the first short-term experiments, the expected decrease in oxygen content was registered due to the test subjects eating ordinary food combined in uncontrolled ratios. The achieved results demonstrated the need to search for ways to correct the GEC value, that is, to produce conditions under which AQ = RQ. Naturally, these quotients are equal only when the AQ is decreased, or the RQ is increased.

Data in the literature bear witness to the fact that the AQ depends on the form of nitrogen nutrients fed to the algae (Myers, 1958; Meleshko, 1967). In systems such as those described in this book where algae utilize human liquid excretions, the source of nitrogen is human urea. This fact deprives us of the chance to correct the balance by diminishing the AQ.

To accommodate human and algal culture gas exchange coefficients, the well-known dependence of human RQ on the biochemical composition of oxidizing food substances was used to reach gas exchange balance.

Earlier, it was noted that the composition of oxidizing substances, or more precisely, the biochemical composition of human food intake, can change the RQ value to one within a wide range of values. The difficulty lies in finding a way to quickly and reliably predict the RQ value produced by the use of one or another ration, and to solve the converse, that is, to contrive a ration that would achieve the desired RQ value.

Carbon dioxide excreted from the body forms due to the oxidation of all food group substances. The largest quantity of carbon dioxide forms when fats and carbohydrates are metabolized since the largest amount of absorbed oxygen is spent on their oxidation. From this it follows that the ratio of oxidized quantities of these substances basically determines the RQ value.

Calculation results demonstrate that the so-called no-protein RQ depends directly on the amount of energy released during carbohydrate oxidation and is expressed in terms of the percentage with respect to the total energy released when these two protein and carbohydrate groups are jointly oxidized (Berkovich, 1964). Measurements showed that a similar RQ dependency exists for the carbohydrate–fat ratio included in the food ration, and this significantly simplified calculating both the RQ value based on the food ration and the food ration composition based on the desired RQ value. These calculations additionally require knowledge of two easily determined values: ration total calorie content and the quantity of protein in the ration. These calculations were performed assuming that the RQ depends only on the carbohydrate–fat ratio since: 1) the part played by proteins in gas exchange is sufficiently small; 2) when only proteins are oxidized the RQ is close to the AQ, and 3) a higher carbohydrate assimilability as compared with that for fats compensates for the fact that proteins tend to reduce the RQ.

Repeated experiments have established that the disagreement between measured RQ values and calculated RQ values does not exceed 1–3%.

Ration composition was calculated using the above approach so that the ratio of proteins to fats to carbohydrates was 1:1.07:3.96. This ration was the mainstay of the test subject diet and produced an RQ for which RQ = AQ = 0.89. Moreover, the biochemical composition of corrective rations No.1 and No.2 were defined (see Table 9.20). The corrective rations were used to manipulate the direction of the oxygen content change in the regenerated atmosphere.

Table 9.20 Ratio of food substances in rations employed in experiments.

Ration	Content ratio			Calculated RQ
	Proteins	Fats	Carbohydrates	
Mainstay ration	1	1.07	3.96	0.89
Corrective ration No.1	1	1.20	2.75	0.85
Corrective ration No.2	1	0.79	4.60	0.93
Accepted physiological norm	1	0.8–1.0	2.5–4.0	

Figure 9.23 illustrates the dynamics of CO_2 and O_2 concentrations that were recorded during the course of one experiment. The graph clearly shows the daily periodicity in the change of CO_2 concentration and the inverse dependence of the O_2 concentration on it. Consequently, the oxygen content in the biological-system air is a function of two variables: RQ and CO_2 concentration. To generate a characteristic for O_2, because it depends only on the RQ value in a regenerated atmosphere, the total oxygen content (TOC) was calculated. TOC is the sum of the system air oxygen content at a given moment in the experiment and the oxygen

content increase that would result if the regenerating link were to assimilate the entire quantity of carbon dioxide found in the system air at that same moment. Curve 3 in Figure 9.23 represents the outcome of these calculations. The trace of this curve demonstrates that: 1) gas exchange equilibrium can be maintained between humans and the photosynthesizing link with respect to oxygen and carbon dioxide simultaneously (the stability of TOC for days 1–15 and days 29–30); and 2) oxygen content change can be directed in system air, and this is manifested by the decrease in TOC for days 16–25 and its increase for days 26–28. The described TOC dynamics were elicited using rations that produce the following RQ values respectively: 0.89, 0.85, and 0.92 (see Table 9.20).

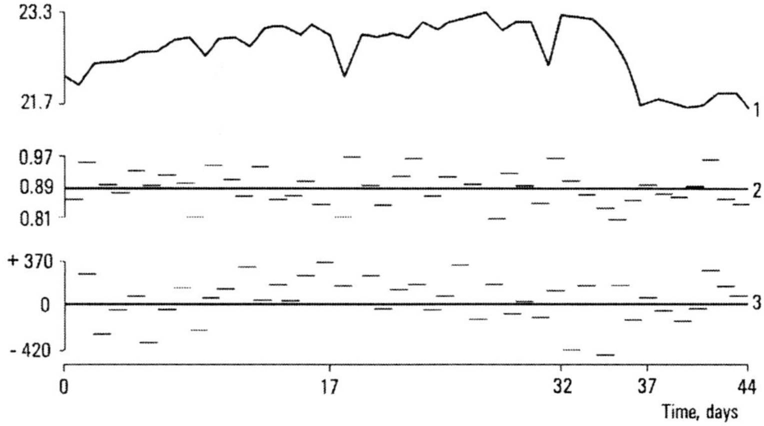

Figure 9.24 Interrelation between diet energy balance, RQ value, and total oxygen content in the sealed cabin air in experiment. 1—total oxygen content, volume %; 2—test subject's RQ; 3—diet energy balance.
(+)—diet energy exceeding test subject's energy expenditure;
(−)—test subject's energy expenditure exceeding energy supplied by food, kcal.

Although the RQ value increased for all test subjects at a consistent 1–2 days after corrective ration No.2 was first used, the RQ value decreased at various times for different test subjects when corrective ration No.1 was applied: within a range of 2–8 days, and for some test subjects it took 16 days to see the decrease. An energy imbalance in the food ration produced some inertia in the research method for correcting the system's air oxygen content. Figure 9.24 clearly illustrates the dependence between human energy expenditures, the energy that came from the food, the RQ value, and the TOC change in the Bios-3 complex. A positive food energy balance ensures that the RQ value exceeds the AQ value, and this in turn makes for a higher TOC. When the food energy balance is negative, the inverse is true. Thus, by manipulating the biochemical composition of the diet within physiologically acceptable norms, we can maintain the balanced gaseous composition of the atmosphere in the closed system.

Microflora of Humans and Habitable Zone in Biological Life Support Systems

Factors Affecting the Formation of Microbial Cenosis

When the macroorganism (a human in our case) is in a normal physiological state, there exists a dynamic equilibrium between it and its microflora, determined by physiological and immunologic peculiarities of the macroorganism, on the one hand, and the species and quantitative composition of microbial associations as well as the diversity of their biochemical activity, on the other hand.

Human Microflora in a Closed Space

With the development of astronautics, the exploration of ocean depths, as well as the Arctic and Antarctic zones of Earth, researchers meet new challenges, microflora of closed spaces being one of them.

Many researchers report that under the influence of extreme conditions—nervous and physical overwork, severe climate, feeding on specially developed diets, etc.—the composition of gut microflora can be shifted (Moore and Holdeman, 1974).

The amount of gut microflora of humans staying in a closed space is reduced (Riely *et al.*, 1966). When the ecological balance is upset, protective mechanisms may be depressed, which may allow the introduction of foreign microflora and the activation of conditionally pathogenic microorganisms always present in the organism (Tashpulatov *et al.*, 1971). For instance, by the end of their stay on the Antarctic Continent members of an expedition had *E. coli* vegetating in their mouth cavities.

After a two-week flight by astronauts, a 50% reduction in the total number of the detected microorganism species was recorded (Taylor *et al.*, 1973). Disturbance of the microflora composition may prove detrimental for astronauts' health and working ability in long-duration space missions (Taylor, 1976; Zaloguev *et al.*, 1971).

When people stay long in a hermetically sealed space (during a year-long experiment), significant shifts in their gut microflora composition are recorded: bifidobacteria and lactic-acid bacilli sharply decrease in numbers.

As for the risk of "microbial shock" that can be experienced by astronauts after a long-duration flight, it has been mentioned by several authors, e.g. Luckey (1966). Meanwhile, a wide variety of microflora in isolated conditions and a decreased ability to inhibit the growth of potentially pathogenic microorganisms may constitute a serious threat of infection to astronauts in long-duration space flights.

Humans staying in isolation for a long time exchange their microflora (Prokhorov *et al.*, 1971). Moreover, it seems that it is mostly pathogenic microflora that is exchanged. Inhabitants of Antarctic polar stations often contract intestinal and other infectious diseases caused by conditionally pathogenic microbial species.

After a 96–140-day flight (*Salyut-6*) it was found that cosmonauts mutually exchanged pathogenic staphylococci. No accommodation of non-pathogenic staphylococci was recorded (Shilov *et al.*, 1979).

Mutual exchange of microorganisms in a closed space creates the problem of biological compatibility. Conditionally pathogenic microbes are exchanged in any group of people, but in a hermetically sealed space such exchange occurs more frequently.

Microflora of Test Subjects in Bios Closed Biological Life Support Systems

Problems of the microflora in a closed ecosystem will be discussed in the next chapter, except for the microflora of humans and their habitat, which is associated with the human so closely that is actually his/her symbiont.

That is why we think it appropriate to discuss the issue of human microflora in the chapter devoted to the human. Human microflora is described here on the basis of the data obtained in the course of several long-duration experiments in Bios-3. The 120-day experiment had an important peculiarity: throughout the experiment the pressure inside the Bios complex was 5–10 mm of water higher than in the outer atmosphere. This prevented the invasion of air microflora from the outside.

In this part we analyze the results of investigations of the test subjects' gut microflora in experiments with the Bios-3 biological–engineering system, lasting 180 and 120 days each.

Experiments with autonomous biological ecosystems yielded a great number of data on the effect exerted by these systems on human gut microflora. Analysis of the literature data given above demonstrates the influence of isolation on gut flora.

In the experiments the state of gut microecology was estimated by the same methods, so we have the right to compare the results. Feces samples for analysis were collected for three days in succession every month.

In the 180-day experiment the highest and the most stable numbers were recorded for bacteroids. The total amount of microorganisms per g of feces also varied little: from 9.1 to 10.2; from 9.6 to 10.4; from 9.2 to 10.4; from 9.4 to 10.4 lg.

For the so-called "transit microflora"—staphylococci and yeasts—wide numerical variations were recorded in the samples analyzed—from 0 to 10^4 cells/g. Such variations in "transit microflora" can also be registered under ordinary conditions in healthy people.

Though in some periods the gut microflora was unstable and tended to simplify itself, the total number of microorganisms in 1 g of the sample was relatively constant for all test subjects. Bacteroids were the most while staphylococci and yeasts the least stable and numerous groups.

In this experiment no shifts were recorded in the gut biocenosis unlike in the hermetically sealed space simulating space flights, isolated Arctic and Antarctic polar stations, and in the previous versions of LSS. No shifts in gut microflora were recorded in the four-month experiment, while in the six-month and twelve-month medical–engineering experiments they were evident. To illustrate, in this experiment the numbers of lactic bacilli and bifidobacteria changed little. In the life support system of the described biological structure, from the third month on, human gut microflora remained rather stable (Figure 9.25).

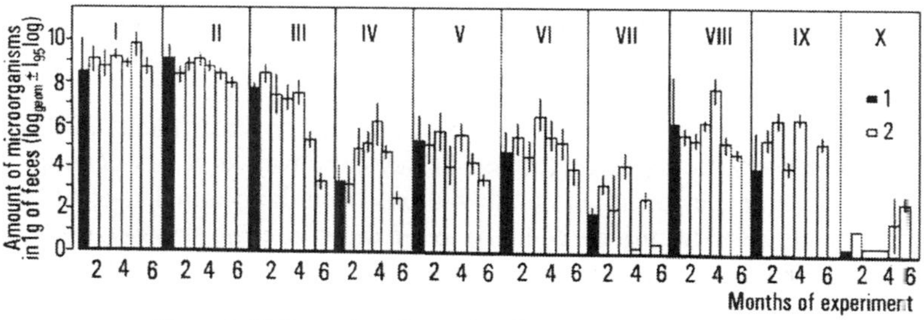

Figure 9.25 Dynamics of gut microflora of the test subject T.
Groups of microorganisms: I—*Bacteroides*; II—*Bifidobacterium*; III—*Bactobacillianaerobic*; IV—*Clostridium perfringens*; V—*Enterobacteria coliaerogenes*; VI—*Enterococcus*; VII—*Staphylococcus*; VIII—*Lactobacilli aerobic*; IX—*Proteus*; X—*Blastomycetes*. 1—results obtained out of experiment; 2—results obtained in experiment.

Common Features of the Sealed Space and Closed Ecological Systems. Differences in Their Influence on Microflora

A closed space (life support systems aboard space crafts and orbital stations, isolated groups of people in Arctic and Antarctic polar stations, the physicochemical life support system) is characterized by restricted exchange with outer environs; small volume; small human team; decreased reservoirs for storage and multiplication of microorganisms; altered, often sterile, diets; and limited personal hygiene. The results are a reduction of the microflora of the intestinal tract, microflora exchange, and a numerical increase of autoflora pathogenic strains.

Biological systems are in many ways essentially different from a closed space, particularly the systems in which the principal environment-forming link is represented by higher plants.

In a biological system, humans consume food and fresh vitamins regenerated in the system; only part of the food is stocked in a lyophilized form. As we have mentioned before, one of the exogenous factors influencing the formation of the gut microbial cenosis is food (its quality and diversity). The ecosystem includes natural reservoirs (phytotrons) for the storage and multiplication of microorganisms. Moreover, they are the microorganisms vegetating on plants in their natural habitat. This must account for the fact that in Bios-3 we did not record negative changes in microflora of the kind described for isolated habitats with no plants.

Our system—Bios—is similar to such conditions in one respect: the Bios is isolated from its surroundings. Special measures were taken to achieve this isolation, and to preclude the introduction of microflora from the outside during the experiment, i.e. gnotobiotic conditions were created. On the other hand, the Bios differs from the above described conditions—humans staying in it contact various plants and eat their fresh produce. It seems that a rather small phytotron can serve to ensure regular self-regeneration of human gut microflora.

Comprehensive medical examinations of test subjects during the half-year experiment and for a long time afterwards revealed neither any worsening of their health nor any deviations of their physiological parameters from the original state. Hence, we can draw a conclusion that the habitat generated in Bios-3 is adequate for human physiological and ecological requirements, and a healthy human can stay in this biological life support system for quite a long time.

That is why later we conducted four- and five-month experiments in Bios-3.

Medical observation was performed in the same way as it has been described for the six-month experiment, and no deviations from the original state of health of the test subjects were recorded either. To avoid repetition, we do not cite the results of the medical examinations in these experiments.

9.3 FOUR-MONTH EXPERIMENT IN BIOS-3 WITH HUMANS AND HIGHER PLANTS, INVOLVING INCINERATION OF INEDIBLE PLANT BIOMASS

The four-month experiment was conducted in the Bios-3 complex. The system consisted of two phytotrons and a living compartment. Their total area was 94.5 m^2, the total volume was 236 m^3. During the first phase (27 days) the crew consisted of three members, for comparison with the previous experiment. During the second (37 days) and third (56 days) phases the crew consisted of two members. The difference between the second and the third phases was that during the third phase a thermocatalytic furnace was used to incinerate inedible plant biomass.

In this experiment, the human environment in the closed system was regenerated only by the photosynthetic activity of higher plants. It was expected that the portion of the vegetable food generated in the system would increase to satisfy 60–70% of the human requirement, or 50% of the total requirement for animal and vegetable food.

As a result of the increased production of edible plant biomass, the production of inedible biomass inevitably increases, too, and the plants must consume more carbon dioxide than the people can exhale, while generating excess oxygen the crew does not need. To maintain the material cycle in the system, part of the inedible biomass had to be oxidized, using excess oxygen and returning to the plants the carbon dioxide they needed. Oxidation was done through incineration in the thermocatalytic furnace introduced into the system. For better microbiological isolation of the closed system, it was pressurized with sterile air, to maintain the internal air pressure in the system 15–30 mm of water higher than the external atmospheric pressure. No direct utilization of human liquid waste by the plants had been attained by the time the described experiment started, so both liquid and solid wastes from the crew were removed from the system. Other human wastes containing moisture and inedible plant biomass were dried within the system. Gray–hygiene water was filtered to remove mechanical impurities and then entered the nutrient solution for the plants. No laundry was done during the experiment. Crew requirements for drinking water were satisfied by the condensate of the plant transpirational moisture after additional treatment by sorption and condensation. That condensate was also used as hygiene water, without additional treatment, but was boiled before use.

System mass transfer between links and with the surrounding environs was monitored by measuring hundreds of parameters. Some analyses required continuous records; others were done once during the experiment. A great number of analyses were performed by the crew within the system, while very many chemical, microbiological, and some other analyses were made outside, which involved the withdrawal of substances from the cycle.

The composition of cultures in the higher plant link, the area occupied by each crop, and the age structure of the conveyer for every crop were selected based on the recommendations of the medicobiological team (Yu.N. Okladnikov). They were aimed to satisfy human requirements for basic vegetable foods. The data obtained in the previous investigations were also taken into account—the growth periods, productivity, biochemical composition, and other peculiarities of behavior displayed by different crops under the conditions of the phytotrons. The structure of the higher plant link for this experiment is presented in Table 9.21.

Table 9.21 Higher plant link in the experiment (species composition, areas occupied, and conveyer age structure).

Crop	Expected edible yield, g/day	Area		Growth period, days	Duration of conveyer "stage", days	Number of conveyer age groups
		m^2	% of total			
Wheat (dry)	250[a]	17.53	44.5	63	9	7
Chufa (dry)	320	13.27	33.7	90	9	10
Carrot	630	3.55	9.0	78	26	3
Radish	235	1.94	4.9	26	13	2
Beet	120	0.80	2.0	60	20	3
Onion	120	1.0	2.5	60	20	3
Dill	30	–	–	20–30	26	3
Kale	30	–	–	20–30	20	3
Potato	78	1.03	2.6	78	26	3
Sorrel	30	0.08	0.2	–	–	–
Cucumber	80	0.20	0.5	60	20	3

[a]of the 500 g/day of wheat grain required by two test subjects, 250 g/day were supposed to be stored.

It was proposed that during the second and third phases of the experiment the daily produce of the edible biomass generated in the phytotrons would provide each crew member with about 35 g of vegetable protein, 30 g of vegetable fat and 215 g of carbohydrates. Besides, 17 g of vegetable protein and 100 g of carbohydrates were to be provided from the store of wheat grain (125 g/day per person) preliminarily grown in the same phytotrons. This was done because of the limited area of the phytotrons. Crew requirements for vegetable fat, including polyunsaturated fatty acids, were completely provided by chufa, first introduced into the composition of the link (see Chapter 5).

The light conditions of the experiment were the same as in the previous one—140–180 W/m^2 PAR. Six-kW water-cooled xenon lamps (WCXL) were used as light sources. Both phytotrons in the Bios-3 system were illuminated continuously, around the clock, except for short (1–3 h) electrical engineering failures. Preventive cleaning of the lamps and the changing of those that failed was done by disconnecting 1 or 2 lamps out of 20, while the others were in operation.

As the wheat was all grown in one phytotron and chufa, requiring higher temperatures, in the other, the temperature of the air in the first phytotron was maintained within the range of 22–25°C, and in the second 24–28°C. The atmospheric CO_2 concentration changed in the course of the experiment but never fell below the level of photosynthesis saturation (0.3%). The relative humidity of the air in the phytotrons was maintained at 70–80%, which could be suitable both for moisture-loving crops (cucumber, radish, beet) and for the crops requiring lower air humidity (wheat, chufa).

The wheat was grown in a subirrigated air culture, chufa and vegetables were grown in hydroponics on expanded clay aggregate. Mineral nutrition was effected with the Knop nutrient medium that did not change in the experiment. By way of compensation for the elements of plant nutrition removed with the biomass, the nutrient medium was replenished 2–4 times during each 24-h period. For 1 kg of dry biomass increment the following amounts of elements were introduced: 79.0 g of KNO_3; 21.6 g of H_3PO_4; 31.6 g of $Ca(NO_3)_2 \cdot 4H_2O$; 25.2 g of $MgSO_4 \cdot 7H_2O$; 38.0 g of HNO_3; 2.0 g of $FeC_6H_5O_7 \cdot H_2O$. The trace elements B, Cu, Mn, Zn, Mo were also introduced as salts with a total weight of less than 1 g per 1 kg of biomass.

The link "higher plants" consumed about 300 l of water a day. The input of water from the link "humans" (condensate of the air conditioning device in the living compartment and the drier, kitchen and hygiene sewage water) amounted to about 18 l/day; the water entering the system from outside (compensating for the water withdrawn for analyses and removed with human wastes) amounted to 2.5–3 l/day. The water requirements of the plants were satisfied within the link, i.e. by the condensate of their own transpirational moisture.

Before the four-month experiment with the closed system started, it took 3.5 months to "warm up" the higher plant conveyer, to bring it to a steady state for all crops, and to assess the correspondence of the conveyer operating parameters to the designed ones.

The main parameters of the state and efficiency of the "higher plant" link ensuring the viability of the system with humans are gas exchange efficiency manifested as daily CO_2 uptake and O_2 emission, and daily productivity in terms of total and edible biomass. Theoretically, under unchanged conditions of the link functioning, the level of daily average productivity of plants, estimated by the final yield, must remain stable owing to the continuous (conveyer) cultivation mode. In a real experiment, the conditions can deviate from the designed ones, affecting the plant condition and, eventually, the yield. It should be borne in mind that the final yield reflects deviations in the living conditions of the plants at different growth phases rather than in the harvesting period. To put it differently, the magnitude of the yield gives delayed information on plant growth conditions, the delay time depending on the length of the crop growth period, the ability of the crop to remedy the damage, etc. Besides, the yield and the derived daily average productivity of the edible biomass are direct indications of the success or failure of the link in vegetable food regeneration during the given period.

The principal producers of both total and edible biomass in the system were wheat and chufa. As is shown in Table 9.22, wheat productivity varied in the experiment. It was most evident at the beginning of the third phase, when wheat productivity for the edible fraction reduced nearly four-fold as compared to the initial and final periods of the experiment. This reduction, however, can be accounted for by some unfavorable conditions for the growth of young wheat plants at the first phase of the experiment.

Table 9.22 Wheat and chufa productivity (on an ovendry mass basis) at different phases of the experiment

Exper. phase	Phase duration, days	Yield for growth period		Productivity, $g/m^2 \cdot day$		Harvest index
		edible	total	edible	total	
		Wheat (growth period—63 days)				
I	27	832	2104	13.2	33.4	39.5
II	37	674	2010	10.7	31.9	33.5
IIIa	30	221	958	3.5	15.2	23.0
IIIb	36	819	2318	13.0	36.8	35.3
average	120	674	1934	10.7	30.7	34.7
		Chufa (growth period—90 days)				
I	27	1737	3186	19.3	35.4	54.5
II	37	2448	4005	27.2	44.5	61.1
III	56	2565	3978	28.5	44.2	64.5
average	120	2340	3807	26.0	42.3	61.5

Observations of the leaf pigment system and visual evaluation of plant conditions showed that at the very start of the experiment the plants were subjected to the inhibiting effect of long wavelength (> 310 nm) ultraviolet radiation that penetrated into the phytotrons together with PAR through heat resistant glass "jackets" of the 6-kW water-cooled xenon lamps. It is noteworthy that before the system was closed, during the conveyer "warming-up" period, the plants in freely ventilated phytotrons did not suffer from ultraviolet radiation. The plants in the hermetically sealed system could be affected not so much by direct ultraviolet rays as by ozone, toxic for plants, which was generated under the effect of ultraviolet radiation in higher concentrations than in the freely ventilated phytotrons. However, this is merely an assumption as no direct measurements of the ozone concentration in the phytotron atmosphere were made. Between days 10 and 12 of the experiment all the "jackets" of the phytotron lamps were changed for those made of ordinary soda–lime glass, which reduced the dose of ultraviolet radiation received by plants and limited the wave length to 325 nm. Within 10–15 days the pigment systems of chufa and the vegetable crops recovered almost completely, and development of their leaves improved. The wheat, though, did not recover entirely. Hence, there were some other reasons for the inhibition, peculiar to this plant or to the method of growing (subirrigated air culture for wheat and hydroponics on expanded clay aggregate for the other crops).

A possible reason for the depression of plants could be the accumulation of toxic substances in the atmosphere, which had been recorded in the earlier half-year experiment (Gitelson *et al.*, 1975). The situation was aggravated by an accident: on day 24 of the experiment the motor of the cooling system in one of the phytotrons burned. The atmosphere was additionally contaminated by carbon monoxide and organics with low oxidation susceptibility. To get rid of these pollutants, the catalytic furnace was switched on and it worked continuously, except for two 3–4 day intervals, up to the end of the experiment. The incineration of organics by the catalytic furnace seemed to produce a positive effect on the condition of the plants, primarily vegetables. The condition of the wheat, however, did not improve radically even two weeks after the furnace was switched on. Further observations and investigations showed that one of the factors depressing wheat was elevated CO_2 concentration in the atmosphere of the phytotron, which at the end of the first phase–beginning of the second reached 2–2.3%. We assumed that elevated CO_2 concentration in the air resulted in increased CO_2 and CO_3^- content of the nutrient medium. It is well known that the bicarbonate ion can disrupt the transport of other ions from the roots to the aboveground part of the plant, causing peculiar chlorosis. That was what we first found on days 15 to 20 of the experiment on the wheat plants which started forming only chlorotic leaves when they were 8–10 days old, while their biomass almost stopped growing. This was all observed under the usual cultivation conditions in non-closed phytotrons at pH levels of about 6.5 with the usual elements, including iron chelates—but at abnormally high levels of CO_2. Experimental conditions did not allow the reduction of atmospheric CO_2 in the system, as it would have involved the depressurization of the whole system. To decrease the CO_2 solution and HCO_3^- accumulation in the nutrient medium, a decision was made to reduce the medium pH by a single injection of nitric, phosphoric and sulfuric acids. After day 40 the medium pH was reduced and maintained at 5.6–5.3. As a consequence, the plants that had formed earlier became greener, and new wheat plants of the next age group formed normally, without chlorosis. Later, to the end of the experiment wheat plants grew normally, and the grain yield was nearly the same (819 g/m^2) as at the start of the experiment (832 g/m^2).

In chufa and vegetables grown in hydroponics on expanded clay aggregate, no "carbon dioxide" chlorosis was recorded although the pH of the medium for these crops was maintained at 6.3–6.8 to the end of the experiment. Throughout the experiment, chufa, the crop first introduced into the higher plant link of the closed system, was more productive than wheat in terms of both total and edible biomass (Table 9.22), and the edible fraction in the chufa yield was larger. It should also be noted that chufa manifested a high tolerance to the peculiar conditions of the closed system (high CO_2 concentrations, toxic substances in the atmosphere, etc.): the productivity of chufa not only remained undiminished but increased by the end of the experiment. All these good points revealed in the experiment, together with the high food value of chufa, suggest that this crop conforms to the main requirements imposed on the higher plants included in the closed system.

The productivity of vegetables in the system did not vary so much as the wheat productivity, so Table 9.23 lists average values for all vegetable crops through the

Table 9.23 Productivity of vegetable crops in the experiment.

Crop	Growth period, days	Yield for growth period, kg/m^2			Productivity, g/m^2/day			Harvest index, %
		Edible biomass		Total dry biomass	Edible biomass		Total dry biomass	
		Fresh	Dry		Fresh	Dry		
Beet[a]	60	6.97	0.94	2.35	116	15.8	39.2	40.2
Carrot	78	11.82	1.59	3.16	152	20.4	40.5	50.3
Radish	30	3.9	0.23	0.68	132	7.7	22.6	34.0
Onion	60	11.4	1.64	1.70	191	27.4	28.3	97.0
Dill[b]	78(30)	1.07	0.11	0.12	14	1.5	1.6	95.0
Kale[b]	60(20)	3.84	0.29	0.30	64	4.9	5.0	97.0
Cucumber	60	23.10	0.93	2.64	385	15.5	44.0	35.0
Potato	78	2.35	0.37	5.40	30	4.8	69.2	6.9
Sorrel	60	30.90	3.22	4.12	515	53.8	68.7	78.9

[a] Edible beet biomass without tops.
[b] Dill was sown as a companion crop between rows of carrot, and kale between rows of beet.

experiment. It is worth noting that productivity in terms of total and edible biomass for such vegetables as carrot, beet, cucumber, onion, and sorrel was not lower than that for the main crops—wheat and chufa. Dill and kale, companion crops for carrot and beet, should not be considered separately when estimating the productivity but rather together with the crops they were companions for. Potato had a high total productivity and an extremely low productivity of edible biomass. The reasons were the continuous illumination hindering tuber formation and elevated substrate temperature unfavorable for tuber formation. The productivity of radish was also rather low; we assume that this can be related to the early maturing of the crop and the relatively slow development of the leaf apparatus in the first half of the growth period.

As compared with earlier experiments (Gitelson *et al.*, 1975), the productivity of beet, carrot, and radish in the system remained at the same level, and for green crops, except dill, it was somewhat higher.

The green crops all contained a large fraction of edible biomass in the yield (harvest index), but they could not contribute much to the crew ration as their biomass was rich in moisture and poor in stored carbohydrates, notably starch. Therefore, the crops prevailing in the structure of the higher plant link were those with quite a low harvest index but making a considerable contribution to the regeneration of carbohydrates—the food components making up the major portion of the ration in weight.

Table 9.24 Daily average yield of biomass in the system phytotrons.

Crop	Area		Edible biomass			Total biomass, dry	
			Fresh, g/day	Dry			
	m^2	% of total		g/day	% of total	g/day	% of total
Wheat	17.53	44.5	–	186.7	27.4	538.1	37.3
Chufa	13.27	33.7	–	345.6	50.7	561.8	39.0
Beet	0.80	2.0	93[a]	12.6	1.8	31.3	2.2
Carrot	3.55	9.0	538	72.3	10.6	143.8	10.0
Radish	1.94	5.0	257	14.9	2.2	43.8	3.0
Onion	1.00	2.5	191	27.4	4.0	28.3	2.0
Dill	–	–	49	5.2	0.8	5.5	0.4
Kale	–	–	51	3.9	0.6	4.0	0.3
Cucumber	0.20	0.5	77	3.1	0.5	8.8	0.6
Potato	1.03	2.6	31	4.9	0.7	71.3	4.9
Sorrel	0.08	0.2	41	4.3	0.6	5.5	0.4
Total	39.40	100.0	1328[b]	681.1	100.0	1442.2	100.0

[a] Beet edible biomass without tops.
[b] The total comprises only vegetables.

The data presented in Table 9.24 can offer a general idea of the actual contribution of each crop to the synthesis of edible and inedible biomass, and, consequently, to the gas exchange of the system. It is noteworthy that the daily edible yield for each crop, except potato, was close to the planned yield (cf. with Table 9.21). This fact alone indicates that plant growth conditions in the closed system "crew–higher plants" were similar to those in the phytotron operating outside the system.

In the experiment the largest portion of total biomass and particularly edible biomass was provided by chufa, though it occupied a smaller area as compared to wheat. As for wheat, it occupied 44.5% of the area but yielded only 37.3% of the total biomass and as little as 27.4% of the edible dry biomass. We have to remember, however, that the potential of wheat had not been realized as the crop was damaged at the first phase of the experiment for the reasons discussed above, and the yield was lower than it could be.

The vegetable crops together occupied 21.8% of the area and yielded 21.9% of the edible dry biomass and 23.7% of the total biomass. Hence, in spite of differences in the productivity of individual vegetable crops, this group of plants as a whole reached the mean levels of biomass synthesis and gas exchange characteristic of this link.

The biochemical composition of the edible part of all crops was regularly analyzed by I.N. Trubachev and colleagues for total carbohydrates and their fractions, proteins, amino acids, lipids and fatty acids, vitamins, mineral components, etc. (see Lisovsky, ed. 1979, Chapter 6). Most parameters of the edible part of plants produced in the experiment differed little from the typical characteristics of the same crops grown in the field. It may be alarming that some crops contained elevated amounts of such trace elements as nickel, chromium, titanium, and aluminum. They were not introduced from the outside but could have come from the structural materials, stainless steel and expanded clay aggregate. During the experiment quantities of some elements (titanium, vanadium, aluminum) found in the edible biomass of certain crops increased by several factors of ten, still remaining within the permissible limitations.

Table 9.25 demonstrates the extent to which the crew was provided with the vegetable food regenerated in the closed system during the third phase of the experiment, when a "full complement" operated in the system—2 test subjects, 2 phytotrons, and a thermocatalytic furnace for the incineration of inedible biomass. The table gives two values: the first is the entire amount of the edible biomass regenerated by plants; the second is the amount of edible biomass actually consumed by the test subjects. The latter value is naturally somewhat lower than the former, since part of the biomass was used as samples for analyses and, what is more important, some products were generated in greater amounts than the dietary norm required. The productivity of the higher plant link was sufficient for the complete satisfaction of the crew requirements for vegetable proteins and fats, essential fatty acids, ascorbic acid, and carotene. The total mass of the vegetable products regenerated in the system (proteins, carbohydrates, fats) was 616 g/day, or 65% of the crew requirements. As we have already explained, according to the plan of the experiment, only part of carbohydrates were regenerated in the system, as the planting area of the phytotrons was limited.

Table 9.25 Provision of food substances generated by plants to the crew at the third phase of the experiment with the closed system, as reported by I.N. Trubachev *et al.* (see Zamknutaya sistema…, 1979, Chapter 6).

Food substances	Crew's physiological requirement, g/day	Provision, % of the requirement	
		Produced by plants	Consumed by crew
Proteins	165.0	44.9	32.1
including vegetable	73.0	101.6	72.6
Carbohydrates	826.0	53.4	38.3
including starch	734.0	41.6	33.1
mono- and disacharides	92.0	85.2	53.3
Ballast substances	46.0	193.3	143.5
Fats	165.0	61.6	43.8
including vegetable	40.0	254.2	180.1
including polyunsaturated amino acids	7.3	147.9	110.9
Organic acids	3.7	394.6	208.1
Vitamins, mg/day:			
ascorbic acid	110.0	175.4	88.4
thiamine	3.2	68.8	50.0
riboflavin	4.0	57.5	35.0
carotene	7.3	578.1	269.9

I.N. Trubachev and colleagues reported that the vegetable products synthesized in the system completely satisfied crew requirements for such mineral food components as phosphorus, magnesium, iron, zinc, manganese, and copper; 60–70% of crew requirements for potassium, sulfur and molybdenum were satisfied; 40% for calcium, and less than 1% for sodium were satisfied. The other dietary minerals were provided by lyophilized food and cooking salt from storage.

Condensates of transpirational moisture from phytotrons were used as sources of drinking and hygiene water.

The phytotron condensate from the system cooling the air flowing into the living compartment was similar to distilled water (total mineralization of 0–1 mg/l, permanganate oxidation of 2–3 mg O_2/l). This condensate was additionally purified using an installation with columns filled with activated carbon and specially selected ionites, sterilized (by 15-min boiling) and slightly mineralized to make it conform to the standard acceptable for drinking water. Water produced this way was used by the crew for cooking and drinking.

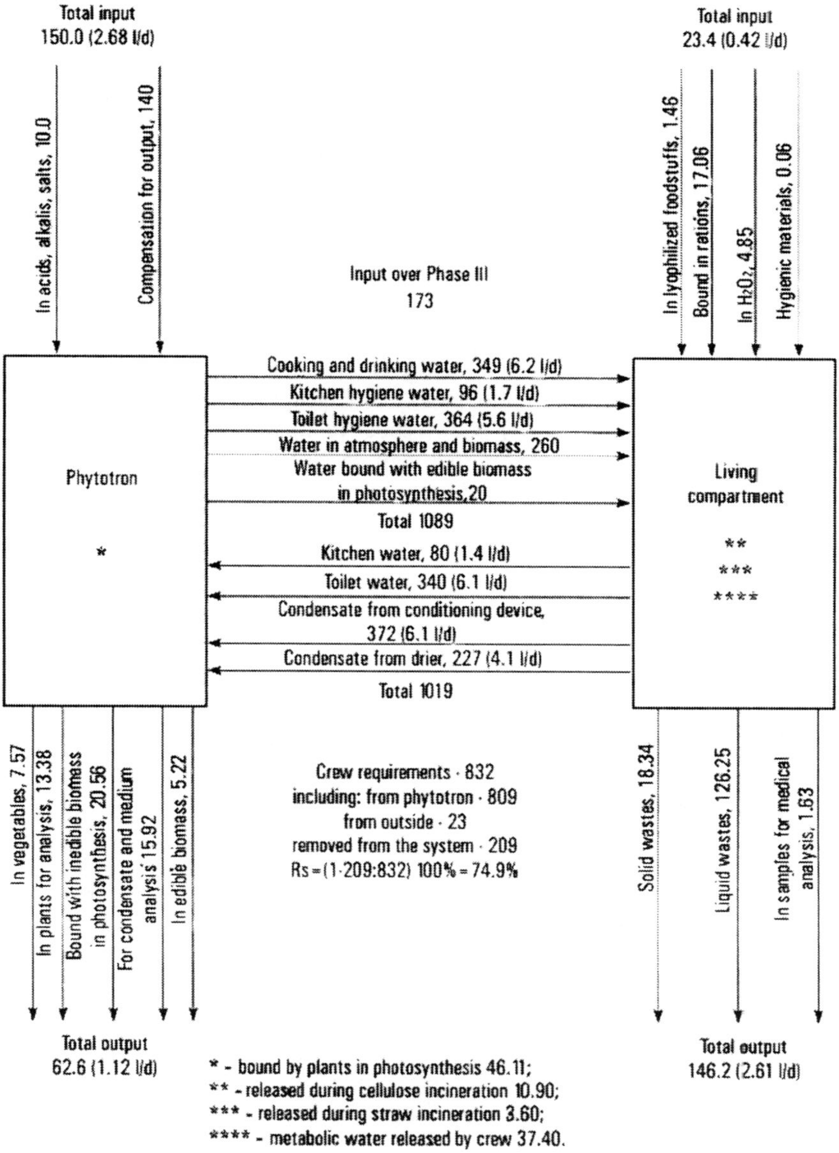

Figure 9.26 Scheme of internal and external water exchange (liters) of the system at Phase III of the experiment (for 56 days).

The phytotron condensate forming in the cooling systems of the phytotrons differed from that described above by somewhat elevated mineralization and permanganate oxidation. After being boiled for 5 min it was used in the habitation module—for washing dishes and vegetables in the kitchen, and for toilet and wet cleanup. This condensate mostly remained inside the phytotrons to replace the water evaporated by plants. The phytotrons also received other condensates of the system (from the drier of inedible biomass, the air conditioning device of the habitation module, etc.) as well as toilet and kitchen waste water.

Water exchange between links and with the surrounding environs is illustrated schematically by the example of the third phase of the experiment in Figure 9.26. The mean daily requirements of the crew for water were mainly satisfied from the sources inside the system (14.45 l), and only 0.42 l/day entered from the outside (the water bound in the rations and contained in materials for hygiene), i.e. sources inside the system satisfied 97.2% of the crew requirements for water.

Daily 3.73 l of water were removed from the system with human wastes, samples for analyses, etc. The input of water in the system with foodstuffs, materials for hygiene, salts for the plants and as clean water to replace different outflows was 3.10 l/day. Water exchange imbalance in the system, estimated by the ratio of the water mass removed to that consumed by the crew, was 25.1%, i.e. the closure of the system (R) with respect to water was 100 – 25.1 = 74.9%. It should be pointed out that more than half of this imbalance was accounted for by liquid wastes which were not used in the system, 2.25 l/day of 3.73 l/day.

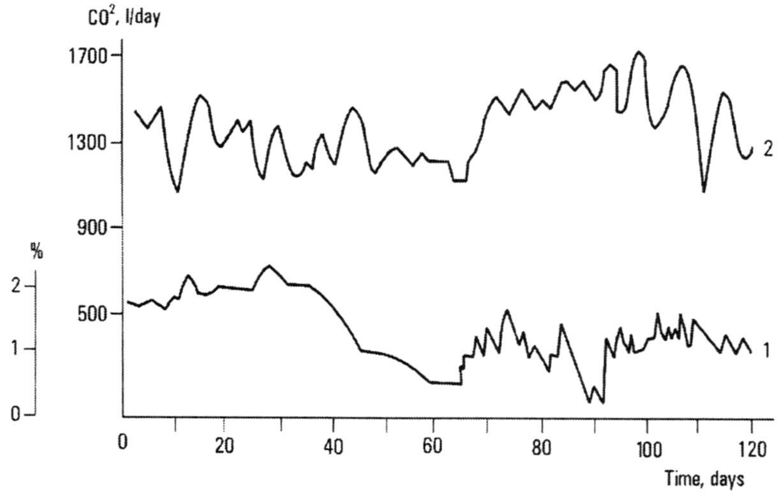

Figure 9.27 CO_2 content of the closed system in volume percent (1), and the rate of CO_2 assimilation by plants, (2).

Gas exchange of the higher plant link was estimated in the experiment by the data obtained from the analyses of CO_2 concentration in the system (Figure 9.27, curve 1), taking into account the average amount of CO_2 respired by test subjects daily, losses due to incomplete sealing of the system, and CO_2 produced in the process of vegetable biomass incineration at the third phase of the experiment. Calculations of the daily CO_2 assimilation by plants (in volume liters not adjusted to 0° and normal pressure), processed by the method of a sliding curve, are plotted in Figure 9.27 (curve 2).

At the start of the experiment daily CO_2 assimilation by plants was 1350–1450 l, while three test subjects daily emitted 1368 l CO_2. Such a rate of plant gas exchange was sufficient for the sustainable existence of the system. The photosynthetic apparatus of plants was, however, damaged by ultraviolet radiation and ozone, therefore the assimilation of carbon dioxide progressively decreased, and the CO_2 concentration in the atmosphere increased to 2.0–2.2%. By reducing the proportion of ultraviolet radiation in the radiant flux of the xenon lamps, CO_2 assimilation was restored, but for a number of reasons described above the condition of the wheat remained unsatisfactory, and the gas exchange of the higher plant link gradually reduced from 1400–1350 to 1200–1150 l/day until the beginning of the third month of the experiment. After day 28 there were only two test subjects in the system. Their mean daily CO_2 emission was 1125 l, i.e. less than the plants assimilated. So the CO_2 concentration of the system atmosphere was gradually reduced to 0.64% (Figure 9.27, curve 1). As the atmospheric CO_2 concentration got lower and the wheat assimilation apparatus improved noticeably, a new stage of gas exchange investigations became possible—the incineration of vegetable biomass to compensate for the difference between the CO_2 emission by test subjects and CO_2 assimilation by plants.

The thermocatalytic furnace for the incineration of vegetable biomass (see Chapter 9, Section 1) was first brought into operation on day 65 of the experiment. To estimate the quality of organic matter oxidation with the furnace, first the crew burned pure cellulose from storage, which was used to imitate straw. Incineration of 600 to 2000 g of cellulose a day, 2–3 days apart, from day 65 to day 76 proved that the working thermocatalytic furnace did not contaminate the atmosphere, oxidized organics completely, and supplied the additional carbon dioxide required by the plants. During this period CO_2 assimilation by the plants (Figure 9.27, curve 2) reached a rather high level, 1400–1500 l/day), and the condition of the plants was visually evaluated as good. On days 78, 79 and 80, dry wheat straw was incinerated: 500, 400, and 300 g/day. As a result of the oxidation of the nitrogen-bearing straw components, short-lived jumps of the atmospheric NO_2 concentration, to 0.5–0.6 mg/m^3, were recorded. Though the recorded NO_2 concentrations were at the level of the maximum concentrations permissible for humans, the plants responded by the yellowing and then drying of leaf tops. Carrot and radish turned out to be the most sensitive to nitrogen dioxide. However, the recorded damage was not lethal.

After straw incineration was discontinued (but cellulose incineration proceeded), the plants gradually restored their rate of CO_2 assimilation. The reversibility of damages done to plants by NO_2 under incineration of small portions of straw is clearly seen in Figure 9.27 (curve 2), where there are dips in the curve of CO_2 assimilation by plants, corresponding to the days when straw was incinerated (days 93–95, 99–101, 109, 115), and the gradual restoration of the gas exchange function during the intervals.

Altogether, during the last 55 days of the experiment 19.69 kg of cellulose and 6.42 kg of straw were incinerated. Regular incineration of the required amounts of straw in the system proved impossible because of plant poisoning. Therefore, first, it is essential to develop such technology of plant biomass oxidation that would sharply reduce or even prevent the release of nitrogen (and maybe sulfur) oxides into the atmosphere of the system; and second, it would be expedient to find for different crops the maximum permissible toxic oxide concentrations resulting from the incineration of inedible plant biomass.

On the whole, gas exchange of the higher plant link of the closed system was intensive enough to regenerate the atmosphere throughout the experiment. Despite some deviations from the prescribed growth conditions, during the last 40–50 days of the experiment carbon dioxide assimilation by plants was higher than at the start of the experiment. This suggests a feasibility of regular gas exchange by the higher plant link when combined with the gas exchange of the human link as well as with the equipment links introduced into the structure of the closed system investigated.

It should be remembered, however, that the CO_2/O_2 ratio is not the same for human and plant gas exchange. The ratio of the CO_2 produced to the O_2 consumed (a respiratory quotient) in the human gas exchange is usually lower than the ratio of the CO_2 consumed to the O_2 evolved for plants (an assimilatory quotient). The difference is chiefly accounted for by a higher fat content of the total plant biomass. As reported by I.N. Trubachev and colleagues (see Zamknutaya sistema..., 1979, Chapter 9), the respiratory quotient of the crew during the experiment was 0.871, while the assimilatory quotient for the same period was 0.928. It should be noted that at the third phase of the experiment the incineration of straw (and cellulose) increased the "respiratory" quotient of the link "humans + furnace" to 0.903 since the "respiratory" quotient of the furnace was 1.004. The imbalance between O_2 and CO_2 in the closed system could be avoided by lowering the plant assimilatory quotient somewhat, introducing more fat-producing plants (chufa, soybean, etc.) or reducing within permissible limits the consumption of fats (essentially animal fats) by the crew, thus increasing the human respiratory quotient. For instance, if at the third phase of the experiment the daily consumption of fats by each crew member had been reduced from 92.7 to 70 g (by introducing into the diet 55 g/day of carbohydrates, which corresponds in caloric content to 22.7 g of fats), the "respiratory" quotient of the link "humans + furnace" would have increased from 0.903 to 0.921, i.e. it would have been nearly the same as the plant assimilatory quotient.

One of the essential characteristics of gas exchange in the closed system is the presence or absence of minor, primarily toxic, components of the atmosphere. Each individual link of the system—humans, plants, technical devices—continually release into the atmosphere various volatiles (carbon monoxide, ammonia, ethylene, alcohols, etc.) which can progressively accumulate in the sealed space. Our investigations show, however, that in a system with a biological material cycle, including all these links, no steady accumulation of minor components of the atmosphere is recorded (Figure 9.28). The amounts of carbon monoxide, ammonia, sulfur dioxide, acetic acid, and some other impurities in the atmosphere of the closed system fluctuated about some average level, within the limits of concentrations permissible for humans (Okladnikov *et al.*, 1979).

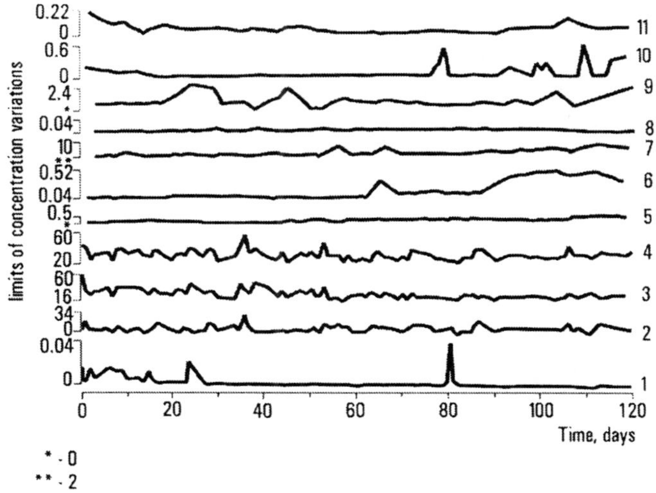

Figure 9.28 Toxic gaseous impurity contents in the air of the living compartment. 1—carbon monoxide, mg/l; 2—total organic substances with high oxidation susceptibility, O_2 mg /m^3; 3—total organic substances with low oxidation susceptibility, O_2 mg /m^3; 4—total organic substances, O_2 mg /m^3; 5—ammonia, mg/m^3; 6—aldehydes, mg/m^3; 7—alcohols, mg/m^3; 8—sulfur dioxide, mg/m^3; 9—acetic acid, mg/m^3; 10—nitrogen dioxide, mg/m^3; 11—hydrogen sulfide, mg/m^3.

These substances could be eliminated within the system through absorption by abiotic components of the system (water from the phytotron cooling system, expanded clay aggregate inhabited by roots), growing plants and associated microflora as well as by the catalytic furnace for oxidizing toxic atmospheric substances, which was brought into operation at the third stage of the experiment.

An attempt to differentiate between the effects of minor atmospheric component absorption by abiotic and biotic components of the closed system (without a catalytic furnace) was made by V.E. Rygalov (Rygalov *et al.*, 1995; Rygalov, 1996). He showed that, in the case of a one-time input of minor impurities into the system, they are absorbed from the atmosphere by biotic components faster than by abiotic ones, the absorption by biotic components being more efficient and often irreversible. The impurities must have been destroyed or even utilized by plants and/or microflora, i.e. the closed ecological system could purify itself via biological components removing atmospheric impurities.

The integrated quantitative characteristics of the system mass exchange at the third phase, when all links, including biomass incineration, interacted, are schematically shown in Figure 9.29. It should be specified that the scheme presents link and system inputs and outputs only of those quantities of substances that participated directly in the processes of biological exchange. Some substances designated in the scheme as inputs (lyophilized foodstuffs, cooking salt, materials for hygiene, mineral components of plant nutrition, etc.) had been stored in Bios before the experiment started, while, for example, inedible biomass, designated as output, was stored in the system in place of the consumed "incoming" foodstuffs. The scheme also leaves out ventilation interlink gas exchange (O_2, N_2, H_2O vapor), the mass of these gases exceeding the amount directly consumed by the humans. This can, for instance, explain why the water input to the human link is approximately 4.5 l/day larger than the water output from this link. This difference was made up of the neglected vaporized water that moved with the air from the crew link to the phytotrons.

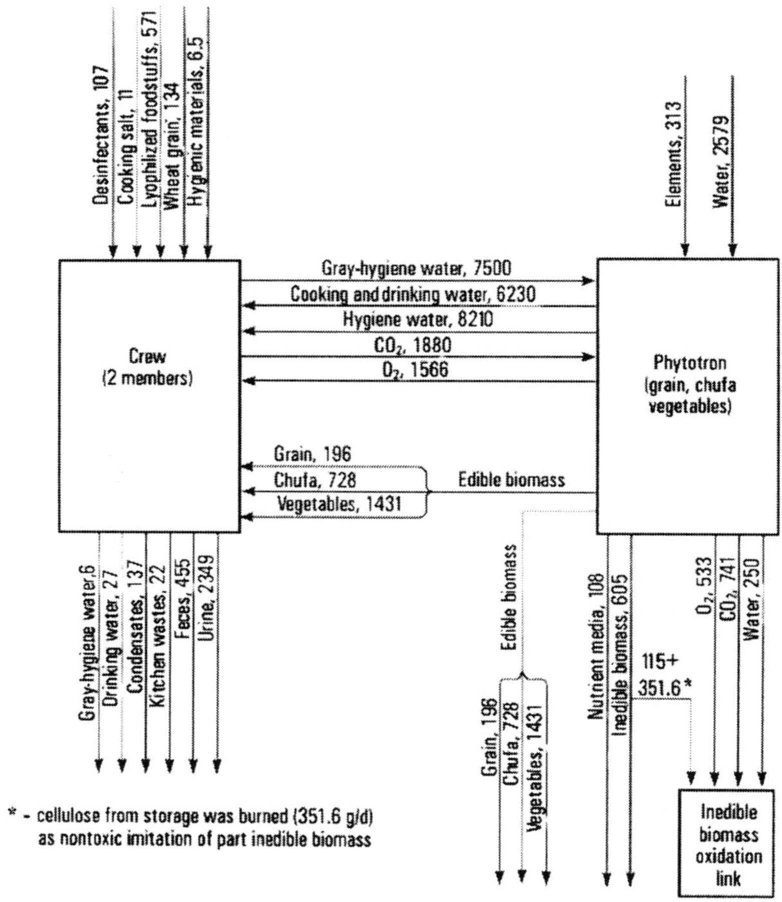

Figure 9.29 Daily average "input" and "output" of substances for the system "human–higher plants" of the experiment (grams).

Adding up the input and output parameters of the links, we can see that the crew link daily received an input of 18361 g of substances regenerated by plants,

Table 9.26 Requirements of the crew and life support system at different phases of experiment, g/day.

Crew requirements	Experiment phase		
	I	II	III
Lyophilied foodstuffs	727.0	587.0	571.0
Vegetables, grain and chufa synthesized within the system	1781.0	2241.0	2355.0
Grain synthesized within the system (ovendry)	280.0	193.0	134.0
Oxygen	2025.0	1675.0	1566.0
Cooking and drinking water	7630.0	5000.0	6230.0
Hygiene water	11407.0	7030.0	8210.0
Hygienic materials	9.8	6.5	6.5
Disinfectants	183.0	113.0	107.0
Cooking salt	16.6	12.2	11.0
Total:	24059.0	16857.7	19190.5
System requirements	**Experiment phase**		
	I	II	III
Lyophilized foodstuffs	727.0	587.0	571.0
Grain synthesized outside the system (ovendry)	280.0	193.0	134.0
Salts, acids for plants	293.0	332.0	313.0
External water (to replace the removed urea and samples for analyses	3705.0	2518.0	2579.0
Hygienic materials	9.8	6.5	6.5
Disinfectants	183.0	113.0	107.0
Cooking salt	16.6	12.2	11.0
Sorbents for purification of drinking water	2.7	2.7	2.7
Total:	5217.1	3764.4	3724.2
The index of system closure, %	78.3	77.7	80.6

while the daily input of substances received from the "outside" was 829 g. Hence, 95.7% of crew requirements were satisfied by substances regenerated within the system. Plants in the phytotrons received 10371 g/day of various substances from the crew and the catalytic furnace, i.e. from the sources within the system, while the input from the "outside" was 2892 g/day. Only 78.2% of plant requirements were provided by the sources within the system. The main reason why large amounts of substances (mostly water) were introduced into the phytotrons from the outside was that all the urine removed from the crew link (2349 g/day) had to be compensated for. If methods were developed to utilize all urine within the system, up to 95.5% of plant requirements could be satisfied from the sources in the system, and the entire material cycle closure could be considerably increased.

Quantitative parameters of the mean daily requirements of the crew and the requirements of the whole system for external substances are given in Table 9.26 for each phase of the experiment. The table also includes the index of closure of the biological material cycle (R_s) as a percentage difference between total crew requirements (M) and system requirements from the external sources (m) relative to the total crew requirements for substances ($M - m/M$ 100%). At all stages of the experiment the index of the system closure was not high—77.7–80.6%, mostly due to the large inputs of water compensating for the removal from the system of crew liquid waste and expenditure of system fluids for analyses.

* * *

Summing up, it should be noted first that throughout the experiment, mass exchange characteristics of the test subjects, their health and psychological condition (investigated as minutely as it is described in Chapter 9, Section 2) remained within the norm. No adverse effects resulting from the participation in the four-month experiment were recorded in the test subjects, who have also been under observation for many years after the experiment.

The basic link for the regeneration of the environment, higher plants, coped with most pre-assigned functions. The plants satisfied all the crew's requirements for oxygen, drinking and hygiene water, various vegetable foodstuffs, and, in part, grain products. The plants utilized all the carbon dioxide produced by the crew and released during the incineration of some part of the inedible biomass, as well as hygiene gray water; they aided in stabilizing minor atmospheric components within safety limits acceptable for humans. Plants showed a capacity for quite fast restoration of photosynthetic productivity after they had been damaged owing to technical hitches in the system (that gave rise to ultraviolet radiation) or volley releases of toxic oxides from the incineration of straw in the system.

It should be pointed out that certain difficulties in conducting the experiment were related to the operation of the link first introduced into the closed system: the furnace for the incineration of inedible plant biomass. Toxic substances released into the atmosphere during incineration reduced the photosynthetic productivity of plants, which brought about the progressive accumulation of carbon dioxide in the atmosphere, limiting the possibilities for further incineration of inedible biomass. Hence, it followed that techniques and equipment for the thermal incineration of biomass should be improved or other methods of biomass oxidation developed ("wet" oxidation, "biological" oxidation, etc.).

Thanks to special measures that lessened the probability of external microflora getting into the system, microflora of both the humans and the higher plant link gradually stabilized (Pankova *et al.*, 1979). It is too early, however, to suggest the possibility of the purposeful formation of microbial landscapes in different structures of the system; it should rather be stated that microflora can adapt spontaneously to the living conditions of any link in the closed system, as they are being optimized.

On the whole, this experiment demonstrated that, under the conditions of the self-regulating closed system it is feasible to attain a rather steady agreement between humans and the higher plant link in terms of the rate and direction of biological exchange processes; some inconsistencies in the exchange processes can be corrected by using a physicochemical link.

Investigation of the closed system in the four-month experiment revealed the possibility of a 78–80% reduction of the pre-stored substances needed for complete human life support and suggested the avenues and possibilities for further perfection of closed ecological systems that include humans.

9.4 FIVE-MONTH EXPERIMENT IN THE LIFE SUPPORT SYSTEM FOR TWO HUMANS. INVESTIGATION OF THE FEASIBILITY OF FULL REGENERATION OF VEGETABLE FOOD AND IMMEDIATE UTILIZATION OF HUMAN LIQUID WASTES BY PLANTS

The results of the experiment with the system "human–higher plants" formed the basis for tackling new problems in the Bios-3 experimental facility. The main objectives were:

1) to increase food regeneration for humans in the closed system to 75–80% of the total food requirement, that is to regenerate the entire vegetable fraction of the human diet;
2) to bring the closure of the water cycle in the ecosystem to 100%, by means of utilizing human liquid waste within the system.

To approach these objectives, some changes had to be made in the Bios-3 facility. A third phytotron was established in the compartment previously occupied by microalgal cultivators. So, the area occupied by plants for two test subjects increased from 39.4 m^2 in the previous experiment to 63 m^2 in the one we are describing.

To reduce the toxicity of the gases released by the incineration of inedible biomass in the system, a new technology for incineration was developed, with a decreased release of nitrogen oxides to the atmosphere. A new installation (a thermocatalytic converter) was made, with three catalysts placed in series. The chief peculiarity of the new technology of biomass incineration was that at the first stage of the process the amount of air used was not sufficient for complete oxidation of the biomass. The major part of the organic nitrogen was oxidized to the molecular state but not to toxic oxides. At the second and third stages combustion products were completely oxidized in catalysts with excess air. The small amount of nitrogen oxides produced was almost completely trapped when gases were scrubbed with water in the scrubber closing the process of biomass incineration. Control experiments with straw incineration in Bios-3 showed that of all the nitrogen contained in the straw 0.2% was released into the atmosphere as nitrogen oxides and 99.8% as molecular nitrogen. An hour after incineration there were no nitrogen oxides in the atmosphere: they had dissolved in the condensate of transpirational moisture and plant nutrient solution. They were also taken up by plants and produced an adverse effect on them, as became apparent in the course of the subsequent long-duration experiment involving humans.

To increase the water closure of the system, for the first time the solid wastes of the test subjects were dried in the specially devised drier provided with the catalyst for removing organic contaminants from the air. Solid wastes were placed into one of the interchangeable cylinders, 1 liter in volume, and dried for 3 days at a temperature of 100°C. Volatile contaminants were subjected to catalytic oxidation at 600°C. Upon cooling, the air entered the phytotrons. No adverse effects on the plants were recorded. The drier conformed to hygiene standards. The dried solid wastes were removed from the system.

With the same aim, to increase the water closure of the system, human liquid waste was used immediately for growing plants. This involved the changing of wheat cultivation technology, as will be discussed below.

To attain the objective of regenerating a greater part of the vegetable fraction of human diet within the system, in this experiment the structure of the higher plant link was altered to increase the quantity of the principal food product, grain, and to compose a more diverse and nutritionally optimal diet. To increase grain production, the total wheat planting area was enlarged by launching the third phytotron. A greater variety of the diet was gained by introducing the crops that had not been used in such systems before—pea, tomato, and kohlrabi. At the same time kale and sorrel were excluded because beet, dill, onion, and kohlrabi were able to supply crew members with salad vegetables to the full. Based on the rated rations of the test subjects and the feasibility of regenerating a particular vegetable product in the closed system, the structure of the higher plant link was worked out. It included 12 crops, each occupying a certain area (Table 9.27).

The properly functioning higher plant link of such a structure was to provide two test subjects with almost all the required carbohydrates, vegetable proteins, fats, and vitamins, simultaneously regenerating atmosphere and water. Such crops as tomato, potato, pea, and cucumber were also to serve as live indicators of the atmospheric pollution by toxic contaminants that could be formed as a result of the periodic incineration of part of the inedible biomass to maintain the cycling of nitrogen and some other biogenic elements.

As a significant portion (two-thirds) of the planting area was allocated to wheat, this crop was chosen to perform the function of utilizing the liquid waste of the test subjects. Wheat has some advantages over other crops. First, the edible part of this plant (grain), unlike chufa and potato tubers, carrot and beetroot, etc., does not contact the nutrient solution and human liquid wastes added to it. This annuls a number of objective and subjective arguments against using native urine as a component of the nutrient solution for food crops. Second, the higher the sodium chloride concentration in the medium, the more intensively inedible wheat biomass (comprising up to 60% of the total biomass) takes it up from the solution. Thus, sodium chloride input to the nutrient solution with liquid wastes, its concentration in the medium, and its removal with inedible biomass can come to a dynamic equilibrium precluding the unlimited build-up in the solution of sodium and chlorine ions regularly coming out of the human organism with urine but unnecessary for plants. Native urine provides plants not only with sodium chloride but also with nitrogen and some other nutrients contained in it in a combined state as urea and other compounds. On the other hand, urine contains numerous minor components (hormones, enzymes, etc.), the action of which is hardly predictable if urine is added to a plant nutrient solution. That is why, prior to the experiment with the system closed, a series of preliminary investigations was performed.

Table 9.27 Calculation of the structure of the higher plant link for a crew of two persons.

Crop, product	Daily crew requirement, g	Expected crop productivity, $g/m^2 \cdot day$	Area to be occupied, m^2
Wheat, grain (ovendry)	520	13	40.0
Chufa, tubercles (ovendry)	234	26	9.0
Pea, seeds (ovendry)	52	13	4.0
Carrot, roots (fresh)	220	160	1.4
Radish, roots (fresh)	110	125	0.9
Beet, roots (fresh)	105	120	0.9
Beet, leaves (fresh)	25	50	–
Kohlrabi, leaves (fresh)	40	50	–
Kohlrabi, stems (fresh)	140	130	1.1
Onion, stalks, bulbs (fresh)	120	170	0.7
Dill, green part (fresh)	20	30	–
Tomato, fruits (fresh)	150	110	1.4
Cucumber, fruits (fresh)	100	250	0.4
Potato, tubers (fresh)	250	80	3.2
Total			63.0

Human liquid wastes had not been used in previous experiments with humans as a source of water, nitrogen, and some other elements for higher plants. Experiments in which urea and sodium chloride, the basic components of liquid waste, were added to the nutrient medium separately, yielded the following results. The replacement of the entire nitrate nitrogen with urea deteriorated the condition of the plants and reduced the yield. The acceptable contribution of urea nitrogen to the total nitrogen nutrition was 70–75%. At a NaCl concentration of 2 g/l and under urea nitrogen nutrition (70% of the total nitrogen), no deterioration in the condition of plants or reduction in their productivity was recorded.

In the experiments with continuous wheat culture, 12 g of NaCl, equivalent to the daily excretion of sodium chloride by a human, was daily added to the nutrient solution for 0.7 kg of synthesized biomass providing the daily human requirement for oxygen. Throughout the experiment (27 days) the NaCl concentration of the solution was maintained at a level of 2 g/l. This was indicative of a balance between the NaCl input and removal with harvested biomass and of tolerance of plants to such level of chloride salinization. Elevated NaCl content (1.2–1.5%) in the total plant biomass was recorded.

Encouraged by the positive results of experiments with the addition of urea and NaCl to the nutrient solution, we made a series of experiments under the same conditions but with native urine containing the same amounts of NaCl and urea. The results of these experiments were negative. In subsequent experiments we used a different procedure of adding liquid waste to the nutrient solution (portionwise, 2–3 times a day) and also changed the approach to mineral nutrition: instead of growing wheat on a mineral solution for the first 36 days and then on pure water (as was the case in previous experiments with closing the systems), plants were grown on nutrient solution throughout the growth period. Under such growth conditions maturing plants had a greater opportunity for passive uptake of salts from the nutrient solution. Two long-duration experiments (150 and 190 days) with conveyer wheat culture carried out in phytotrons in an open system showed that the introduction of native urine into the nutrient solution did not lead to a significant decrease in the wheat yield, if the diurnal quantity of the urine of one human was added to 200–250 l of nutrient solution needed for wheat planting of 16–18 m^2 capable of synthesizing 600–700 g of dry wheat biomass a day. In the devised experiment the complete utilization of two test subjects' liquid wastes might be attained by including 40 m^2 of wheat planting into the higher plant link. In the above long-duration experiments the sodium chloride concentration in the nutrient solution reached 1.5–2.0 g/l and stabilized within this range. The sodium chloride content of the total wheat biomass reached 1.3% and the total amount removed with 0.7 kg of daily biomass increment was approximately the same as the daily amount of sodium chloride supplied with the liquid waste of one human. Thus, a way was found for plants to directly utilize human liquid wastes in the closed system "human–higher plants," which provided prerequisites for significantly increasing water cycle closure and, consequently, the total closure of the system.

As compared to earlier experiments in the closed system (Section 9.3), in this experiment the overall structure of the higher plant link and its operating conditions were changed so

a) to augment the amount of vegetable food regenerated in the system per person and to improve its quality by including a greater variety of products (tomato fruits, green and ripe peas);
b) to increase the closure of the water cycle by including human liquid wastes into biological cycling through the direct utilization of them by higher plants (wheat);
c) to obtain various test objects for visual monitoring of the occurrence in the atmosphere of contaminants toxic for plants and of the degree to which they affect plants.

Other conditions of the higher plant link functioning had been substantiated and tested in earlier experiments.

All plants were grown under the light of 6-kW xenon lamps, the incident photosynthetically active radiation (PAR) intensity being 130–170 W/m^2 in the first and second phytotrons and 120–140 W/m^2 in the third. All crops were irradiated continuously, except potato and kohlrabi that were shaded for 8 h a day and were grown under a 16L:8D photoperiod.

The air temperature in the phytotrons was maintained at 22–24°C, the air humidity within 70–80%. The atmospheric CO_2 concentration varied in the course of the experiment between 0.5 and 1.8%.

Wheat was grown in subirrigated culture; chufa and all vegetable crops were grown hydroponically on expanded clay aggregate.

The starting nutrient medium was the Knop solution. It was corrected 2–4 times a day to replenish the elements taken up by plants. The addition contained the following quantities of salts and acids calculated for a kilogram of dry biomass: KNO_3—79.0 g; H_3PO_4—21.6 g; $Ca(NO_3)_2 \cdot 4\ H_2O$—31.6 g; $MgSO_4 \cdot 7\ H_2O$—25.2 g; HNO_3—38.0 g; $FeC_6H_5O_7 \cdot H_2O$—2.0 g. Besides, such trace elements as B, Cu, Mn, Zn, and Mo were added in salts, weighing altogether less than 1 g per 1 kg biomass. Due to such corrections element concentrations in the nutrient solution could be kept at the prescribed levels and the pH level within the range of 5.8–6.5.

If the productivity of the higher plant link in the system is in a steady state, the nutrient medium composition can be corrected as calculated, without chemical checking. In the real experiment, however, samples of the nutrient solution had to be regularly taken for analysis for two main reasons: 1) plant productivity and harvest composition might vary; and 2) it was proposed that native urine should be added to the wheat nutrient solution, the elemental composition of which could vary within a certain range. In the described experiment macroelement concentrations in the wheat nutrient medium were measured once a week, and in the vegetable nutrient medium once every two weeks; trace element concentrations were measured twice during the experiment. Depending on the results of the analyses, "correction of the correction" was performed when needed.

In the closed system including humans, plants received water from the following sources: the condensate of transpirational moisture in the phytotrons and the moisture exuded by humans in the living compartment; hygiene wastewater after taking a shower, washing, etc. (upon rough filtering); and human liquid waste as an additional source of water for wheat.

Preparation of the crop conveyer started 27 days before the crew closed the system. Vegetable crops were planted according to the schedule of conveyer formation. Wheat was sown simultaneously on 20 m^2, half of the total wheat planting area. The objective was to effect photosynthetic gas exchange in the system until the planned development of the wheat conveyer could ensure complete CO_2 absorption within the system. As the conveyer developed, part of the extra wheat, sown before the experiment started, was harvested in the vegetative state.

The experiment with two test subjects was launched when the higher plant conveyer, not fully realized yet, could already maintain gas exchange in the system "human–plants."

The conditions under which plants were grown in the phytotron were kept close to the designed ones. Most environmental parameters varied little throughout the experiment. In most cases the composition and pH of the nutrient solution as well as the atmospheric CO_2 concentration were changed following the program of the experiment. Sometimes, though, the CO_2 concentration changed as a result of the deterioration in the photosynthetic activity of plants caused by toxic contaminants released into the atmosphere owing to faulty operation of physicochemical links of the system.

Beginning on the first days of the experiment the following devices were tested in the system: a drier for solid wastes (volatile organics being oxidized in the catalytic converter operating continuously) and a thermal incinerator for burning inedible plant biomass. On day 47 the test subjects began to add their liquid wastes to the wheat nutrient solution.

During the test of the solid waste drier, intended for the reintroduction of water contained in solid wastes to the material cycle of the system, no volatile products were recorded as building up to toxic levels in the atmosphere of the system. The tests caused no noticeable changes either in the external features of the plants or in their photosynthetic rate. So a decision was made that it would be expedient to use this device throughout the experiment.

During the first days of the experiment, the plants in the system were able to consume all of the carbon dioxide produced by the test subjects. So, in the first week experiments were started in which, first, inedible plant biomass and then cellulose were incinerated portionwise, 0.0 to 2.4 kg a day, averaging 0.76 kg/day for the experiment. As soon as several hours after the first portion of inedible biomass was burned, the plants exhibited signs of gas poisoning: leaf folding in pea, potato and carrot; tendril twisting in pea; etc. During the days that followed, when biomass or cellulose was burned repeatedly, more sever the damage to plants was done: potato and pea ceased to grow and began to cast their lower leaves; pea and cucumber stopped flowering and fruiting altogether; cucumber and carrot leaves as well as chufa leaf tips turned partly yellow and covered with necrotic blotches; in wheat the height growth was retarded and leaves changed their spatial orientation; etc. The damage done to the plants in many respects resembled that done by ethylene, but unfortunately, we did not try to determine it in the atmosphere. As a result, the photosynthetic productivity of plants was impaired, and at harvest the yield of the edible biomass was lower than expected. It should be noted that the detrimental effect increased with the length of the growth period. For instance, the yield of radish roots in this phase of the experiment was 3.6 kg/m^2—rather than the expected 3.7; beet root yield was 4.6 kg/m^2 rather than 7.2; the yield of dry chufa tubers was 0.8 rather than 2.3 kg/m^2. The wheat planted first, which had been repeatedly subjected to the action of the toxic products of biomass combustion, formed almost sterile ears.

The crew examined the incinerator and remedied faults, interrupting incineration for periods from 5 to 20 days. Owing to this, the atmospheric conditions of plant growth were partially normalized, so the photosynthetic function of the plants intensified, reaching 70–80% of the designed value during the third–fourth months of the experiment. But the production of edible biomass was more strongly affected by the atmospheric toxicity than the total biomass yield was. Besides, the after effect of atmospheric toxicity on the productivity of crops with prolonged growth periods showed up within 40–70 days after the plants were actually subjected to the action of such an atmosphere. Therefore, only at the end of the experiment did it become possible to close the food cycle. On day 110 a new catalyst for biomass oxidation was installed, and the adverse effect of the toxic atmosphere on plants, which persisted till the end of the third month of the experiment, gradually became less pronounced.

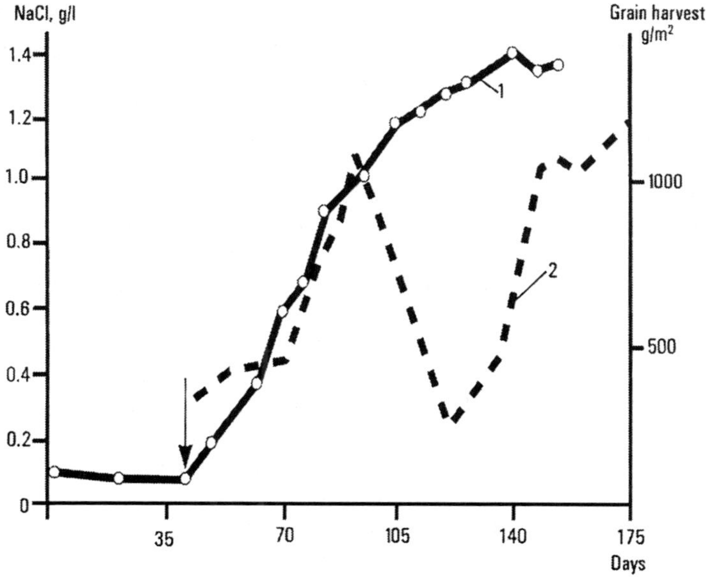

Figure 9.30 Sodium chloride accumulation in the nutrient medium for wheat (1). ↓ is the time the human liquid wastes began to be introduced. Grain harvest of wheat (2).

Much earlier, on day 47, when the wheat conveyer was structurally complete, the test subjects' liquid wastes were first used as a component of the wheat nutrient medium. Every day the urine of the two test subjects was all—1.5 to 2.5 l, excluding samples for analysis—added to the wheat nutrient solution, the total volume of which was 450–500 l. During the 103 days, from day 47 to the end of the experiment, 210.7 l of native urine were introduced into the solution. First, the daily system requirement for water from the outside or storage became 2.046 l lower, which significantly, by 15.7%, increased the closure of the system as a whole. Second, since plants utilized the mineral components of the urine, smaller amounts of salts, acids, and alkalis were needed to prepare the wheat nutrient media: the expenditure of P and Ca was reduced by 10–12%, K and S by 20–26%, and N by 70%.

The addition of liquid wastes to the wheat nutrient media resulted in a gradual build-up of sodium chloride in the medium, both elements of it being unnecessary for plants in the quantities that can daily come with human liquid wastes (Figure 9.30). Concurrently sodium chloride content in wheat biomass began to increase, too, primarily due to passive consumption by maturing plants. As a result, sodium chloride inflow to the nutrient solution from the test subjects and an increasing outflow with plant biomass were approaching ever closer to an equilibrium. During the first four-week period of adding urine to the solution, the NaCl concentration in the medium became 512 mg/l higher, during the second four-week period 603 mg/l higher, during the third only 136 mg/l higher, and during the last four-week period it increased by mere 68 mg/l, i.e. the input of NaCl with urine and output with harvested wheat biomass actually approached the balance.

The sodium chloride content of wheat biomass (on an ovendry basis) grew from trace levels (hundredth fractions of percent) during the control period to 3.50–4.05% in roots, 1.12–1.37% in shoots, but only 0.01% in grain. Such "salinization" of the wheat nutrient medium and biomass produced little effect on the wheat productivity. Once the adverse effect of toxic atmospheric pollutants was eliminated. It was in the period when the levels of sodium chloride content in the nutrient medium and biomass were close to equilibrium that wheat daily average productivity and the overall grain yield in the system reached the designed levels. Hence, we can assert that the tested method of utilization of test subjects' liquid wastes by wheat plants is quite acceptable as part of the general system of biological methods for the regeneration of an environment suitable for humans.

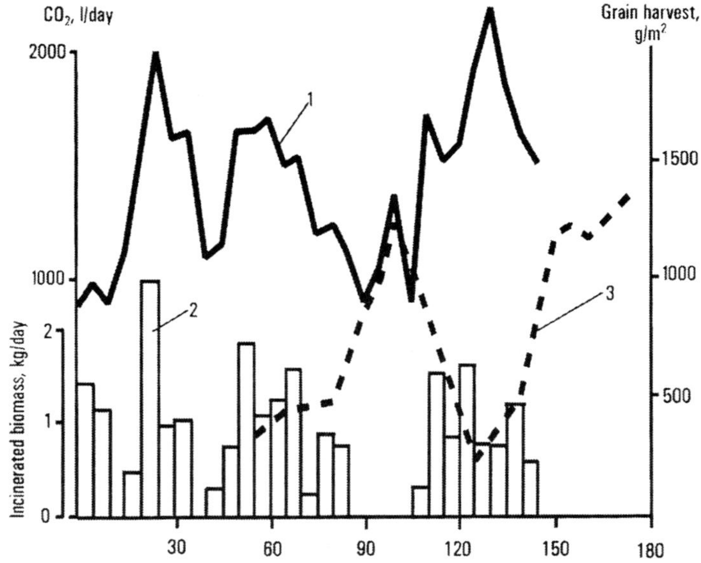

Figure 9.31 Daily average CO_2 uptake by the higher plant link (1), daily average amount of incinerated cellulose or inedible plant biomass (2), average wheat grain harvest (g/m^2) for the growth period in the first and second phytotrons over the course of the experiment (3.)

For a number of reasons, in the experiment the functioning of the higher plant link as a whole was unstable, but the link revealed a great capacity for repairing and restored the required level of productivity by the end of the experiment.

The character of the changes recorded during the experiment in the total photosynthetic plant productivity (on the basis of the consumed carbon dioxide) and grain productivity of the principal crop, wheat, can be seen in Figure 9.31. During the first three months and at the beginning of the fifth month, plant photosynthetic rate was lowered considerably by the effect of atmospheric contaminants formed in the course of inedible biomass incineration in the system, the process being not quite optimized. The consequences of this effect could be seen in the grain productivity of wheat (and in edible biomass productivity of most vegetable crops) being almost as long as the end of the experiment. Wheat restored its productivity to the designed level only at the very end of the experiment. However, this fact suggests that the closure of the system "human–higher plants," including a number of new elements of closing, such as the drying of inedible biomass and solid wastes as well as the immediate utilization of urine in the nutrient solution, cannot be obstacles to the proper functioning of the higher plant link, unless the technical devices interfere with the operation of the biological components of the system. The incineration of inedible biomass may be regarded as the only exception to this.

How close the actual plant productivity and the total daily average yield of the harvested edible biomass in the experiment are to the target parameters can be derived from the data listed in Tables 9.28 and 9.29. Testing the technology of thermocatalytic oxidation of inedible biomass was accompanied by the release of toxic substances, primarily nitrogen oxides, into the gaseous environment of the phytotrons. As a result, the productivity of all the crops grown in the phytotrons was reduced. Thus there is no point in calculating how fully the crew was provided with vegetable food during the experiment, as the resulting data would not reflect the true potentials of the system. Moreover, the harvest of the last month of the experiment was also lower than that potentially attainable: for the most part it was formed during the third and fourth months under unfavorable conditions created by the still imperfect technology of thermal oxidation of biomass.

During the fifth month of the experiment almost all crops were recovering their productivity. Only by the end of the experiment and during the following month, crops whose growth period did not exceed 70 days steadily retained their productivity at a target level. Later maturing crops (chufa, pea, and tomato) had not completed the process of recovery yet. With this in mind, we included in the above-mentioned tables the yield harvested in the first month after the test subjects left the Bios-3. In this period the overall edible biomass yield of all crops essentially corresponded to the test subjects' requirements for vegetable food (Table 9.30). Thus it has been demonstrated that the higher plant link consisting of the tested set of crops occupying the mentioned areas, cultivated under the discussed conditions, and not subjected to the effect of toxic volatiles from the incinerated inedible biomass can completely utilize the test subjects' gaseous and liquid wastes, as well as regenerate the atmosphere, water, and the whole vegetable part of the test subjects' diet.

The microflora of the higher plant link has been investigated by a group of microbiologists headed by L. Tirranen, using methods described in the monograph by Gitelson, Manukovsky, Pankova, *et al.* (1981). It was shown that prolonged use of unchanged nutrient solution brings about an increase in the numbers of aerobic bacteria, ammonificators, and urobacteria due to some accumulation in the medium of readily oxidized organic substances. The effect of system closure also made itself evident in the larger total numbers of microorganisms in the plant nutrient medium. Among them were *E. coli*, *Proteus* and fungal diaspores. The addition of human liquid wastes to the wheat nutrient solution increased the number of microorganisms contributing to the transformation of nitrogenous compounds. Increased numbers of indicator groups of microorganisms were also recorded when the state of plants was impaired by the faulty operation of the catalytic incinerator for burning the inedible fraction of the plant biomass. Once the action of this factor was overcome, the state of the plants gradually improved and their productivity increased. Simultaneously the disturbed microbial cenoses restored themselves in the plant nutrient solution. For the 40 days after the system was opened some groups of microorganisms (urobacteria, denitrificators, and *E. coli*) reduced their numbers in the wheat nutrient solution. No distinct changes in the microflora were recorded in the nutrient solution of chufa and vegetable crops grown on expanded clay aggregate.

Throughout the experiment the crew was fully provided with oxygen and water recycled in the closed system. The edible plant biomass grown in the phytotrons

Table 9. 28 Daily average crop productivity (in terms of edible biomass).

Crop	Growth period, days			Productivity, g/m^2 · day		
	Expected	Last month of exper.	First month after exper.	Expected	Last month of exper.	First month after exper.
Wheat (dry)	63	72	67	13	6.4	12.5
Chufa (dry)	90	90	86	26	10.7	14.3
Pea (dry)	90	90	90	13	6.6	6.4
Radish (fresh)	30	30	30	125	157	296
Carrot (fresh)	80	80	80	160	94	197
Beet (fresh)	60	60	60	120	136	147
Potato (fresh)	90	90	90	80	8.2	4.2
Tomato (fresh)	120	120	120	110	78	73
Cucumber (fresh)	90	128	98	250	272	693
Kohlrabi (fresh)	45	45	45	130	60	150
Onion (fresh)	60	60	60	170	81	183

Table 9.29 Daily average yield of edible biomass in the Bios-3 system.

Crop	Crop area, m^2		Edible biomass, g/day		
	Expected	Exper.	Expected	Last month of exper.	First month after exper.
Wheat (dry)	40.0	39.6	520	255	495
Chufa (dry)	9.0	8.4	234	90	120
Pea (dry)	4.0	4.0	52	26	26
Radish (fresh)	0.9	0.9	110	142	266
Carrot (fresh)	1.4	1.2	220	112	236
Beet (fresh)	0.9	0.9	105	123	132
Potato (fresh)	3.2	4.8	250	40	21
Tomato (fresh)	1.4	1.2	150	94	87
Cucumber (fresh)	0.4	0.4	100	109	277
Kohlrabi (fresh)	1.1	1.0	140	66	164
Onion (fresh)	0.7	0.6	120	49	110
Dill (fresh)	–	–	–	6	–

Table 9.30 Food regeneration by plants in the system during the first month after experiment.

Food components	Crew's monthly requirement	Vegetable food yield during the first month after experiment	Regenerated food fraction of crew's total requirement
Ovendry substance, g	27716	21484	77.5
Including:			
Proteins, g	4369	2851	65.2
Lipids, g	4679	1652	35.3
Carbohydrates, g	16030	15290	95.4
Calorific value, kcal	119699	83609	69.8

was all used as food by the test subjects. For the reasons mentioned above, though, plant productivity varied widely in the course of the experiment, often dropping below the planned one. That is why the fraction of the foodstuffs produced outside the system ranged between the planned 22.5% of the total ration to 66.5%, when the plant productivity was reduced (days 60 to 90 and 110 to 140, see Figure 9.31.). Around-the-clock medical observation throughout the experiment and regular registration of some parameters characterizing the main functions of the organism (for details see Chapter 10) proved that each of the measured parameters was stable and within the normal physiological range. Though some plant species were seriously damaged by the temporary contamination of the atmosphere caused by the operation of the incinerator burning the inedible plant biomass, it did not affect the condition of the test subjects. Neither complaints about worsening of health nor diseases were recorded during the experiment.

It is therefore concluded that there is no evidence, either subjective or objective, about harm done to humans by their prolonged stay in the system with plants performing biological regeneration of the chief components of their habitat.

Generally speaking, in this experiment several leaps were made toward increasing the closure of the ecosystem with a biological material cycling.

Planting area enlarged to 31.5 m^2 per person and irradiated around-the-clock at the average of 130–140 W/m^2 PAR afforded (barring technical hindrances) full regeneration of all the vegetable food required by a human (to 77% on the dry mass basis and 70% of the total caloric value of the ration).

A drier for test subjects' solid wastes included in the system and all native liquid wastes of the crew added to the wheat nutrient solution made it possible to provide immediate requirements of biological links throughout the experiment only with the water cycling in the system, without even a liter of water introduced from the outside. Limited amounts of water were supplied only to replace samples of liquid media regularly removed from the system for analysis (nutrient media of each phytotron, urine, condensates, etc.). Excluding water withdrawal for investigation purposes, the biological water cycle was 100% closed.

Excess (inedible) plant biomass and, naturally, excess oxygen unnecessary for humans were brought back to the biological cycle of the system as CO_2 produced by thermal oxidation. During the experiment 113.5 kg of dry plant biomass was burned. However, technical imperfections of the thermocatalytic converter for biomass incineration led to the formation of a small amount of toxic contaminants (primarily nitrogen oxides) that had no adverse effect on the test subjects but reduced the photosynthetic productivity of plants, which sometimes upset the course of the experiment. The physicochemical recycling of the deadlock product of the higher plant link proved to be an acceptable way. However, a search should be continued for different, more advantageous energetically, ways of utilization of inedible biomass, for example through biological oxidation by fungi, herbivorous fish, etc., which could simultaneously provide humans with more food products.

The overall closure of the system determined by the relationship between substances needed to provide human life support in the systems and substances that must be introduced into the system (or pre-stored in it) can be derived from the data given in Table 9.31:

$$R_s = 1 - \frac{598.2}{13010.5} = 0.954, \text{i.e.}\, 95.4\%$$

Thus, thanks to the development of some new technologies, in this experiment the highest closure was attained for a human life support system basing on the biological principle of habitat regeneration.

Table 9.31 Requirements of the crew and the ecosystem including the crew during the experiment with normal plant productivity, g/d[a].

Crew's material requirements		Ecosystem material requirements	
1. foodstuffs (without water)	924	1. foodstuffs	208
2. oxygen	1220	2. chemical substances as nutrient solution components	350
3. drinking water	5133	3. hygiene materials	9.5
4. hygiene water	5696	4. sorbents for water purification	2.7
5. hygiene materials	9.5	5. cooking salt	28
6. cooking salt	28		
Total	13010.5		598.2

[a] Amounts withdrawn from the system for analysis are not included in the table.

* * *

The Bios-3 made possible an experiment with humans staying for prolonged periods in the environment regenerated by photosynthetic processes. The overall time humans stayed within the system was 2 years, the longest experiment involving humans lasting 6 months. The biological life support system was inhabited by 21 persons, 1–3 at a time. So the overall time of human life in the biologically regenerated environment amounted to 49 man/months.

Meticulous medical examination, involving the determination of numerous parameters, during the experiment and one month after it revealed no changes in the medical state of the test subjects. Some test subjects have been members of 2–3 crews for 1.5–10 years. The maximum time one person stayed in the system was 13 months. After the experiments the test subjects—researchers of the Institute—proceeded with their research work, presented dissertations, gave birth to children, i.e. lived ordinary lives, the same as other people. This was a unique opportunity to trace the remote after-effects of the experiment. Post-experimental observation of the test subjects for 10–15 years revealed no diseases that could be confidently associated with their staying inside the Bios. Hence, we can state with assurance that in the Bios of the tested design, with biological regeneration of atmosphere, water, and vegetable fraction of the human diet, the living conditions are adequate to meet the principal human requirements, that is the ecological ones.

CHAPTER 10

Microbial Life in a Closed Ecosystem[1]

A high reproductive rate of microorganisms and the diversity of their enzymes make the microbial population the most labile element of life in closed ecosystems. That is why comprehensive investigation of microflora was one of the main objectives of our work in creating the Bios.

The part played by microflora in closed ecosystems is significant and its living conditions peculiar. Photosynthesizing microorganisms can be primary producers of biomass, regenerators of atmosphere and water, and consumers of human exometabolites.

Heterotrophic microflora—bacteria, yeasts, actinomyces, and fungi—can act in the ecosystem in two ways. Possessing a powerful enzymatic apparatus, it can destroy the exometabolites of other organisms, humans, animals, and plants, to connect consuming and producing links, thus closing the system metabolically. Under controlled cultivation this function can be intensified sharply and performed quite effectively in very small volumes. These microorganisms make up *functional* microflora.

Another aspect of the problem is that microflora is not specially cultivated in the closed ecosystem but is inevitably present in it as microbial associations concomitant with animal and plant species included in the system. These microorganisms can perform a positive function as, say, producers of vitamins to which the host is auxotrophic. They may also exhibit antibiotic properties towards pathogenic flora. In a small ecosystem, where individual macroorganisms (humans) are very few, there is a danger of sacrificing this normal flora, accidental loss of species with unique metabolic functions. Owing to bacteriological isolation the balance cannot be restored by the introduction of the lost species from the environment.

The balance can also be upset by changes in the immune status of the human, mutation process, and microevolution within a confined and isolated population of unicellular organisms.

High rates of reproduction and mutation of both functional and concomitant microflora constitute a potential threat that the direction and rates of material exchange processes in the biocenosis would be changed, and that a balanced material cycle in the closed system would be disturbed. It is in this link that the small ecosystem is the least stable and controllable. On the other hand, microbial populations, having a short life cycle and no supracellular homeostasis mechanisms inherent in multicellular animals and plants, can serve as very sensitive indicators of the state of the system as a whole.

---

[1] This chapter is an outline of the monograph "Microbiological Problems of Closed Ecological Systems" by I.I. Gitelson, N.S. Manukovsky, I.M. Pankova, M.N. Posadskaya, L.A. Somova, and L.S. Tirranen published in 1981.

The behavior of the microbiota is less controllable and predictable than that of the principal species.

There is a risk of disturbing the balance between the immune protection of the human organism and microorganisms inhabiting its intestine under the conditions of a closed system. It is not unlikely that under such specific conditions microevolutionary processes might soon result in the emergence of new forms of microorganisms, which could happen to be pathogenic or upset the balanced material exchange in the ecosystem. That is why, in a special gnotobiotic experiment in the Bios, we set ourselves a task of tracing the dynamics of microflora under nearly absolute isolation, which is extremely hard to attain on Earth but which becomes a natural condition for an ecosystem outside of Earth. To this end the sealed cabin was continuously pressurized with sterile air, and during four months investigations were made to trace the dynamics of the microflora of humans, plants and their environment including the atmosphere. A pressurization system was used consisting of a membrane compressor, gas cylinders, a gas meter, a microbiological filter, and a pressure regulator controlled with diaphragm pickups installed in the habitation module of the system. As a result of pressurization the pressure inside this module was maintained 15–30 mm of water above the external level, which prevented external microflora from entering the Bios during the experiment.

10.1 MICROFLORA OF THE MICROALGAL CULTIVATOR

Eley and Myers (1964) were the first to attempt to create a closed system. Their experiments with the system "microalgae–mammals" lasted 82 days. The heterotrophic link in this system was represented by a mouse, and the function of the bioregeneration of the atmosphere was performed by pure *Chlorella* culture. As Odum (1971) pointed out, the creation of such simplified systems gives rise to a number of difficulties: colonization by foreign microorganisms, the build-up of toxic metabolites of algae, the presence of mutant algae with decreased metabolism.

Preliminary experiments, which investigated the admissibility of introducing human and animal liquid and solid metabolites into media for green algae (Golueke and Oswald, 1963; Rerberg *et al.*, 1965), yielded positive results. However, it was shown (Rerberg and Vorobyeva, 1964; Oswald and Golueke, 1964) that pure *Chlorella* culture could not assimilate the organic matter of human metabolites. Algae can grow for long periods and the medium can be mineralized only under nonsterile conditions, with bacteria being actively involved.

Microflora of the Microalgal Cultivator

Microbial suspension cultivated in our system is not pure microalgal culture but rather a biocenosis of algae in association with dozens of bacterial species exchanging metabolites in a very intricate metabolic network. Trophic relations between algae and the totality of associated bacterial species can be described as symbiotic. To illustrate, algae cannot assimilate directly the organic matter of human urine, but they can utilize the biogenous elements contained in it after it has been mineralized by bacteria. In the same manner bacteria mineralize the organic substances excreted by algae, and algae consume biogenous elements excreted by bacteria during mineralization.

Organic matter in the algal–bacterial cultivator of a multi-link biological life support system, besides metabolites of the cenosis proper, includes various metabolites entering it from other links. Human liquid metabolites consist of fully oxidized organic substances, some of which can be utilized by algae directly, e.g. urea; others first have to be mineralized by bacteria: uric and hippuric acids, creatine, creatinin. The microbial fermenter processing human solid wastes supplies organic matter the composition of which differs from the composition of organic matter present in the microalgal link.

General Characterization of Microflora of Algal–Bacterial Cenosis as a Link of Biological Life Support Systems

Bacteria of the microalgal link can be subdivided into *autochthonous*, constituting their own microflora that is well adapted to the ecological conditions of the algal–bacterial cenosis, and *allochthonous*, which enter the microalgal link from other links of the ecosystem in the course of inter-link water and gas exchange.

Organic matter concentration in the closed water exchange of the life support system differs from that in natural water bodies, amounting to 1–3 g/l, sometimes to 5–6 g/l, i.e. approaching and even exceeding the organic matter concentration in nutrient media, such as beef-extract agar and peptone agar.

The number of bacterial cells grown on peptone agar and medium with centrifugate amounted to $1 \cdot 10^9$ cells/ml and more, which corresponded to 1–2% of algal biomass.

More than 4000 bacterial strains belonging to 12 families and 17 genera were extracted from the microalgal link on peptone agar.

Fungal flora included representatives of 10 genera belonging to the family *Moniliacea*. Yeast-like fungi of the family *Sacchromycetaceae*, genera *Rhodotorula* and *Candida* were extracted sporadically.

Bacteria of the genus *Pseudomonas* dominated. Bacteria of the genera *Flavobacterium* and *Achromobacter* also occurred frequently.

Among bacteria persistently accompanying *Chlorella*, a numerous group of bacteria close to corynebacteria (of the genus *Arthrobacter*) stood out. In the microalgal link, the counts of this group in different experiments varied from $130 \cdot 10^6$ to $1.5 \cdot 10^9$ cells/ml of the medium.

In the algal–bacterial cenosis of a closed ecological system (CES), *Chlorella* is permanently accompanied by bacteria of the genus *Caulobacter*, earlier found in algal cultures and often promoting lysis of *Chlorella* cells.

In the microalgal link the number of “stalked” bacteria per 1 ml of the medium is small, but in the experiments with the higher plant link included in the system it increases to several million/ml. The reason may be that they are supplied by the higher plant link in inter-link water exchange.

The bacterial population of the algal–bacterial cenosis becomes more diverse when a microbial cultivator and a phytotron are included in the system. The algal cultivator receives bacteria of the genera *Enterobacter*, *Escherichia*, *Alcaligenes* with the fluid from the microbial cultivator.

When the phytotron is added to the system, the biocenosis of the microalgal link receives microorganisms typical of epiflora and rhizospheric microflora, such as *Achromobacter* agile, *Achromobact.* sp., *Enterobacter cloacae*, *Proteus vulgaris*, *Brevibacterium liquefaciens*, and *Alcaligenes aquamarines*.

These microorganisms may be classed with allochthonous microflora, representatives of which are not dominant in the biocenosis of the algal cultivator. As soon as water exchange stops, they are eliminated from the biocenosis of the algal cultivator or die out.

The number of fungal spores in the culture liquid constitutes just a small fraction of the total number of microorganisms. The main source supplying fungi in the CES is the higher plant link. Conditions in the phytotron are highly favorable for the multiplication of fungi.

Chlorella was permanently accompanied in the system by fungi of the genus *Aspergillus*, *Penicillium* and *Mucor* were recorded somewhat less frequently.

With the higher plant link in the system, the fungi characteristic of plant microflora (*Fusarium*, *Botrytis*, *Cladosporium*, *Cephalosporium*) are detected in the microalgal cultivator.

Transformation of Nitrogenous Compounds by the Algal–Bacterial Cenosis

As has already been mentioned, to the microalgal nutrient solution are added human liquid wastes containing organic substances in the form of urea, uric and hippuric acids, creatine, creatinine, various amino acids, and an insignificant amount of mineral nitrogen as ammonium salts. In the course of water exchange between the links of the system, nitrates, nitrites, ammonia nitrogen, polypeptides, and proteins also enter the nutrient medium.

Ammonium salts and urea are the preferred forms of nitrogen for *Chlorella*. No build-up of any compounds of this element was recorded in the solution.

The principal ammonificators of protein substances in the biocenosis of the microalgal link are aerobic bacteria. The number of ammonificators amounts to $1 \cdot 10^9$ cells/ml, which makes up 50–80% of the total bacterial count in the biocenosis. The species composition of this group is rather diverse and includes representatives of 8 genera: *Pseudomonas* (6 species), *Enterobacter* (2 species), *Achromobacter* (2 species), *Flavobacteriaum* (2 species), *Micrococcus* (2 species), and the genera *Caulobacter* and *Arthrobacter* are represented by one species either.

Urea contained in the culture medium is not only directly utilized by *Chlorella* but can also be subjected to bacterial transformation to ammonia. The number of uric bacteria ranged between 250 and 400 million cells/ml. They were represented by *Ps. dacunhae*, *Ps. caudata*, *Flavobacterium fucatum*, *Achromobacter* sp., *Alcalirenes aquamarines*, *Micrococcus* sp., and *Arthrobacter* sp.

The mineralization of the products of human nitrogen exchange is effected by the following species of microorganisms decomposing:

uric acid
Ps. dacunhae
Ps. putrefaciens
Pseudomonas sp.
Nocardia sp.
Arthrobacter sp.

hippuric acid
Ps. dacunhae
Ps. ambigua
Pseudomonas sp., 1
Pseudomonas sp., 2
Pseudomonas sp., 3
Staphylococcus epidermidis
Arthrobacter sp.

creatine
Ps. dacunhae
Staphylococcus epidermidis
Micrococcus sp.
Nocardia sp.

creatinine
Ps. dacunhae
Ps. putrefaciens
Ps. caudata
Achromobacter agile
Achromobacter sp.
Nocardia restricta

Nitrifying bacteria

In the biocenosis of the microalgal link the following bacterial species can oxidize ammonium salts: *Ps. cerevisia*, *Alcaligenes aquamarines*, *Arthrobacter* sp., and *Flavobacterium* sp.; nitrites and nitrates: *Ps. putrefaciens*, *Ps. caudata*, *Bacteroides furcosus*, *Flavobacterium vreve*, *Vibrio* sp., *Micrococcus* sp., *Proactinomyces* sp., and *Actinomyces* sp.

The total number of the former is approximately $0.025 \cdot 10^6$ cells/ml, and the number of the latter ranges from 7.5 to $200 \cdot 10^6$ cells/ml. Though the intensity of these processes is low, ammonium oxidation can make a significant contribution to nitrate formation in the medium owing to large numbers of nitrifying heterotrophs.

Denitrifying bacteria

In the microalgal link the group of true denitrificators reducing nitrates to free nitrogen includes *Ps. fluorescens*, *Ps. aeruginosa*, *Ps.* sp., *Agrobacterium radiobacter*, *Arthrobacter* sp., and some strains of *Ps. dacunhae.*

Most bacteria of the genera *Pseudomonas, Achromobacter*, *Micrococcus*, and *Flavobacterium* isolated from the algal–bacterial cenosis possess the capacity for reducing nitrates to nitrites.

Bacteria fixing free nitrogen

The same bacterial species, most of which are permanent companions of *Chlorella*, are involved in the ammonification, denitrification, and oligonitrophilic processes.

Due to the involvement of these microorganisms in the microbiological processes of transformation of nitrogenous compounds, the nitrogen content of the microalgal link and the biocenosis are maintained in a steady state.

Microorganisms Involved in the Transformation of Sulfur Compounds

The pattern of water exchange in the CES presupposed the introduction of unprocessed human liquid wastes to the microalgal link. So, besides nitrogen compounds, the *Chlorella* nutrient solution also received different sulfur compounds.

The daily amount of urine from one human contains on the average 3 g of sulfur including 2.1 g of sulfates, 0.3 g of sulfur-coupling, and 0.6 g of organic sulfur.

We detected in the biocenosis the following groups of bacteria involved in sulfur cycling:

- putrefactive bacteria, decomposing proteic substances to yield hydrogen sulfide;
- sulfate-reducing bacteria, reducing sulfate to H_2S;
- purple bacteria and thionic bacteria, oxidizing hydrogen sulfide, elementary sulfur, and incompletely oxidized sulfur compounds (thiosulfates) to sulfates.

The results of analyses showed putrefactive bacteria to constitute 15–25% of the total bacterial number. This group includes *Ps. fluorescens*, *Ps. denitrificans*, *Ps. putrefaciens*, *Enterobacter aerogenes*, *Proteus vulgaris*, *E. coli*, *Enterobacter cloacae*, *Achromobacter* sp., *Artheobacter* sp., *Bacillus subtilis*, *Bacillus* sp., and *Micrococcus* sp.

Thus, in the algal–bacterial cenosis the sulfur cycling involves many species of microorganisms, both heterotrophic and autotrophic, the transformation of sulfur compounds being one of the basic features of their metabolism. The presence of these bacterial species and the absence of the build-up of any sulfur forms or loss of this element in the microalgal link are indicative of the balanced processes of sulfur transformation in the algal–bacterial cenosis.

Microorganisms Involved in the Transformation of Inaccessible Phosphorus Compounds

Investigations were made to estimate the ability of microorganisms of the biocenosis of the microalgal link to decompose organophosphates and transform insoluble inorganic phosphates to soluble forms.

It was found that *Ps. fluorescens*, *Ps. dacunhae*, *Pseudomonas* sp., and *Agrobacterium radiobacter* possess this ability.

Cellulose-Fermenting Microorganisms

One of the essential functions of a biocenoses is the hydrolysis of cellulose—one of the least easily mineralized biopolymers. The culture medium in the microalgal link is continuously enriched with cellulose. The cellulose content of the nutrient solution recirculated many times is 0.5–2% of the organic fraction of the medium (Trubachev *et al.*, 1969). In nature, cellulose is decomposed by aerobic and anaerobic bacteria as well as by myxobacteria and fungi. In our biocenosis the range of microorganisms having this property is much narrower. There are no anaerobic cellulose-fermenting bacteria in the aerated cultivator. Cellulose-fermenting bacteria were represented only by myxobacteria of the genera *Cytophaga* and *Sporocytophaga*, which could be regarded as autochthonous microflora, and *Cellvibrio*, isolated in the experiment with the microbial cultivator in the system, which belongs to allochthonous microorganisms. The number of bacteria did not exceed 10 cells/ml.

Microorganisms Decomposing Humic Substances

Elevated humate content in the microalgal culture medium is directly related to the length of periods of medium recirculation in the algal cultivator and water exchange in the system. It can be suggested that if the period of medium recirculation in the closed water loop of the CES increased, humate content would get higher.

A group of bacteria of the genera *Pseudomonas*, *Nocardia*, *Micrococcus*, *Bacillus*, and *Actinomyces*, decomposing humic substances, was isolated on the medium containing carbon only in the form of sodium humate doped with 1% peptone. A mixed culture breaks down humates more energetically than do the pure bacterial species comprising it.

Such elements as nitrogen, phosphorus, sulfur, etc. are neither lost nor accumulated in the system, which proves the efficiency of the microbiological processes of matter transformation by the algal–bacterial cenosis. These processes perform the main stabilizing function in the algal–bacterial cenosis.

The Effect of Cultivation Conditions on Biocenosis Stability

In the described closed ecological systems, the parametric regulation of biosynthesis keeps the principal parameters within the range of values optimal for the producing species. Such stability of parameters results in steady-state *Chlorella* photosynthesis and, consequently, a steady level of bacterial numbers. In the course of water exchange other links of the system supply the algal cultivator not only with organic matter, allochthonous to the organic matter formed in the microalgal link, but also with a great number of microorganisms constituting allochthonous microflora.

However, under these conditions, instead of the expected sharp increase in the total number of bacteria in the biocenosis, an overall numerical reduction of microorganisms accompanied by an enrichment of species composition was recorded (Figure 10.1).

An illustration of this is also the death of *E. coli*, entering from the microbial cultivator, and of epiphytic microflora, primarily spores of microscopic fungi supplied by the higher plant link (Fig. 10.2).

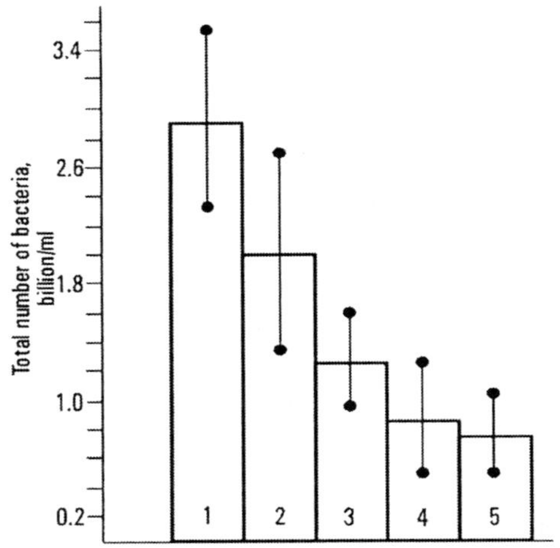

Figure 10.1 Total number of bacteria in the microalgal link in different variants of the ecosystem. 1—two-link (human–microalgae); 2—three-link (human–microalgae–microbial cultivator); 3—three-link (human–microalgae–phytotron with vegetables, 30 days); 4—three-link (human–microalgae–phytotron with wheat, 30 days); 5—four-link (human–microalgae–microbial cultivator–phytotron, 90 days). The straight line segments show the confidence range.

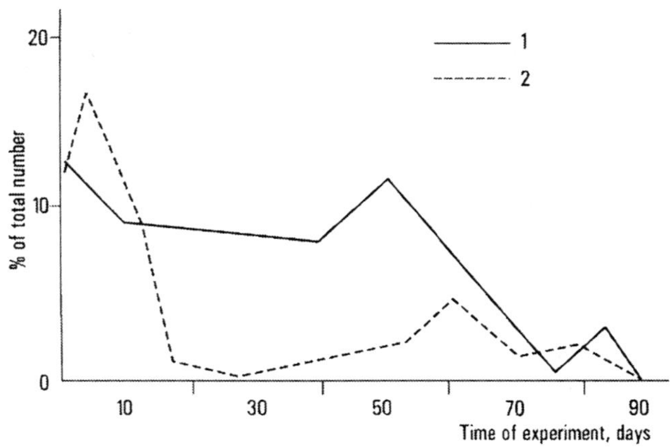

Figure 10.2 Dynamics of *E. coli* (1) and fungi (2) in the microalgal link in the 90-day experiment with the four-link system.

Analysis of the specific structure, physiological groups, and trophic relations has shown that the naturally formed biocenosis used as the bioregenerative link in the CES contains a great many bacterial and fungal species, having numerous enzymatic systems. The *Chlorella* culture medium contains macroelements and trace elements, products of algal and bacterial metabolism, lysed cells, amino acids, carbohydrates, proteins, fats, cellulose, humic substances, etc. Against this background develops the algal–bacterial cenosis, in which the bacterial component numerically approaches the primary producer.

The transformation of nitrogen, sulfur, phosphorus, and iron involves a wide range of nonspecific heterotrophic and specific microorganisms. Bacteria, in the process of their life activity, decompose compounds difficult to oxidize, making them accessible to *Chlorella*. This is the principal environment-forming function of microflora. The cultivation of algae and bacteria together precludes an inhibitory effect of metabolites thanks to an ecological balance between these groups of microorganisms, which is of crucial importance in effecting cyclic water exchange in the system.

Biocorrosive Properties of Microorganisms in the Algal–Bacterial Cenosis

The results of our investigations demonstrate that some species of microorganisms in the algal–bacterial cenosis are potentially hazardous for the existence of the ecosystem. They can damage metal units and parts of installations, causing a release of trace elements in amounts several times higher than the optimal concentrations. In designing installations, particularly those intended for long-duration usage, this probability should be taken into consideration.

10.2 MICROBIOLOGICAL OXIDATION OF ORGANIC WASTES IN A CLOSED ECOLOGICAL SYSTEM

One of the most promising techniques of organic matter decomposition is aerobic microbiological oxidation, which is widely used in aerobic sewage treatment. It is based on the oxidizing activity of aerobic microbial associations, the composition of which depends on the chemical composition of sewage and which can break organic substances down to inorganic compounds.

The biomass of the principal physiological groups of bacteria performing oxidation amounts to 80–90%. The biomass of associated bacteria and other organisms does not exceed 10–20%.

The major active mass of sludge contains bacteria of the genera *Pseudomonas*, *Bacterium*, *Bacillus*, *Micrococcus*, *Corynebacterium*, *Sarcina*, *Sacharmyces*, and *Mycobacterium*.

Cellulose and other natural polymers, such as pectic substances and lignin, are rather resistant to the action of microorganisms. Slow mineralization of these compounds is favorable for humic substance formation.

Lignin biological degradation takes the longest time. The following fungi were recorded as active decomposers of lignin: *Fusarium lactes*, *F. nivate*, *Trichoderma lignorum*, *Alternaria tenius*, and *Stemphylium batroysum.*

Formation of Microbial Cenoses Oxidizing Human Solid Wastes and Wheat Straw

To investigate the microbiological oxidation of ecosystem wastes, microbial cenoses oxidizing human solid wastes and wheat straw were created. We have worked out the technology of batch microbiological oxidation that made possible nearly complete oxidation of ecosystem wastes. We have also looked into ways to optimize the process and to utilize oxidation products in the human closed ecological system

It was not our objective to regenerate clear drinkable water. We were ready to content ourselves with oxidizing organic matter to soluble mineral and organic compounds. The resulting mineral compounds were intended for use by the autotrophic links, and organic soluble substances for utilization by the algal–bacterial cenosis. Human solid wastes contain various organic substances, mostly those that are easily oxidized by microorganisms. One third of the inedible fraction of plants, mostly wheat straw, is lignin, hydrolysis of which by microorganisms is difficult to effect. To oxidize these wastes it is necessary to form microbial cenoses able to oxidize multicomponent organic substrates.

An important stage in working out the technology of mineralization of human solid wastes and wheat straw is the formation of adapted microbial biocenoses, the activity of which ultimately determines the efficiency of the method. It takes not only initial microflora and substrate but also certain cultivation tactics to form a biocenosis. In our work this was the consecutive transfer of microorganisms, after a certain cultivation period, to the fresh portion of the substrate.

In forming a microbial association capable of oxidizing human solid wastes, the basic substrate itself served as a source of microflora.

The microbial biocenosis capable of oxidizing wheat straw was formed by a similar scheme. Sources of microflora were forest soils and the contents of a cow's rumen.

Formation of the microbial biocenosis was considered to be completed if during one cycle at least half of the introduced organic substances were oxidized, and the sludge increment amounted to 0.4 g/l.

Experimental Installation and Process Control

A setup for the mineralization of solid wastes is a vertical flat-parallel cuvette with a 2-cm-thick inner space (Figure 10.3). The full internal volume of the cuvette is 5.5 l. In the lower part of the cultivator, in the outer wall made of acrylic plastic, there are bubbling openings.

A setup for straw mineralization is a vertical cylinder 100 cm high and 10 cm in diameter (Figure 10.4).

Microbiological Oxidation of Human Solid Wastes

Having investigated the dynamics of the oxidation of solid waste organics in the batch microbial cultivator, with the chemical oxygen demand of 7.56 g O_2/l at 22°C, we found the optimal duration of a mineralization cycle to be 8 h (Figure 10.5). Within this period the chemical oxygen demand of the medium was reduced by 84%, 16% of them used to synthesize microbial cells, and 68% oxidized to carbon dioxide and water. The soluble fraction contained up to 16% of organic substances.

Microbiological Oxidation of Wheat Straw

Straw oxidation was performed using the devised recovery technology. With this technology, straw oxidation could be conducted at a rate of 10 g/l · day. In this process 93.5% of organic substances decomposed to the end products CO_2 and H_2O, 4.6% went into soluble compounds, and 2.0% entered into the composition of microbial cells.

To oxidize 400 g of wheat straw, a daily Bios yield satisfying the daily human requirements for bread, the capacity of the cultivator must be only 40 l.

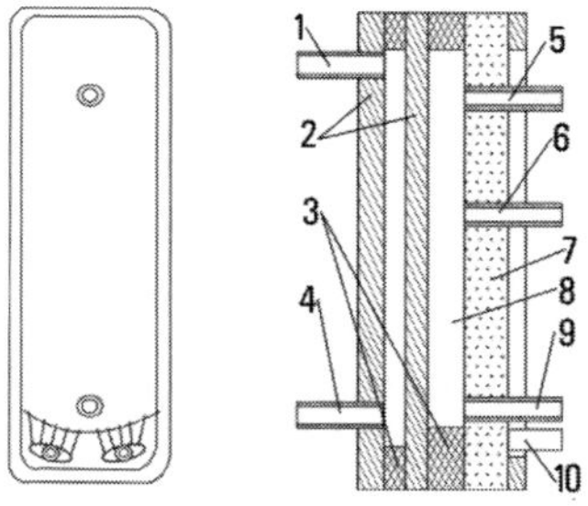

Figure 10.3 Microbial cultivator for oxidizing solid wastes. 1, 4—thermostat inlet and outlet water pipes; 2—walls of water jacket; 3—rubber gaskets; 5—gas and foam outlet to separator; 6—cultivator inlet suspension pipe; 7—cultivator transparent wall; 8—cultivator working volume; 9—cultivator outlet suspension pipe; 10—air inlet pipes.

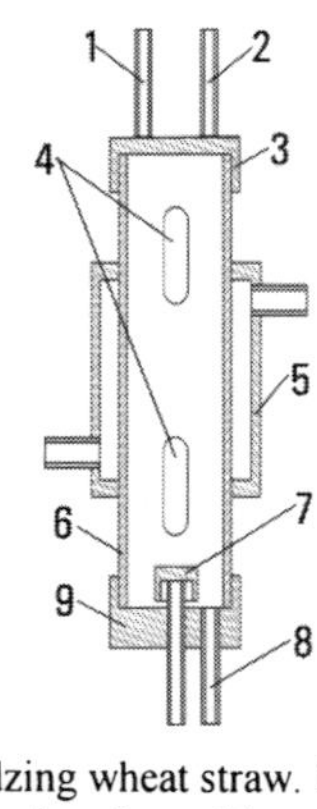

Figure 10.4 Microbial cultivator for oxidzing wheat straw. 1—inlet pipe for feeding suspension to foam separator; 2—cultivator inlet suspension pipe; 3—cultivator covers; 4—observation ports, 5—thermostabilization jacket; 6—cultivator casing; 7—bubbler; 8—cultivator outlet suspension pipe.

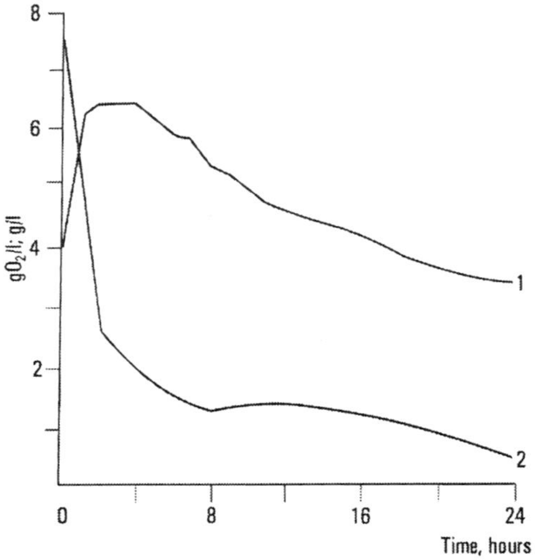

Figure 10.5 Dynamics of oxidation of human solid wastes. 1—microbial biomass concentration, g/l; 2—chemical oxygen demand of the medium, gO_2/l.

Microbiological Characterization of Biocenoses Contributing to the Oxidation of Human Solid Wastes and Straw

The bacterial cenosis oxidizing human and *Chlorella* metabolites contained bacteria, fungi, actinomyces, yeasts and protozoans—ciliate infusoria and flagellates.

In the bacterial component of the cenosis microorganisms of the genera *Pseudomonas* and *Bacterium* prevailed, making up 80% of the total microbial population. With a rise in temperature, an increase was recorded in the relative numbers of gram-negative bacilli of the genera *Achromobacter*, *Flavobacterium*, and coccal flora (Figure 10.6A).

The composition of bacterial flora in the biocenosis oxidizing straw differed little from that described above (Figure 10.6B).

Besides *Penicillium*, *Aspergillus*, and *Mucor*, usually occurring in the biocenosis oxidizing solid wastes, the fungal flora contained the microscopic fungi characteristic of wheat—*Fusarium* and *Trichoderme*—which must have been introduced with the straw. The dynamics of the oxidation of organic compounds is shown in Figure 10.7: a—for human solid wastes, b—for straw.

With an initial dry matter concentration in the medium of 90 g/l and the rate of substrate decomposition 10 g/l per day, to process 500 g of daily inedible wastes the microbial cultivator has to be 45 l in volume. The mass exchange of the mineralization link, with the process lasting 4 days, is schematically represented in Figure 10.8. Using this technology, 450 g of organic substances can be oxidized daily to carbon dioxide and water in the microbial cultivator.

Thus, microbiological oxidation is a practicable way of decomposing organic wastes produced in the closed ecological system. Unfortunately, this process yields end products; some of them cannot be oxidized by microorganisms, with others the process of oxidation is too slow. To return to the material cycle, these products have to undergo physicochemical oxidation. It should be recognized that the link of physicochemical correction will always be an essential part of closed ecological systems.

10.3 MICROFLORA OF THE HIGHER PLANT LINK IN THE CLOSED ECOSYSTEM

In hydroponics as well as in soil, the main and active factor contributing to the generation of microflora is higher plants, mostly their root exudate.

Microbial cenoses of wheat close-to-root and root zones as well as phyllospheres were investigated by the wash-off method, in the phases of germination, booting, earing, and waxy ripeness. By the same method radish and chufa were investigated in the phases of germination, root formation (tubercles for chufa), and harvest maturity. Microflora of the substrate (expanded clay aggregate), of seeds of vegetable and cereal crops, and of chufa tubercles was also analyzed.

Phytopathogenic properties of isolated microorganisms were estimated by the ability to form necrotic zones on seedling stems and leaves around the points of injections with the bacterial cultures under investigation.

The bioassay method was used to determine the toxicity of the nutrient solution, by the growth rates of wheat or radish seedlings.

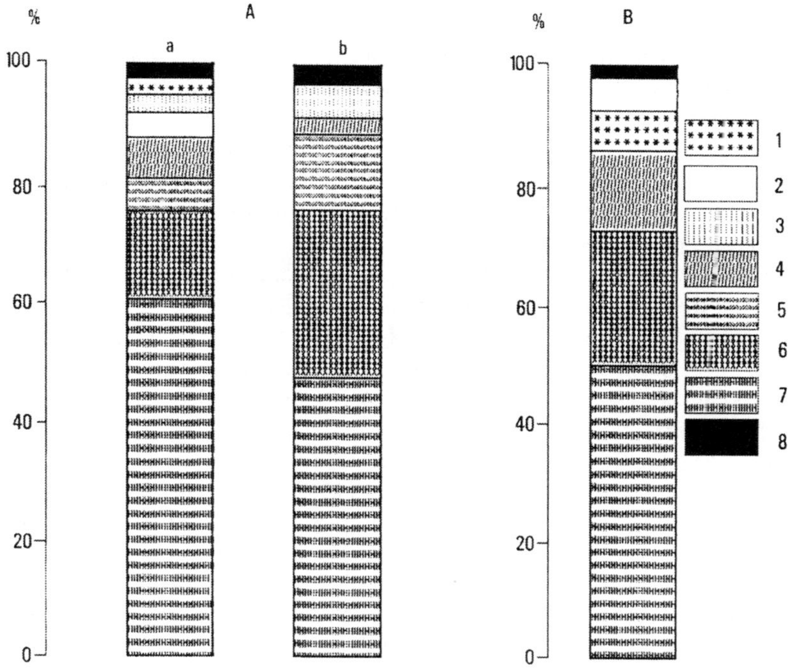

Figure 10.6 Composition of biocenosis oxidizing organic compounds. A—human solid wastes: total number of microorganisms, million/ml: a—1100 at 22°C, b—at 37°C. B—straw: total number 3600 million/ml: 1—*Sarcina*; 2—*Mycobacterium*; 3—*Micrococcus*; 4—*Bacillus*; 5—*Flavobacterium*; 6—*Achromobacter*; 7—*Pseudomonas*; 8—other.

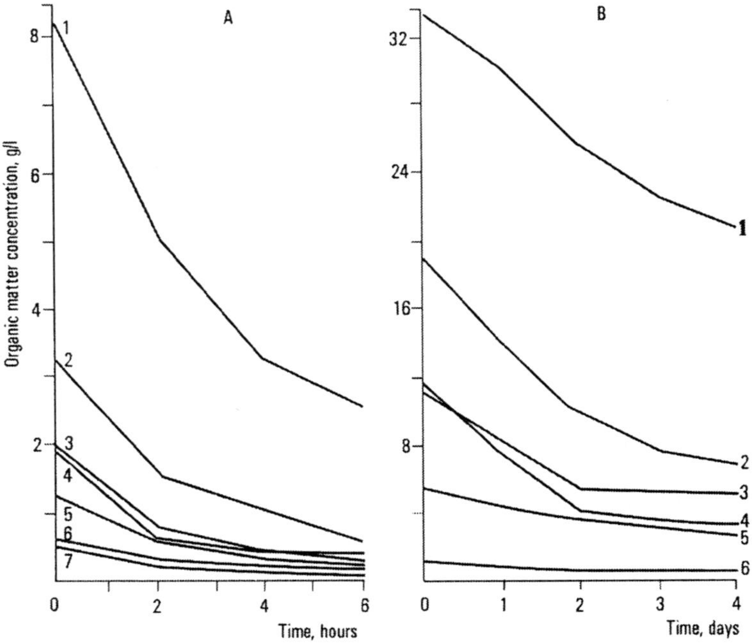

Figure 10.7 Dynamics of the oxidation of organic compounds. A—human solid wastes: 1—crude protein, 2—humus-like substances, 3—carbohydrates, 4—cellulose, 5—hemicellulose, 6—starch, 7—lipids; B—straw: 1—lignin, 2—cellulose, 3—crude protein, 4—carbohydrates, 5—hemicellulose, 6—starch.

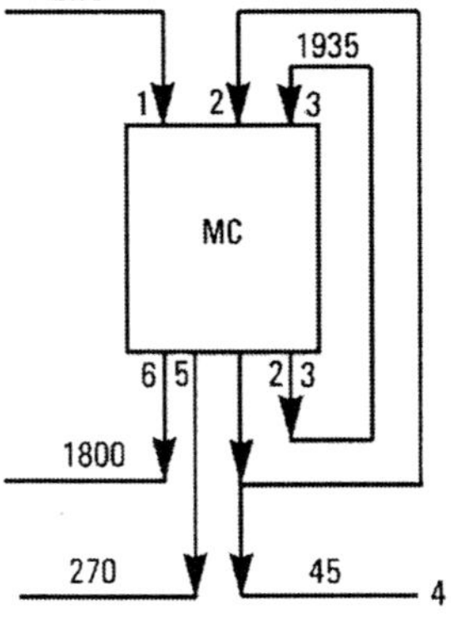

Figure 10.8 Version of mass exchange (grams) under oxidation of ecosystem wastes designed for 45-l cultivator. 1—ecosystem wastes; 2—microbial biomass; 3—reentering residue (part of wastes that has not been completely oxidized in a cycle); 4—biomass increment; 5—soluble fraction; 6—part oxidized to CO_2 and H_2O.

Microflora of the Cereal Crop—Wheat

In many experiments with the closed ecosystem the principal component of the higher plant link was spring wheat grown from seed to harvest maturity.

The highest density of the microbial population is recorded in the close-to-root zone, the lowest—in grain, phyllosphere, and the nutrient solution.

The microflora of the freshly harvested wheat grain grown in the phytotron in and out of the closed system differed from the microflora of the control (field-grown) seeds. Experimental plants carried 8–10 times fewer microorganisms, their composition was less diverse, with diaspores of mold fungi predominating, making up 99% of the total numbers of microorganisms. Microflora composition was not cultivar dependent. It was recorded that the numbers of diaspores of microscopic fungi grew in four generations of phytotron-grown plants; after that the proportions of fungi and bacteria remained essentially the same. The peculiarities of the experimental grain microflora had no adverse effect on the germination and emergence of seedlings.

Whatever the growth conditions or plant age, the group of non-sporiferous bacteria prevailed. The most abundant were bacteria of the genera *Pseudomonas*, *Erwinia*, *Eneterobacter*, *Flavobacterium*, *Nocardia*, *Achromobacter*, and *Mycobacterium*. Bacteria of the genera *Bacillus*, *Micrococcus*, *Xanthomonas*, and *Promyxobacterium* were recorded in smaller amounts. Bacteria of the genera *Vibrio*, *Streptococcus*, *Sarcina*, and *Mycococcus* were isolated sporadically.

Microorganisms proved to be sensitive indicators of changes in the nutrient solution and the state of the plants.

Microflora of Vegetable Polycultures

The higher plant link included a "greenhouse" with an uneven-aged vegetable polyculture.

We made microbiological analyses of the solutions and expanded clay aggregate used there. Of the nine CES vegetable crops—table beet, carrot, dill, turnip, kale, radish, onion, cucumber, and sorrel—the microbial cenosis of radish was selected for investigation. This crop had the shortest growth period, and we were able to detect compositional and numerical changes in the microbial population of several generations. Microflora of the same objects grown outside the system was used as control.

The area most heavily populated by microorganisms is the close-to-root zone, the least—phyllosphere. On the expanded clay aggregate there were 40–60 times more microbes than in the nutrient solution. In the cultivation of vegetables under artificial conditions, the substrate adsorbs various substances and microorganisms.

Possible Reasons for Inhibition of Plant Biosynthesis in Closed System Cultivation

When plants were grown on the unchanged solution in experiments in the open system, no significant changes in culture growth and state were recorded by visual observation. However, by the end of four months of repeated use of the solution radish productivity was reduced by a factor of 1.3 and wheat productivity was almost halved. When the unchanged solution was used for a long time by the vegetable polyculture, the yield was less affected than it was in the monoculture grown under the same conditions.

The impaired state of plants grown on the nutrient solution with wastewater can be accounted for by a build-up in the medium of substances toxic for crops—plant metabolic products, microorganisms, and organic substances introduced with wastewater. All groups of microorganisms increased numerically, and changes were recorded in the microbial system associated with higher plants. A particularly dramatic increase was recorded in the bacterial population—in the *E. coli* group, *Proteus*, and lactic-acid yeasts. Species diversity became narrower, and the prevalence of bacterial and fungal species potentially harmful to humans increased. After the supply of wastewater to the nutrient solution was discontinued, the organic substance content and the numbers of microbes decreased, while their species diversity got wider.

We found (Tirranen *et al.*, 1979) that some gaseous metabolites produced by microorganisms affect wheat growth. The growth of seedling roots was inhibited more than the stem growth was.

Thus, the results of our investigations suggest that plant growth inhibition and productivity reduction in the closed system must be caused by a number of factors acting together. They are: prolonged use of the solution, the addition of wastewater, and closure of the gas cycling in the system atmosphere.

Bearing this in mind, in the subsequent experiment with a CES measures were taken to eliminate unfavorable factors causing the inhibition of plant biosynthesis and changing the microbial system in the higher plant link. A catalytic converter to oxidize organic contaminants and microorganisms in the system atmosphere was added to the system. Besides, air and plants in the phytotron were subjected to UV-radiation treatment.

As a result of these measures, in this experiment, neither a progressive deterioration of the state of the plants nor a microbial population explosion was recorded.

Contribution of Microflora of the Higher Plant Link to the Microbial Landscape of the CES

The ultimate microbial landscape of a closed ecosystem depends on the composition of the microbial population of the conjugate links and the possibility of exchanging microflora between links. In the CES, the microflora distribution and formation occurs mostly by aerogenic, contact and alimentary ways.

The higher plant link is quantitatively and qualitatively richer in microorganisms than the other links of the system.

The microflora of higher plants can be taken to other links of the system not only with flows of air and water, but also with the crops grown in the CES, while harvesting them and using them as food.

When the higher plant link became part of a closed system, the microflora of the other links became more diverse and the occurrence of fungal diaspores more frequent.

In the algal cultivator the microorganism species introduced from the higher plant link were not dominant, but diaspores of mold fungi occurred more frequently. In the microbial cultivator a similar situation was observed.

Bacteria of the genera *Erwinia*, *Achromobacter*, *Pseudomonas*, and other bacteria belonging to the microflora of the higher plant link were detected on the internal surfaces of the sealed cabin.

With the air inflow from the phytotron, the fungi of the genus *Penicillium* began to prevail in the air of the sealed cabin. Those were the fungi dominating on plants, particularly on wheat stems and grain, as well as in the air of the higher plant compartment.

The effect of human microflora on the microbial landscape of the higher-plant link was mostly exhibited in the increased numbers of *E. coli*, *Proteus* groups, yeasts, and yeast-like fungi in the wheat nutrient solution with wastewater.

The microbial landscape of the habitation module is largely formed under the influence of the higher-plant-link microflora.

To reduce the crew's immediate contact with the microbial community of the higher-plant link, the test subjects, while working in the phytotron, wore protective outer garments and respirators filtering off microorganisms.

These investigations demonstrated that the formation of the microbial community in a closed ecosystem is subjected to the influence of the specific conditions in it.

Human microflora in the closed ecosystem has been described in Chapter 9, Section 2, dealing with the issue of humans in the CES.

CHAPTER 11

Theoretical Analysis of Closed Ecological Systems[1]

PROBLEMS OF MATHEMATICAL MODELING OF CES

Development of Closed Ecological Systems (CES's) is driven by two main needs: 1) a need to understand the mechanisms of the Earth's biosphere stability in order to overcome global ecological crisis and 2) a need to ensure safe human life under unfavorable or dangerous environmental conditions by means of Life Support Systems (LSS's). Therefore, experimental CES's come to the fore as both experimental models of the biosphere and space Biological LSS's (BLSS's). The necessity of using experimental models is well known—it is very dangerous and practically impossible to conduct experiments with the Earth's biosphere and it is very expensive and time-consuming to conduct comprehensive testing of full-scale experimental LSS prototypes. So the two lines of investigation are closely related since any CES is essentially a life supporting system—an LSS (not only human life) based, preferably, on biological processes of environment regeneration. In this chapter we will consider CES's from the engineering point of view and will concentrate on the problems of developing CES's for practical applications.

An experimental CES as a research tool must satisfy a number of demands arising from the direct purposes of these two lines of research.

The purpose of the LSS line (first of all for space applications) is to create an LSS that:

1) supports life, i.e. it has to: i) be reliable; ii) provide a necessary quality of life; and iii) be autonomous or self-sustainable—or nearly so;
2) is manufacturable, i.e. it has to: i) be easy reproducible; and ii) be designable in accordance with a plan;
3) has minimal mass, volume, and energy consumption.

Since a long-term LSS has to be self-sustainable or autonomous, closed ecological systems are their research analogs. The purpose of studies of Earth's biosphere is the reliable forecast of the biosphere's state after the unprecedented anthropogenous attack. The present approach is based on building enormous mathematical models adjusted to real observations on the basis of retrospective analysis

[1] This chapter was written by S.I. Bartsev, V.V. Mezhevikin and V.A. Okhonin.

of the available data and the comparison of short-term model forecasts. Obviously, this retrospect covers too short a period in comparison with typical time periods during which the biosphere state changes. Practically, it is necessary to wait tens and hundreds of years to be able to compare the results of a model's forecast with appreciable biospheric changes. The speed of approaching the announced goal is evidently too slow for practical purposes. Moreover, since the adequacy of the biosphere's models is not verified, it is quite possible that another set of parameters necessary for hitting the target will have to be measured and registered.

Native Earth's biosphere is considered to be essentially closed. It means that there are no essential losses or inflows of chemical elements from/to the system of chemical and biochemical reactions running on Earth. Therefore an adequate experimental or theoretical model of the biosphere has to satisfy an additional constraint: closure. So an experimental CES can be regarded as a challenge for specialists in biosphere modeling and to their approach and theories. Table 11.1 displays the correspondence between space and terrestrial applications of CES's and biosphere's aspects of CES research.

Table 11.1 Similar demands on LSS's and CES's.

	BLSS (application aspect)	CES (research aspect)
1	Reliability (safety)	Mechanisms of instability
2	Autonomy (self-sustenance)	Closure
3	Manufacturability	Reproducibility (reliability of results)
4	Designability for an order	Predictability via CES composition
5	Minimal mass and volume	Experiment acceleration (via decreasing relaxation times)

Let us consider these items.

1. A reliable LSS can be designed only after comprehensive investigation of the mechanisms of instability. The same is true for an understanding of the mechanisms of biosphere stability. A comprehensive investigation needs a great amount of data, which cannot be obtained from observing the biosphere and even a full-scale LSS prototype over a certain time. Moreover, the testing of stability includes the thorough investigation of the dynamics of critical regimes, which is impossible in the case of the experimental LSS with humans.
2. Autonomy or self-sustenance ensures the long-duration operation of LSS's without any external supplies and implies a high level of closure. Closure also means the absence of matter deadlocks in CES's. An investigation of deadlock dynamics requires long experiments since the rate of deadlock formation is comparable with the duration of LSS exploitation.

3. Manufacturability, i.e. an ability to mass produce, means that after the development of the design and operation regimes of a given type of LSS's, the creation of any copy of the given LSS will not require any additional investigations. It must become possible after achieving the complete reproducibility of experiments with LSS's. By the way, the reproducibility of experiments is the most significant requirement (and the merit) of the scientific approach. In the field of artificial ecosystems, experimental reproducibility is not achieved, and therefore nobody can consider the present unique CES's (Bios-3, CELSS and CEEF) as the objects of scientific investigation—they are now artifacts only. To develop this area of knowledge to the level of science, a great number of new experiments will be required.
4. An ability to design and build an LSS in accordance with an order without any considerable additional experimental work means that there is a formal description of the operational parameters of the LSS on the basis of a description of its components. To reach such a level of understanding, experiments with different CES's and their adequate (general) theoretical description are necessary.
5. All previous items become achievable if the rate of CES experiments is accelerated many times. This may be possible after an intensification of material cycling inside a CES. For example, in the Bios-3 facility (Kirensky *et al.*, 1971; Gitelson *et al.*, 1989), one oxygen atom took part in the respiration process approximately once a month. If we consider Bios-3 as a crude model of the biosphere and compare the rates of material exchange, we can say that the time in this model runs ten thousand times faster than in Earth's biosphere. But for practical applications this level of acceleration is not sufficient. The bulk rate of chemical transformations in a manned LSS is limited by the intensity of human metabolism.

To develop and design a CES for practical applications, it is necessary to evaluate both its reliability and its exploitation features (closure level, the mass of involved components, energy consumption, etc.). Certainly, the term "stability" in this context is more common, but to emphasize an engineering orientation of this chapter we will often use the term "reliability" even when mentioning Earth's biosphere. So this section is devoted to the consideration of the problem that seems to be extremely significant for designing a long-term space mission, and our life on planet Earth is an example of such a space mission. To ensure the safety of the crew is undoubtedly the most important function of the space system with humans. The crew's safety essentially depends upon LSS's reliability. On the other hand, the closure level is a very important parameter both for artificial biospheres (CES) and for Earth's biosphere since it determines the rate of gradual shifting from an initial state of a CES to some border of instability.

So there are obvious reasons to increase the LSS's reliability and to prefer for practical use the most reliable LSS of all. As for our biosphere the question concerns the strategy of overcoming the global ecological crisis. Any work at LSS's reliability requires *measuring* it. But the usual ways of reliability evaluation applicable for engineering systems are not quite applicable here due to the peculiarity of an LSS intended for long-term use.

The main restriction is related to the evaluation of the Mean Time Between Failure (MTBF) of the LSS. For example, if the probability of LSS failure per year is assumed to be 0.01 then (if the exponential model of the failure probability is used) the adequate duration of a test is more than 100 years for one full-scale LSS prototype or approximately 1 year for 100 prototypes. A single LSS experiment gives almost nothing in terms of the problem and is only an episode in the personal biography of the investigators.

Mathematical models that generally give a possibility to predict the system state cannot be used here since the process of model adjusting requires the same amount of experimental data. So it is not so easy to build a BLSS for long-term missions, but it is much more difficult to test it.

Possible Approach to Reliable LSS Design

The reliability of an LSS can be estimated if the reasons for LSS instability are known. A thorough investigation of the critical regimes of LSS's functioning is needed. But it is not realistic to test the reliability of several full-scale LSS prototypes during a time equal to the expected duration of LSS's real exploitation. Obviously, it is necessary to accelerate the test processes. We could not essentially increase the intensity of the mass exchange processes inside a full-scale LSS since it is limited by the human metabolic rate.

A possible way to solve the problem is to use small (micro- and meso-) unmanned CES's, where highly intensive processes take place (Kovrov and Fishtein, 1985; Folsom and Hanson, 1986). But there is a problem in using the obtained data for full-scale LSS design. As a possibility to solve this problem or essentially decrease its significance, the approach of ecological similarity (scaling) is proposed (Bartsev and Okhonin, 1994; Bartsev *et al.*, 1996).

The idea of this approach is based on an analogy with that used in aircraft construction. At the beginning a designer draws a general draft of a new aircraft and makes approximate estimates of future aircraft parameters. Then, in accordance with the aerodynamical principle of similarity, a small model of the aircraft is built and then tested in the wind tunnel. During these experiments all the necessary modifications of the model are made and its parameters achieve the desired values. The principle of similarity expressed in the so-called Numbers (Mach, Reynolds, etc.) allows the conducting of a reverse procedure and the modifying of the original, i.e. a full-scale aircraft prototype, in accordance with the model modify-cation. Only after this stage is a pilot aircraft built and real flight tests begun.

The process of LSS creation should look like aircraft building. In this case the role of a small aircraft model will be performed by micro-LSS's of small size that contain plants and animals possessing high rates of mass exchange and transformation. Unfortunately, the experience of the aerodynamical similarity approach can not be used directly in LSS design since there are no equations that describe ecological systems exactly as the Navier&Stokes equation does. The objective is to develop the principles of ecological similarity, which can be applied to real LSS's without a detailed mathematical description of them. A possible structure of this approach is shown in Figure 11.1.

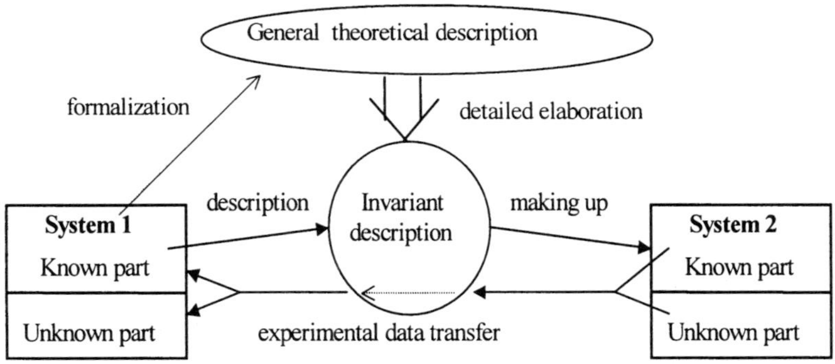

Figure 11.1 Methodological pattern of the approach to ecological similarity.

Here "System 1" is a full-scale prototype and "System 2" is an experimental micro-CES. It is very important that there are two parts of any biological system, the known and unknown part. This makes the situation basically different from that in aerodynamics since it does not allow the construction of an accurate model of a CES. Really, if there is a complete mathematical description of a CES, it is easy now to conduct computational experiments and obtain data on reliability. But since there is an unknown part of CES properties, computational experiments are not comprehensive and reliable. The ecological similarity approach is largely based on the hope that two systems with similar or identical (at the level of CES's invariant description) known parts are similar at the level of unknown properties due to their biochemical, genetic, and, in general, biological similarity. In addition, the approach promises to develop the principles of an adequate description of CES's by generalizing the experience obtained with small fast CES's.

The type of general theoretical description is determined by the level of accuracy, available data and goals of the CES designer. So the chosen level of theoretical consideration and a specificity of "System 1" determine the invariant description of "System 1" and those similar to it. This description allows the obtainance of the description of properties that the experimental CES ("System 2") must comply with. The experimental data obtained with "System 2" can be transferred by the reversible application of the invariant description to "System 1." For example, the area of safe operation or dynamics range of "System 2" can be transformed into those of "System 1."

Direct Use of Approximate Equations

A possible way is to write some general equations that can describe a given CES. Then, on the basis of these equations, some "numbers" can be determined. These "numbers" allow the selection or combining of CES's with equal "numbers" and, consequently, with similar behavior including the similarity of reliable area. As an example let us consider a model describing a CES, which is completely closed with respect to atmosphere. A real prototype of such CES's is the Bios-3 facility (Gitelson *et al.*, 1989; Kirensky *et al.*, 1971). The simplest version of the Bios-3 description having a practical value can look like the following system (Equation 11.1):

$$\begin{cases} \dot{x}_1 = \gamma y_1 \dfrac{v_1 x_2}{K_1 + x_2} + \delta y_2 \dfrac{v_2 x_2}{K_2 + x_2} - \mu V_0 \\ \dot{x}_2 = -\gamma R_1 y_1 \dfrac{v_1 x_2}{K_1 + x_2} - \delta R_2 y_2 \dfrac{v_2 x_2}{K_2 + x_2} + \mu R V_0 \\ \dot{y}_1 = y_1 \dfrac{v_1 x_2}{K_1 + x_2} - q V_0 \\ \dot{y}_2 = y_2 \dfrac{v_2 x_2}{K_2 + x_2} - (1-q) V_0, \end{cases} \tag{11.1}$$

where x_1 is oxygen concentration; x_2 is carbon dioxide concentration; y_1 is wheat biomass "concentration" (first plant species); y_2 is chufa biomass "concentration" (second plant species); V_0 is the rate of human ("customer" component) metabolism; v_i is the maximum growth rate of plants under optimal illumination; γ is a coefficient relating the rate of wheat growth to the rate of oxygen production; δ is the same for chufa; μ is a coefficient relating the rate of human metabolism to the rate of oxygen consumption; R_1 is an assimilatory quotient for plants; R is a human respiratory quotient; q is the part of human metabolism supported by wheat photosynthesis.

If $\mu = \gamma q + \delta(1-q)$ and $\mu R = R_1\gamma q + R_2\delta(1-q)$ take place, then the CES can achieve complete balance with respect to atmospheric composition. The invariant description of systems having behavior similar to the given CES behavior can be obtained by introducing new variables:

$$\varepsilon_1 = \frac{x_1}{\widetilde{x}_1}, \quad \varepsilon_2 = \frac{x_2}{\widetilde{x}_2}, \quad \eta_1 = \frac{y_1 v_1}{V_0 q}, \quad \eta_2 = \frac{y_2 v_2}{V_0(1-q)},$$

where $\widetilde{x}_1$ are the concentrations of oxygen and carbon dioxide, which are normal for the "customer" of a CES.

The invariant description of the Bios-3 facility at a given level of accuracy is:

$$\begin{cases} \dot{\eta}_1 = v_1 (\eta_1 \dfrac{\varepsilon_2}{N_1 + \varepsilon_2} - 1) \\ \dot{\eta}_2 = v_2 (\eta_2 \dfrac{\varepsilon_2}{N_2 + \varepsilon_2} - 1) \\ \dot{\varepsilon}_1 = N_4 \dot{\eta}_1 + N_5 \dot{\eta}_2 \\ \dot{\varepsilon}_2 = -N_4 N_6 \dot{\eta}_1 - N_5 N_6 \dot{\eta}_2, \end{cases} \tag{11.2}$$

where

$$N_1 = \frac{K_1}{\tilde{x}_1}, \quad N_2 = \frac{K_2}{\tilde{x}_2}, \quad N_3 = \frac{v_1}{v_2}, \quad N_4 = \frac{\gamma V_0 q}{v_1 \tilde{x}_1},$$
$$N_5 = \frac{\delta V_0 (1-q)}{v_2 \tilde{x}_1}, \quad N_6 = R_1 \frac{\tilde{x}_1}{\tilde{x}_2}, \quad N_7 = R_2 \frac{\tilde{x}_1}{\tilde{x}_2}$$

are the "numbers" that must be equal for systems similar to the Bios-3.

So it is possible to build accelerated small CES's including some analogs of human and plant components. If all the numbers presented above are equal, then this small analog will have the same dynamics and, therefore, the same (after scaling) areas of reliable operation.

Generally speaking, this approach provides a possibility to test our understanding of CES's by greatly accelerating the experiments in which 20-days dynamics will be equivalent to 400 days of full-scale CES experiments. Since these fast CES's are really small and cheap, it is possible to conduct mass-scale experiments. It will be possible to obtain enough experimental points to evaluate the accuracy of the description and to improve it, which is absolutely unreal for a full-scale prototype model.

Sometimes the consideration of partial cases of similarity can be fruitful in CES design and evaluation. However, if these cases were discussed, this chapter would become far to long.

Peculiarity of CES's as a Model Object. Inverse Principle of Gauze

There are some limits on the minimal number of species in a CES necessary for complete closure of the ecosystem with respect to the given number of chemical elements. Let the CES contain N substances of n_i concentrations where $i = 1,..N$ and M species numbered from 1 to M. Let j-th species produce V_i^j of i-th substance per mass unit. Then taking into account L mass conservation laws:

$$\sum_{i=1}^{N} C_\alpha^i V_i^j = 0\,, \qquad \alpha = 1 \ldots\ldots L\,, \tag{11.3}$$

where C shows how much of α-th element is in the unit of i-th substance, it can be shown that for long-term existence of the CES the following inequality must be satisfied:

$$M > N - L. \tag{11.4}$$

This inequality shows that a CES has to contain a sufficient number of species to be closed. Moreover, the more the mass conservation laws L—the smaller number of species has to be involved for providing closure with respect to N substances. Expression (11.4) can be compared with the well-known Gauze principle, which says that for autoregulation of M species whose growth rates depend upon N substance concentration, the following inequality must be satisfied:

$$M < N. \tag{11.5}$$

So expressions (11.4) and (11.5) show that if autoregulation is the only control mechanism in a CES, then the number of species must conform to the combined constraint:

$$N > M > N - L.$$

Due to the structural similarity of expressions (11.4) and (11.5) expression (11.4) can be called the *inverse principle of Gauze.*

Influence of Closure on the Accuracy of Predicting CES's State

It is possible to select three aspects of an a priori forecast of ecological system parameters: the forecast of the stationary state of medium variables (O_2, CO_2, H_2O, etc.); the forecast of the stationary state of organism numbers; and evolution forecast. It is generally possible to consider all these three levels separately since the relaxation times are essentially different for every level.

Let us evaluate the error of the forecast of the stationary state of medium variables. The evaluations presented below could be applied to artificial biospheres like CELSS and Bios-3 (Gitelson *et al.*, 1989) best of all. Since for CELSS it is possible to assume the conditions of ideal mixing and the stability of external parameter values, the dynamics of the ecosystem state variables are sufficiently well described by the system of ordinary differential equations:

$$\dot{x}_i = \overline{K}_i(\vec{x}, \vec{\beta}), \tag{11.6}$$

where β is the vector of parameters of organisms involved in the ecosystem and $\overline{K}_i(\vec{x}, \vec{\beta})$ are some functions obtained from experimental data, which are taken as the evaluation of general set features. However, in every ecosystem, the parameters of the functions differ from those of the general set due to inevitable fluctuations, thereby resulting in differences between the experimental stationary concentrations and those expected from the solution of the system:

$$\overline{K}_i(\vec{x}, \vec{\beta}) = 0.$$

The task is to estimate the dispersion of the stationary concentrations and the dispersion of the mass cycling rate in a cycle.

For simplicity, let us consider a closed system in which every trophic level is represented by only one species. In addition, let us assume a simple rate dependence on substrate concentrations, i.e. the absence of regulatory 'allosteric' interactions.

After necessary calculations taking into account the closure of the system ($\sum_i x_i = \text{const}$), the formula for the dispersion of a stationary substance concentration is obtained:

$$D(x_i) = \frac{b_i}{m_i a_i^2}\left(1 + \frac{2}{m_i a_i \sum_{j=1}^{n} \frac{1}{m_j a_j}}\right) + \frac{1}{m_i^2 a_i^2} D(f), \tag{11.7}$$

where $D(f) = \dfrac{\sum_{i=1}^{n} \dfrac{b_i}{m_i a_i^2}}{\left(\sum_{i=1}^{n} \dfrac{1}{m_i a_i}\right)^2}$ is the dispersion of the substance cycling rate;

where $a_i = \dfrac{v_i}{x_i}$; $b_i = \sum_{\varepsilon i} \left(\dfrac{v_i}{\beta_{\varepsilon i}^{i}}\right)^2 D(\beta_{\varepsilon i}^{i})$,

where $D(\beta_{\varepsilon i}^{i})$ is the dispersion of the ε-th parameter of the *i*-th species due to inevitable fluctuations of different origins, which are not considered here; v_i is the rate of transformation of the *i*-th substance by the *i*-th species; m_i is the number of organisms of the *i*-th species.

Analysis of Equation 11.7 allows us to obtain some general statistical properties of CES's:

1. Closure reduces the value of dispersion in comparison with the open case. For the simplest CES's consisting of two species, dispersion is only half of that for one species under open conditions.
2. Dispersion of the CES's state cannot be more than that of the stage with maximal value (for open conditions).

In conclusion, the forecast of an ecosystem stationary state is generally not precise, and the error could achieve large values: therefore, artificial biospheres will need 'soft' control (adjusting, initial and permanent regulation).

CES Applicable Statistical Data Processing

Experimental measuring of any parameter always involves an error. A usual statistical approach of evaluating some significant parameters (arithmetic mean value, dispersion, etc.) and their accuracy is based on the assumption of the independence of these parameters. However, inside a CES mass conservation laws make these parameters not independent. So any data obtained in a closed system have to be processed taking into account these constraints. The approach of the maximum likelihood makes it possible to obtain the necessary equations for calculating the adequate expected value. Here only the formula for the expected value evaluation in the case of one conservation law is presented:

$$\mu_i = \bar{x}_i + s_{\bar{x}_i}^2 \frac{(1 - \sum \bar{x}_i)}{\sum s_{\bar{x}_i}^2} \text{ (for} \sum \mu_i = 1\text{)}, \tag{11.8}$$

where μ_i is the expected value of an *i*-th substance; $\bar{x}_i$ is the arithmetic mean value of an *i*-th substance; $s_{\bar{x}_i}^2$ is the standard deviation of the arithmetic mean value of an *i*-th substance. The case of several mass conservation laws requires the solution of a system of linear equations and has no elegant analytical form.

Project-Oriented Levels of CES Model Description

In this section we will consider the problem of the adequate choice of the LSS type for a given practical application. There are many versions of possible LSS compositions and there is a continuum of operation regimes. Since there are not enough experimental data on different LSS features (only few variants of LSS's have been experimentally tested), the problem of a priori choice of the most promising LSS composition is essential for the concentration of financial support and designer efforts. So the question of an adequate criterion of the LSS choice arises. The level of closure is a very important parameter closely related to the time of autonomous CES existence without external interference. Besides, closure determines the total mass of a LSS, which is extremely important for space applications. As complete closure cannot be achieved in practice, the fractional closure (K) can be defined as:

$$K = (1 - \Delta\dot{m}/\Delta M), \tag{11.9}$$

where K is the closure coefficient; Δm is the mass input to the CES in unit time; ΔM is the mass recycled through the "customer" regeneration trophic level of the CES per unit time.

It is important that Equation 11.9 may only be used to determine the closure of any CES when $\Delta m_{in} = \Delta m_{out}$ and when within the sealed ecosystem a stationary state is reached. If this requirement is not met, then there are deadlocks or deficiency inside the system, and their build-up causes CES destruction. So the term CES in this chapter always presupposes that this requirement is met.

If we compare the two most investigated BLSS—Bios-3 (Kirensky, *et al.*, 1971) and Biosphere 2 (Nelson and Dempster, 1995) we can see big differences that are related not only to the size of the facilities. Bios-3 was designed as a machine with biological units and the designers had to think about the "proper" operation of them. Authors of Biosphere 2 relied on self-organization of the constructed ecological systems and introduced about four thousand organisms into it (Allen, 1991). In Bios-3 the steady state (excluding microflora content) was recorded during the whole period of experiments, while in Biosphere 2 the steady state was not achieved and therefore it was a sealed but not closed ecological system.

Analysis of Bios-3 and Biosphere 2 paradigms and results of the experiments allowed us to select three levels of CES closing technology. Their brief description is presented in Table 11.2. At present the first level is practically achieved, and material balance and the rate of possible deadlock formation *under constant conditions* can be calculated.

Formally the problem at the first level is to find the minimum of an evaluation function with respect to the chosen criterion, subject to constraints imposed by the law of matter conservation, human metabolic requirements, LSS components' normal operating conditions, etc. The first level of closure technology is sufficient to make comparative engineering estimates of different CES's, which allows essential reducing of the set of possible CES variants.

The second level of closure technologies includes fine adjustments of component parameters, relative to each component, and an assessment of system stability. At this level, the dynamics of mass flows can be described by differential equations.

The third level accounts for non-specific interactions between CES components (including physical and chemical ones). For example, it is possible that micro quantities of organic volatile compounds emitted from one plant species or human organism (for example, ethylene, ammonia, and growth hormones) can affect the growth and development of other plant species or can poison the physicochemical catalytic unit. These effects can not be predicted theoretically and therefore, the predominant role of experimental investigations is the specificity of this level.

The purpose of this section is to illustrate the first level of closure technology, which is an obligatory step toward the development of optimal LSS design.

The biological CES's considered here as examples include those based on hydrogen-oxidizing bacteria (*Alcaligenes eutrophus* (former name *Hydrogenomonas eutrophus*))—HB LSS's, microalgae (*Chlorella* genus)—MA LSS's, higher plants (wheat, chufa, vegetables)—HP LSS's, higher plants and mushrooms (oyster mushroom *Pleurotus streatus*)—HPM LSS's, and higher plants with physicochemical incineration of wastes—HPI LSS's.

Calculations are conducted with the data on CES component parameters obtained in experiments under closed conditions, where functioning parameters of CES components differ from the open case (Bartsev *et al.*, 1996). Even with such rigorously prescribed CES component parameters it is possible to assemble many CES versions with a different degree of closure or complementarity to the human. But the optimal CES of the given type can be only one for the given conditions and with respect to the given criterion, and the problem of choosing the best CES type for further design is reduced to the comparison of a small set of optimal CES's.

Table 11.2 Levels of ecological closing technology.

Levels of closure technology	Types of formal description	Required data	Target criteria of LSS's
First (static) (sufficient for comparative engineering evaluations of LSS's)	Systems of algebraic equations of stationary flows with constraints following from the law of matter conservation	Parameters of LSS components obtained in experimentally determined optimum	LSS mass, mass of external supplies, closure coefficient
Second (dynamic) (gives necessary requirements to the control system)	Differential equation systems describing material flows in LSS's	Working characteristics of components (relation of parameters to conditions)	Reliability, stability, dynamics range of an LSS as a whole
Third ("allosteric") (determines specific species composition of BLSS's)	Formal description is not known. Investigation at the level of experiments with a laboratory CES	Non-specific interactions among LSS components	Quality of human life

Chosen Criterion of Optimization

In cosmonautics the minimum mass criterion and the energy criterion admitting simple conversion to mass units are accepted. In this section the criterion of "integrated" or "accumulated" mass is used for optimization and for comparison:

$$m_{\text{int}} = m_{\text{LSS}} + \Delta m \cdot T, \tag{11.10}$$

where m_{LSS} is the mass of LSS components; Δm is the mass of necessary supplies from the outside; T is LSS operation time.

The answer to the question—what is to be taken into account to estimate m_{LSS} and Δm—depends on the chosen framework of consideration. To make up for the incomplete data on LSS conditions and designs, we introduce two levels of LSS description.

Frameworks of LSS Consideration

The most definite at present is the framework of the mass of LSS components immediately involved in the mass exchange process at a given moment regardless of the mass of the maintenance equipment. We call this framework level the Ideal Design Solution (IDS) level, corresponding to the real achieved and tested conditions of the component operation, but the mass and other parameters of the design are not taken into account. This framework complies with the limit LSS characteristics, whatever the equipment design solution, it can serve as a guideline for the designers and indicates the principal constraints for one or another type of an LSS.

The next framework level is the Moderately Optimistic Design (MOD), which is based on LSS parameter estimates taken from general consideration and by analogy. This is quite a subjective estimation level, but it is the level that has to serve as the basic one to compare different LSS's for a given space mission. This framework encompasses the mass of equipment of the LSS itself, the mass of energy-supply and heat-rejection systems maintaining LSS operation, the addition to the spacecraft mass associated with increasing the volume of the living quarters by the LSS volume and so on. In this case Δm should include not only the mass of air, water and food, but also the mass of spare parts to fix the equipment. In this section only IDS case calculations are presented. Both cases can be found in Bartsev *et al.* (1996).

Constraints on Optimization

The CES optimization procedure is conducted under constraints imposed by the conservation of matter law, human metabolic requirements, CES components' normal operating conditions, etc.

Here the constraint system corresponds to the closure technology employed to devise LSS's of the Bios series (Kirensky *et al.*, 1971; Gitelson *et al.*, 1989). The four chosen elements (C, O, H, N) define the set of substances whose transformations are under control: oxygen, carbon dioxide, water, proteins, fats, carbohydrates, and nitrogen-containing chemical compounds. The total biomass of CES components is considered as a combination of these substances. The constraints restrict the values of adjustable variables: the composition of the human diet (amount of proteins, fats and carbohydrates), the mass of biological components, the part of utilized edible biomass (for higher plants and mushrooms), and the mass of incinerated components (for systems with incineration).

In this chapter three different cases are considered:

1. The case of a "normal" diet, when some amounts of animal proteins and fats are considered to be obligatory components of human food and have to be prestored or supplied from the outside. The range for proteins is (100–160 g/day); for fats—(50–100 g/day). The minimal amount of animal proteins supplied from the outside is 50 g/day and animal fats—25 g/day. In addition, CSR (coefficient of salt recycling) is 0.5, which is close to the obtained level of incinerated product utilization (Bubenheim and Wignarajan, 1995). So this case can be considered as real and almost ready to be implemented.
2. The case of a "normal" diet with CSR = 1.0. Other constraints are the same as in item 1.

3. The case of a completely vegetarian diet with CSR = 1.0. The range for proteins is (50–60 g/day); for fats—(30–100 g/day).

IDS Framework Computational Experiments

The results of IDS LSS structure optimization with respect to integrated mass and for different operation times are shown in Figures 11.2–11.4.

The dynamics of higher-plant–mushroom LSS mass is not presented in the figures since the optimization program eliminated mushroom species from the system. Data on microalgae LSS's are presented in all figures for comparison. In Figures 11.2–11.4 every mark shows the point where the structure of a higher-plant LSS undergoes some changes. Inside the intervals the structure is constant.

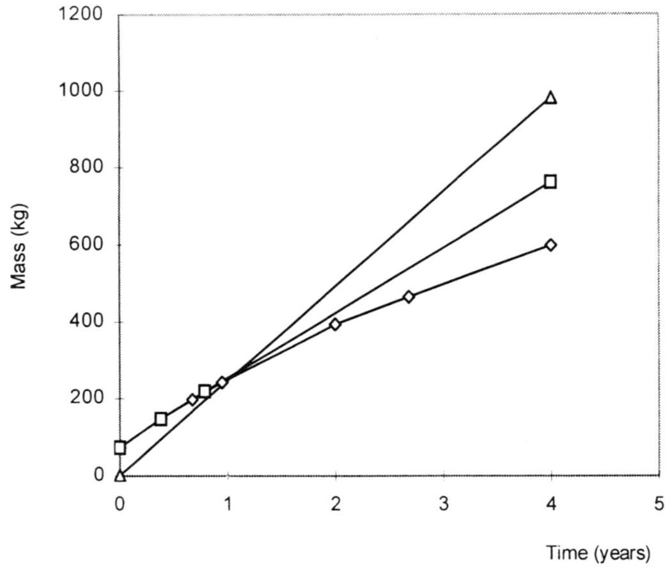

Figure 11.2 Integrated mass of a BLSS ("normal" diet, Coefficient of Salt Recycling (CSR) = 0.5, Ideal Design Solution (IDS) level is considered). In the figure □ denotes an LSS based on higher plants (HP); ◊—an LSS based on higher plants with incineration (HPI); Δ—an LSS based on microalgae (MA).

It can be seen that an optimized biological LSS is promising for long-term space applications. The higher-plant LSS's with or without incineration have no rivals when the duration of the space mission is longer than one year. The horizontal portion of the HPI LSS line in Figure 11.4 presents the state of the LSS with complete closure when there is no need for external supplies. It is certainly true only in the scope of the chosen accuracy and structure of constraints. For example, this system can be imbalanced with respect to phosphorus, but taking phosphorus into account requires additional experimental data, which are not available yet. So the calculation scheme of this kind can be used as an experimental data organizer and planner displaying key points of LSS assessment. But the Ideal Design Solution case shows only the upper level of system properties. To get evaluations closer to a real design, a Moderately Optimistic Design (MOD) framework of consideration has to be used.

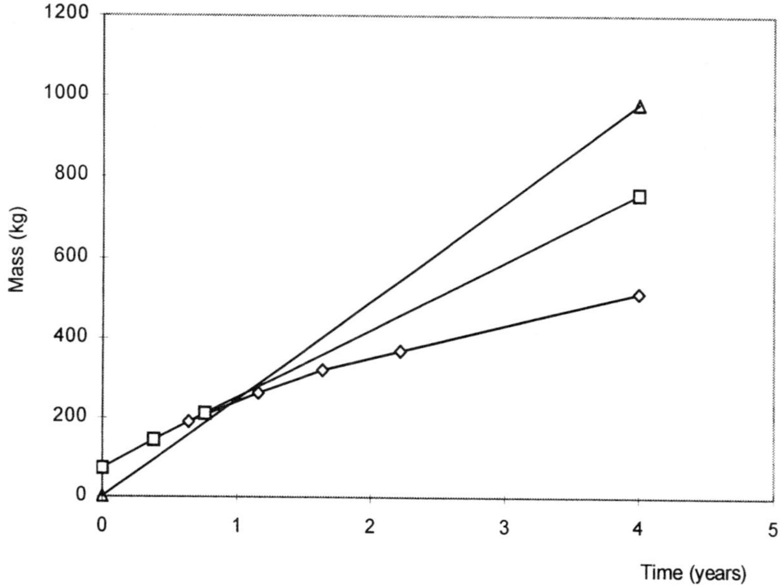

Figure 11.3 Integrated mass of a BLSS ("normal" diet, CSR = 1.0, IDS level). In the figure □ denotes an HP LSS; ◊—an HPI LSS; Δ—an MA LSS.

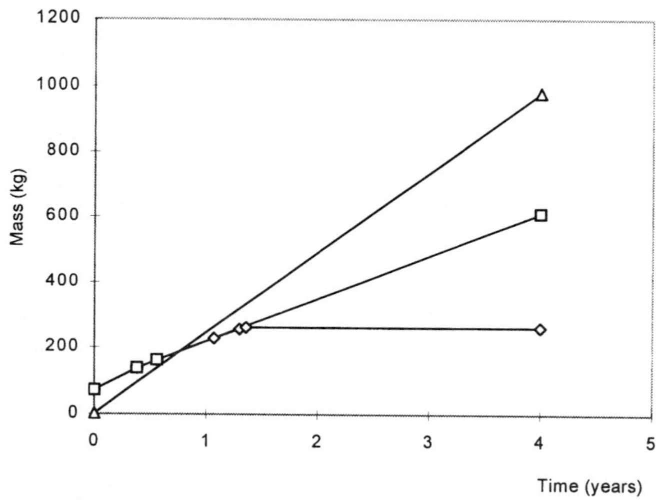

Figure 11.4 Integrated mass of a BLSS (vegetarian diet, CSR = 1.0, IDS level). In the figure □ denotes an HP LSS; ◊—an HPI LSS; Δ—an MA LSS.

Calculations show that the patterns of the dynamics of LSS structures are the same for IDS and MOD cases, and accounting for the equipment provides only shift diagrams over a long time. At the MOD level of consideration, the optimized LSS with higher plants is better than the optimized microalgae LSS, when the space mission lasts more than approximately 5 years for a "normal" diet and about 3 years for a vegetarian diet.

The Structure of Higher-Plant LSS Phytocenoses

Computational experiments show that the value of the coefficient of salt recycling (csr) does not affect the structures, but only shifts them in time.

Wheat/Chufa (Sedge-Nut) compositions for different constraints are shown in Figures 11.5 and 11.6.

Computational experiments demonstrate that the optimal structure of the higher-plant LSS (species ratio, their mass) depends on their operation time. It was shown that mushrooms of the considered species are eliminated from LSS's with higher plants at any time of operation.

In the case of a vegetarian diet the closure can reach 100% for this level of elaboration. But a high level of closure can not be the goal of LSS design since different exploitation times determine the appropriate LSS configuration with respect to the criteria valuable for practical applications. With respect to integrated mass, LSS's on the basis of higher plants are preferable if the duration of LSS operation is more than 3 years for a more or less optimistic design.

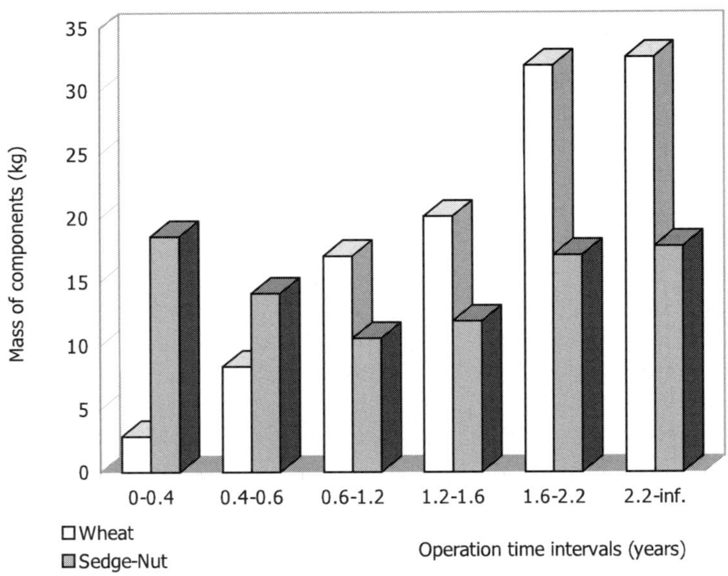

Figure 11.5 Optimal structure of an LSS based on higher plants with incineration ("normal" diet, Coefficient of Salt Recycling = 1.0, Ideal Design Solution level).

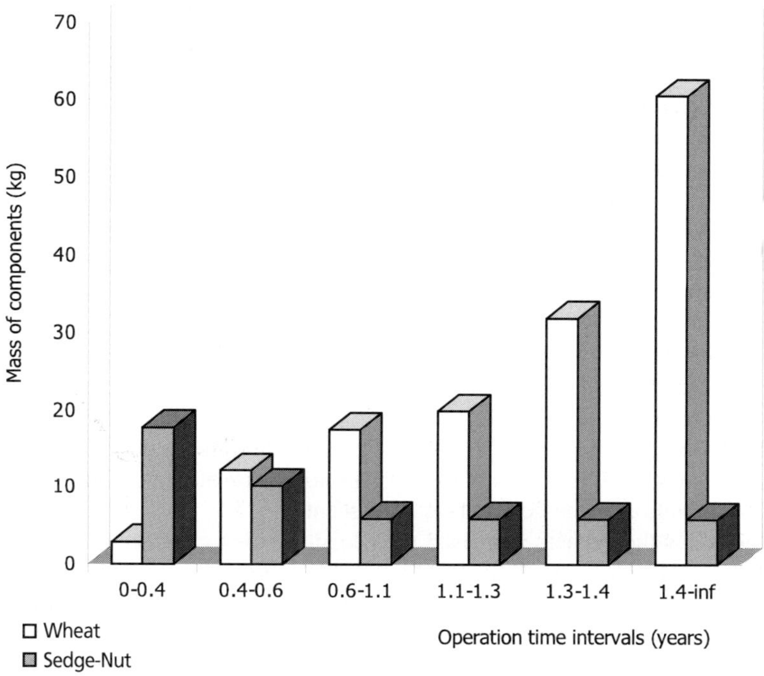

Figure 11.6 Optimal structure of an HPI LSS (vegetarian diet, CSR = 1.0, IDS level).

Integral Criteria of CES Type Selection for Practical Applications

Distinct and clear criteria for evaluating and comparing different LSS's are necessary to improve LSS's and to provide a solid foundation for the choice of an LSS that will meet the requirements of a specific mission.

These criteria need to be expressed as numerical values so that different LSS's could be arranged on a single axis. It is also important for these criteria to be objective, or at least explicit, to allow meaningful discussion.

The statement that the principal and only purpose of a space LSS is to provide for the safety of humans and preserve their health and working ability may be considered trivial. Then, the basic LSS parameter is the reliable functioning and sustaining of such quality of life so as not to make it harmful for human health. However, in the literature the parameters mentioned above are seldom considered.

This situation can be explained by the difficulties of the accurate calculation of the reliability of complex systems and the evaluation of the quality of life in an LSS. Now in cosmonautics the criterion of minimum mass is accepted. In the scope of this criterion power consumption can be accounted via simple conversion of power units into mass ones and following minimization of summarized mass.

Nevertheless, the mass optimality criteria of the space systems are secondary. They were established in the course of the rise and development of cosmonautics with mass as the dominating factor of the reliability of a spacecraft: "The weight of a life support system linearly increasing the launch weight of a spacecraft, the rocket weight and the thrust rapidly increase non-linearly, enormously increasing the number of rocket constituents and systems and decreasing the reliability" (Voronin and Polivoda, 1967). Thus, the mass optimality criterion is a transformed form of the initial, basic criteria related to the reliability of a spacecraft on the whole.

Now, when long-term missions are planned, it is reasonable to return back to a criterion defining the success of space missions on the whole—an integral criterion of the reliability of a spacecraft and a life support system in compliance with the chosen scenario of a space mission.

Certainly, even at present, calculating reliability of a spacecraft is a complicated problem, and certain indices have to be chosen by phenomenological estimates or expert judgments. However, a formalization of the reliability criterion could help in disclosing gaps in our knowledge and retaining the integrity of space mission design. The formalization gives an ability to outline clearly those areas where the intuition of the designers is still essential.

Nowadays the carriers increased their load-carrying capacity, and the LSS operation time essentially increased. But during a long-term and far mission, almost any failure of an LSS leads to catastrophe. All this makes the physical mass minimization criterion inadequate for many cases, disorienting the designers. The criterion of integral reliability is in good agreement both with humanity and the purely practical requirements of space flight success.

Approach

At first, three main indices can be selected for a space LSS to meet its objectives: reliability, mass, and quality of life. Of the possible evaluation criteria, the criterion of integral reliability is proposed. This criterion incorporates the above mentioned indices. These indices can be related only if the space mission is considered as a whole.

The reliability of the overall space mission can be more simply measured by the probability of a catastrophe, which consists of:

probability of LSS failure;

probability of a failure due to a fatal human error;

probability of fatal malfunction of the spacecraft engineering systems.

The general form of the criterion of integral reliability is expressed by the condition:

$$P_{LSS}(m) + P_{QL}(T) + P_{GMS}(m) \rightarrow \min , \qquad (11.11)$$

where m is the mass of the LSS including the mass of spare parts and devices providing for the reliability;

$P_{LSS}(m)$ is the mass-dependent probability of a catastrophe due to LSS failure (which decreases, in general, with increase in mass of an LSS);

$P_{QL}(T)$ is the probability of a fatal error of the crew, which depends on the quality of life and time T of stay in the LSS. P_{QL} varies with different LSS types;

$P_{GMS}(m)$ is the probability of a catastrophe due to the spacecraft engineering systems failure, which increases with LSS mass (here, index "GMS" means General Mission Support).

From the above expression and its constituent functions, it is apparent that there must exist an LSS mass optimum with respect to mission reliability. The first and second terms of expression (11.11) depend only on the type and design of the given LSS. The specific form of the third term depends on the mission scenario and will be considered after the first two.

Estimation of Reliability of an LSS Itself

When estimating LSS reliability, redundancy systems should be taken into account. But Biological LSS's are peculiar due to their self-repair ability, which allows BLSS's to restore quickly after the gravest damage, in principle, even from only a few cells or seeds that retain their ability to reproduce. Long-standing experience in Bios-3 proved this experimentally. The causes of all its failures were engineering devices only, not the living organisms. Even when they were damaged due to engineering faults, they restored themselves after the technical breakdowns were cleared. Certainly spacecraft must have some emergency supply of oxygen, water, and food for the crew to survive in an emergency.

To enhance the reliability of the system on the whole, it is advisable to divide the regeneration component into several identical and self-sustained sections affording boosting, i.e. fast increase of productivity or regeneration rate. Compartmentalization implies the circulation of water and air in several loosely correlated loops. Compartments function simultaneously in the mode optimal with respect to their efficiency. When one of the compartments is out of order, the others are boosted and sustain the life of the crew while the damaged section is being fixed (10 days for higher plant LSS's, for microalgal ones and those with hydrogen-oxidizing bacteria—much shorter). The failure intensity of the LSS itself decreases in this case, and its estimate has the form (Bartsev and Okhonin, 1999):

$$\lambda_{LSS} = e^{-t_{CR}\lambda_\tau (N-N_0+1)} \frac{N!}{(N_0-1)!(N-N_0)!} \frac{\lambda_s^{N-N_0+1}\lambda_\tau^{N_0}}{(\lambda_s+\lambda_\tau)^N} \qquad (11.12)$$
$$N_0 = N/f\ (f \geq 1);\ N_0 = N\ (f=0),$$

where N is the number of self-sustained sections of the regeneration component; f is the permissible boost (how many times productivity of a component can be increased); $\lambda_\tau = 1/t_r$, t_r is the time required to repair (restore) a component; t_{CR} is the crew existence time when all LSS regeneration components fail; λ_s is the failure intensity of one section of a regeneration component.

The causes of a BLSS failure can be: an emergency; leaking of substances that are toxic for this component; or the failure of the mechanical parts of the regeneration component. An approximate estimate for this λ_s is based on the "standard" intensity level accepted for Space Transportation System Orbiter (Lewis, 1989)—10^{-4} h^{-1}.

In addition to the catastrophes due to LSS failure, account should be made of the possible threat to human health and the life of the LSS itself. If the technological cycle of an LSS involves explosives (λ_{exp}), flammable (λ_{flam}) or toxic (λ_{tox}) substances, then the associated risk should be taken into consideration.

Summarizing the above and assuming that the intensity of failures of different units of the system is small (which is true for space equipment), we have an equation to calculate the probability of an LSS accident:

$$P_{LSS} = (\lambda_{LSS} + \lambda_{exp} + \lambda_{flam} + \lambda_{tox}) \cdot T. \quad (11.13)$$

Quality of Life Criterion and Reliability

There is always an other-than-zero probability of a fatal human mistake in controlling the vehicle, landing module, nuclear reactor, or other vital device. It is difficult to estimate the probability of such a mistake, but it is apparent that this probability increases with impaired human health. Human health, workability and, eventually, probability of a fatal mistake depend upon a very important parameter of LSS's—Quality of Life (QL). So there is a way to transform this not easily formalized criterion into a probability of catastrophe, which is a numerical presentation of crew safety.

It was shown (Walkinshaw, ed., 1990) that under closed air cycling which takes place in highly sealed, so-called "energy efficient" buildings a symptom of the Sick Building Syndrome (SBS) appears. By estimates given in Robertson *et al.* (1990), and Zweers *et al.* (1990), elimination of a SBS can reduce absence from work due to illness by 30% and essentially decrease complaints of poor health. Seemingly, the cause of SBS is a set of not completely identified factors, yet, it should be noted that in spite of complete pressurization the crew of Bios-3 did not have complaints typical for SBS; what is more, they preferred to spend time in the phytotron compartment (Gitelson and Okladnikov, 1994). It cannot be ruled out that the higher plants may be the most realistic remedy for reducing the probability of the crew's fatal mistakes.

The probability of a fatal human mistake when health is good can be estimated by means of the published data on the statistics of air catastrophes through human fault. Human "failure" intensity is rather large—once in 30–100 years. We assumed that with health impaired (chronic fatigue, somnolence, headache, etc.), human "failure" intensity linearly increases with time.

But since there are no data on the dependency of human-mistake (fault) rate on the time spent under harmful conditions, a simple linear function is used. Quantitatively this function can be expressed as follows:

$$\lambda_{QL} = \lambda_{hm} + \lambda_{hm} \cdot (1 - R_{LSS}) \cdot dA \cdot T_{mis}, \qquad (11.14)$$

where λ_{hm} is the "fault" rate of fatal mistakes of the crew under normal conditions; R_{LSS} is associated with SBS estimation of QL for the given LSS type (with respect to higher plants it is estimated as 0.7, microalgae—0.1, hydrogen bacteria—0); dA is dys-adaptation coefficient showing the increment of fatal mistakes rate due to duration of living under LSS conditions; T_{mis} is the duration of space mission. So $P_{QL}(T) = \lambda_{QL} \cdot T$ is a quadratic function of T.

Reliability and Mass Criteria

Space systems are characterized by having a close relationship between reliability and mass, but the form of this relationship expressed in $P_{GMS}(m)$ depends on the mission scenario. In this chapter three main types of scenarios are considered. The simplest is the scenario when a spacecraft is launched from a planet.

Launch/landing mission. Let us consider a spacecraft consisting of a LSS unit (payload) and a launcher. It appears obvious that in the general case the spacecraft catastrophe probability increases with the payload mass, i.e. the probability of an accident of the carrier at launch becomes a function of the LSS mass. Possible causes of increasing the catastrophe probability due to mass increases are the prolongation of the initial low stable stage of launching, additional engines load, shift of mass center, etc. A possible general pattern of this dependency is shown in Figure 11.7.

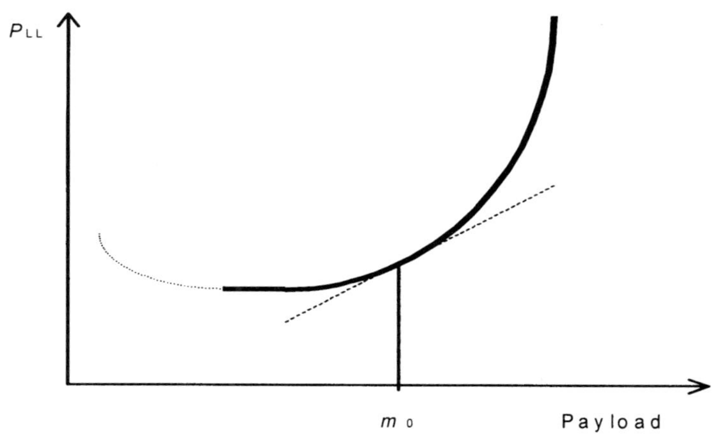

Figure 11.7 Hypothetical curve describing probability of launcher catastrophe versus payload.

Here and below the catastrophe of spacecraft at launch and landing is denoted as P_{LL}. In this type of space mission scenario $P_{GMS} = P_{LL}$. The dotted part of the curve corresponds to the possible growth of the accident probability due to the loss of spacecraft balance. It may look unusual, but the usual way of payload mass determination, when rocket engineers establish the upper level of a payload, practically corresponds to the pattern shown in Figure 11.8, which is obviously wrong.

Practically, it is not necessary to know the curve as a whole—it is quite sufficient to know the part of it near the planned payload mass. In this case it is possible to use the linear approximation:

$$P_{LL}(m) = P_{LL}(m_0) + K \cdot \Delta m, \tag{11.15}$$

where $K = P_{LL}/m$. Parameter K can be derived from mass-scale experiments, which is, however, not realistic, or via expert evaluation:

$$K = P_{LL}(m_0)/\Delta m, \tag{11.16}$$

where Δm is the value of additional payload at which the probability of an accident doubles for the launcher.

If coefficient K is known, an LSS can be optimized regardless of the optimization problems of other spacecraft units; this coefficient takes into account the effect of LSS mass on spacecraft reliability as a whole.

Figure 11.8 Graph of implicitly accepted dependence of a launcher catastrophe probability vs. payload in the case of a fixed prescribed permissible payload mass.

At the dawn of the development of cosmonautics, the small power of the engines prohibited a high admissible excess of overload Δm; the probability of launcher catastrophe increased with the load (K was big). Proceeding from the statistics of successful space flights over the period of 1961–1993 (NASA, 1994) it is possible to estimate the probability of a launch–landing catastrophe as $P_{LL} = 0.02$. This estimation is close to the estimate of the probability of a space shuttle disaster made by R. Feynman in his "Appendix F: Personal Observations on the Reliability of the Shuttle" (Feynman, 1989).

Then, taking for illustration $\Delta m = 1000\,\text{kg}$ and m_{LSS} = 300 kg (Voronin and Polivoda, 1967), then, $K = 0.02\ T^{1}$. Because of their simplicity and small flight time P_{LSS} of the life support systems of the first spacecraft was << 0.02.

The calculation of P_{LL} by Equation 11.15 is applicable only when a spacecraft starts its mission from Earth with an LSS aboard. If a spacecraft is assembled in orbit to begin its mission, the formula for catastrophe probability is different. Analogous is the situation for estimating LSS failure for an orbital station and lunar base at a permanent orbit. These two cases are considered by Bartsev *et al.* (1996b). Here only the results of optimal LSS structure for the lunar base and Mars mission are presented.

Computational Experiments

LSS parameters have been calculated as per 1 human basis with 2800 kcal/day exchange. For all LSS's the following values of parameters were accepted: the "fault" rate of fatal mistakes of the crew—10^{-6} h^{-1}; dys-adaptation coefficient *(dA)*—1 (1/year); engineering units MTBF—10^{-4} h^{-1} (shuttle level) (Lewis, 1989).

Computational experiments presented in the chapter were carried out over LSS's of the following types: a system with hydrogen-oxidizing bacteria (H_2B); a system with unicellular algae (microalgae) (MA); a system with higher plants (wheat, chufa, vegetables) (HP); a system with higher plants and incineration (HPI), and, for a Mars mission, a Bosch physicochemical system. These systems are described in detail in the literature (Kirensky *et al.*, 1971; Huttenbach *et al.*, 1988; Gitelson *et al.*, 1989; Mezhevikin *et al.*, 1994). The criterion of integral reliability (Equation 11.11) was used for the optimization procedure.

Comparison of a Lunar Base LSS with Respect to Integral Reliability

There are several concepts of lunar bases. However, with the given level of estimates the distinctions of all these versions are insignificant for the reasons mentioned in the previous section. Time peculiarities of solar illumination on the Moon are important for natural illumination only. Artificial light and Sun-independent energy sources eliminate some difficulties in using a photosynthetic BLSS.

Tables 11.3 and 11.4 show the estimates of the total catastrophe probability depending on the type of an LSS used. Some parameters of LSS's and their service are also presented. A relatively high probability of catastrophe in the case of H_2B LSS's is due to the presence of an explosive mixture of H_2 and O_2. Perhaps there may be a special design of hydrogen-oxidizing bacteria cultivator that might reduce the danger of explosion. In this case the probability of catastrophe would be more acceptable.

Table 11.3 Lunar base (manned transport spacecraft version).

LSS type →	H_2B	MA	HP	HPI
Total catastrophe probability (1/year)	0.502	0.061	0.036	0.034
The number of LSS section	1	4	14	14
Crew replacement frequency (1/year)	0.66	0.63	0.36	0.36
LSS mass (kg)	113	328	890	1937
Total mass intake (kg/year)	267	256	174	116

Table 11.4 Lunar base (unmanned transport spacecraft version).

LSS type →	H_2B	MA	HP	HPI
Total catastrophe probability (1/year)	0.493	0.052	0.031	0.030
The number of LSS sections	1	5	14	16
Crew replacement frequency (1/year)	0.66	0.63	0.36	0.36
LSS mass (kg)	113	331	890	2007
Total mass intake (kg/year)	267	256	174	116

Comparison of a Martian-Mission LSS with Respect to Integral Reliability

Assume the Martian mission to be employing a nuclear rocket, flight time to Mars to be 5.5 months (T_{to} = 0.46 year), the time at Mars orbit to be 3 months (T_{orb} = 0.25 year) (Manned Mars Mission Study group, 1988). Assume that a part of the crew and the LSS used at "Earth orbit–Mars orbit" route does not land on the Mars surface. To calculate the integral reliability, it is necessary to know the part of the propellant (with respect to the payload delivered) the engine spends to place the spacecraft in Mars orbit and start from this orbit.

For illustration assume that braking at Mars and acceleration from Mars spends a mass equal to half of the accelerated mass.

The results indicate that for the Lunar base the most preferable LSS types are those with higher plants. The results for the Mars mission scenario are not so unequivocal: physicochemical LSS's and higher plant LSS's are nearly equivalent (Tables 11.5 and 11.6). It is necessary to note that presented results are used only for an illustration and practically applied results can be obtained after the thorough verification and expertise of all the parameters involved.

The requirements of reliability being very high, such features of higher-plant LSS's as habitual surroundings for humans, self-restoration of basic elements, simplicity and absence of overloading auxiliary engineering elements can make a higher-plant LSS absolutely advantageous in reliability, comfort, and even in such an insignificant criterion as cargo traffic minimization, if the LSS's are intended for stations located anywhere.

Table 11.5 Mars spacecraft LSS (solar irradiation as power source).

LSS type →	Bosch	MA	HP
Total catastrophe probability during flight	0.59	0.064	0.057
The number of LSS sections	1	3	7
LSS mass (kg)	50	325	772

Table 11.6 Mars spacecraft LSS (nuclear reactor as power source).

LSS type →	Bosch	MA	HP
Total catastrophe probability during flight	0.58	0.054	0.050
The number of LSS sections	1	3	7
LSS mass (kg)	50	123	529

For the space programs in the decades to come, the type of system closure tested in Bios series systems appears quite sufficient—further closure in terms of food would do little to increase the integral reliability.

This chapter, as a whole, can be considered as an illustration of some of the key problems of LSS and CES development. The authors tried to show some approaches, which seem to be promising for the problem solution. However, our main goal is to attract the attention of investigators to key problems of CES research in order to provide the effective space and terrestrial applications of it.

CONCLUSION

Creation of Closed Ecological Systems: Results, Problems and Prospects

Bios-3 was the first to make possible experimentation with a long-term human presence in a closed environment regenerated by photosynthetic processes. The atmosphere and water were regenerated completely while food regeneration amounted to 93%. Bios operated for a total of about a year, and continuous experiment with humans in the closed system lasted 6 months. Some test subjects took part in the experiment 2–3 times; the longest time one human spent inside the system was 13 months.

Thorough medical–physiological observation involving the measuring of numerous parameters during the experiments and one month after their completion revealed no changes in the test subjects' health condition. After the experiments the test subjects, researchers at the Institute of Biophysics, resumed their research work, presented dissertations, gave birth to children, i.e. lived the lives of ordinary people. Post-experimental observation during 10–15 years showed that none of them had any diseases that could reliably be associated with their stay in the Bios. Hence, we can state with assurance that in the biological closed life support system of the tested design, with biological regeneration of atmosphere, water, and the vegetable fraction of the human diet, the living conditions were adequate to meet the principal human requirements.

As experiments have shown, a life support system based on a biological material exchange is fully realizable and contains possibilities for further improvement. This kind of closed human ecosystem can become a model not only of an earthly noosphere, but also of daughter noospheres that can be used to help humanity render space inhabitable without threatening other bodies of the solar system through the intrusion of earthly substances and biological autocatalytic processes. As such, it will allow humans to exist in space or on other solar bodies, while requiring only the input of energy and not allowing the release of metabolites into the surroundings.

We cannot envision a time when a complete closure of exchange will become necessary for practical purposes under conditions anywhere on Earth. But partially closed technologies for regenerating the atmosphere, water, and vegetable nutrients can radically enhance the quality of life in extreme regions—in the Arctic, Antarctica, in deserts, and in high mountain settlements. Another aspect of closed life support technologies is that they enable the minimizing of the environmental pollution that results from human waste products and those of their domestic animals.

It is not at all unlikely that these technologies will find applications on Earth before human settlements appear on Mars or the Moon. The transition to essentially closed nonpolluting, non-deadlock life support technologies will be a noticeable step on the path toward the sustainable development of the noosphere.

Synthesis of closed ecological systems is a new field of research. A vigorous process is going on—shaping of ideas, selection of objects, refinement of methods, and definition of terms.

It would be well to define the niche occupied by the relatively new branch of ecology, studying the biosphere, in ecology as a whole.

If ecology is defined as the study of the relationship between organisms and their surroundings, it is naturally subdivided into hierarchic levels differing in the complexity of the ecological system organization:

- elementary ecosystem—relationship of a single individual to its habitat (autecology);
- population ecosystem—relationship of one-species population to its surroundings (population ecology);
- biocenosis, or community in the American literature—a multispecies ecosystem consisting of many (at least two) species interrelated by trophic, reproductive and signal links (synecology).

A term more commonly used in Russian scientific literature is "biogeocenosis" coined by Sukachev (1944). It emphasizes that a cenosis obligatorily includes not only a living component—"bio"—but also a nonliving one—"geo," to which the living component is related through material exchange.

Ecosystems of increasing complexity, from the elementary biocenosis of microorganisms to the multicomponent ecosystems, covering vast territories and water areas, are all at the same level in terms of organization. Their basis is the energy and matter flowing through them.

The next and the highest level of ecosystems in terms of organization is the biosphere as a whole. The biosphere differs from all ecosystems constituting it by a unique quality—nearly complete closure of the material cycle. So the biosphere is endowed with specific properties justifying detachment of this ecosystem as the object of a separate branch of ecology—biosphere science. In modern literature it is called *biospherics* although *biospherology* would be a more logically correct term, while *biospherics* could only be applied to the part dealing with the synthesis of manmade ecosystems closed to a high degree, the experimental investigation of these systems, and modeling, both physical and mathematical.

Present-day science and, subsequently, society realize more and more that humanity's technological potential exerts a rapidly increasing effect (mostly destructive) on the closure of the material cycle in the biosphere. Thus, even today the biosphere experiences a profound influence of humanity, and this influence will grow stronger.

Therefore, there are ample grounds to regard the current epoch in the history of the biosphere as a transition period of its evolution from the stochastically balanced existence to that controlled by the human race—evolution to the noosphere.

The emerging branch of science with roots in biospherology but growing into the humanities may be termed *noospherology*, and the part of it concerned with synthesis of manmade noospheres and their control may be given the name *noospherics*. In essence, the experimental systems we have described in our book are just the first fruits of noospherics.

Figure 1 is an extremely schematized image of our conception of the relationship between biospherology and noospherology as fundamental sciences, their applied branches—biospherics and noospherics—together with fruits growing on them and buds promising to give fruits in the future.

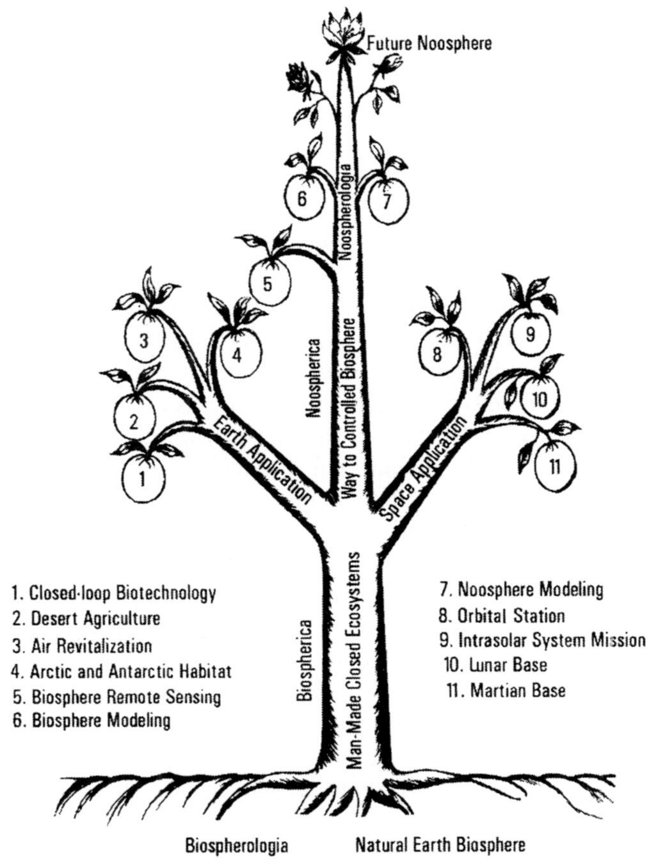

Figure 1 Tree of prospective Earth and space application of manmade closed ecological system.

Methodologically this field of science is rather specific since the object itself—closed ecological systems controlled by humans—does not exist in nature. It is likely that the ideal absolutely closed biological material cycle may be unattainable, but theoretical analysis of the problem shows that this ideal can be approached very closely. Progressing along this path experimentally, researchers, as the reader could see in this book, have not as yet encountered any insurmountable obstacles in closing the material exchange in small ecosystems. In terms of mass exchange, the closure has reached 80–95%, but full closing of the trophic structure is still ahead, and we should expect that each new step will bring increasingly serious difficulties.

Thus, a long distance has been covered on the way to creating closed biosphere-like biological systems within less than half a century. The first "naive" experiments made by Myers (1954) with a mouse under a glass bell, respiring with the oxygen produced by algae, resembles Priestley's classic experiment. The fate of the mouse that died of suffocation in Myers' experiment sadly repeats the history—Priestley's famous experiment. However, as soon as 20 years later people live, breathe and feed for months and months in closed biological life support systems.

Let us, on the basis of the experience that has been accumulated by now, consider the avenues for the further advancement of the described experimental ecosystem directed toward a more complete closure.

Two kinds of obstacles to the further closure of the system and the prolongation of its lifetime can be foreseen: technical and biological obstacles.

We have named the technical problems first. Although they do not seem fundamental, our experience with the system, particularly with the autonomous version, proves that it is the issue of technical reliability of the system that determines to what extent it can be really autonomous. There is a fundamental difference between the live and technical parts of the system with respect to reliability. An organism is a self-sustained system, and a population of organisms is even self-regenerating. Technically speaking, the live part of the biological–technical system not only can be repaired but also can repair itself—it continuously renews itself according to the intrinsic program. This program is the gene pool, and, consequently, the reliability of its restoration depends on the number of autonomously acting genomes. For instance, in the algal reactor of our system it repeats itself in 10^{13} autonomous cells. In the case of some damage, one living cell would be enough for the system to restore itself to its original size. You can find experimental support for this statement in Chapter 6.

In present-day engineering, the functions of operation and restoration are performed by different devices. Machines cannot restore themselves, and their survival is provided either by redundancy or by maintainability. That is why in the biological–technical system, where live and technical parts are as close as organs in a single organism, the overall lifetime is determined by the most short-lived element—and that is the technical part. For many years we have been dealing with experimental closed systems and do not have a single record of system failure for biological reasons. All damages were the fault of technical devices.

A biologist-reader may argue that bacteria, insects, or other organisms that multiply at a high rate can quickly increase their numbers in the ecosystem and cause a catastrophe. This possibility cannot be excluded a priori. However, no new species can invade the closed system, just because it is *closed*, and the species inhabiting it are at equilibrium, limiting each other. Mutants can certainly emerge there; moreover, the rate of mutation must be the same in the closed system and in the natural environment, but a mutant would hardly find its eco-niche in the closed system, where the conditions remain unchanged and under control. For many years we have been conducting experiments in closed ecosystems, but have not dealt with any spontaneous biological catastrophe. General biology and the evolution theory would benefit from an investigation of the possibility and probability of such biological catastrophes in closed ecosystems. This can be done using the material of high statistical power: hermetically sealed microcosms, closed ecosystems consisting of unicellular organisms.

The second technical aspect is the scheme of controlling a small closed ecosystem.

Manmade closed ecosystems differ from their prototype, the biosphere, by the principle of control. The sustainability of the biosphere is ensured by its biological diversity creating an intricate network of metabolic paths with fail-safe redundant functions, by buffer stocks of bio-inert mass, and by the huge size of the planet itself. Such a system, produced and polished by evolution, is sustainable due to stochastic control. By contrast, in a small manmade closed ecosystem all these factors become ineffective: its diversity and size are not sufficient for stochastic mechanisms to operate successfully; it also lacks history, during which stabilizing selection could take effect.

A deterministic control system is a prerequisite of the sustainable existence of a small, closed, ecological system. If we look for a parallel in living nature, the closest will be the central nervous system, controlling a multicellular organism. The human introduces into the manmade ecosystem a regulation system and a scheme of control. This is the fundamental difference between manmade ecosystems and any natural ecosystem, and this is also a condition of its sustainability.

Human intellect must be involved not only in creating a small closed ecosystem but also in maintaining it in a sustainable state. That is why we have to recognize that by the principle of control manmade ecosystems are rather models of the noosphere, the future state of the natural biosphere.

The main result of our experiments is proof that an internally-governed biological life support system can be realized and maintained in a sustainable state within the bounds of a tightly limited space on the condition that it is obligatorily controlled by the human.

A problem arises of devising equipment whose vitality would correspond to the reliability of the biological part of the system. An ideal solution would be to invent self-repairable devices by analogy with living organisms, but this can hardly be the destiny of the nearest generations of devices. Now it must be concluded that of the two parts of the biological–technical union it is the technical part that is unreliable, and this factor will limit the lifetime of autonomous life support systems. We are emphasizing this conclusion because it is in conflict with the opinion popular among engineers, that machinery is more reliable than a fragile and unreliable living body.

Here it would be appropriate to sum up a long discussion between the proponents of physicochemical and biological life support systems. Experience shows that their development is convergent—the future is with hybrid biological–physico-chemical systems. However, this hybrid cannot be a mere summation of similar functions of the current systems.

It is our opinion that in the hybrid system the main function of the biological link will be to provide humans with regenerated products, while the physico-chemical link will maintain the operation of the biological link and recycle dead end products of the live part of the ecosystem.

Let us look at the problems of the biological part of the closed ecosystem with the human. Two principal problems stand out here. First, it is further closing of the system trophic structure. The main problem is to match the food inlet of the human link and the biomass regenerated within the system. The difficulties here are two-fold. To begin with, it has not been clearly defined yet what is an "optimal human diet," whether we know all the compounds required by a human, their quantities and proportions, whether they can replace each other, and what the minimum required diversity is. It is highly probable that the so-called balanced diet is not optimal to follow for a long time as it can result in the detraining of adaptation mechanisms. However, the most important question is if the traditional sources of human food are irreplaceable. There are ample grounds to think that the range of food animals and plants, which is narrow compared to the species diversity of the biosphere, has been selected by humans in the course of history rather for economic and technological reasons than for the unique biochemical properties of these food sources. Acknowledgement of this idea opens up a lot of opportunities for detecting new food sources among animals and plants, more fit to be intensively cultivated in a manmade ecosystem than traditional agricultural objects.

On the other hand, our reasoning has to be based on the fact that the human diet, which has formed empirically, is adequate for people's requirements. This is one of the physiological and psychological stereotypes that must be approached extremely carefully, bearing in mind a lag in the organism's response to changes in diet.

The second aspect of the food problem is the impossibility of selecting one species, a metabolic antipode to the human, that would regenerate all the food required by the human, feeding on human excreta.

We have to cultivate in the system multispecies biocenoses, which complicates its structure. The size and energetics of the system are particularly affected by the augmentation of trophic levels. To synthesize 50 g of animal protein a day, the phototrophic link must increase twice, i.e. the expenditure of the system is equal to that needed to produce the other 500 g of the primary biomass for the human. Naturally, researchers strive to make the system trophic structure as simple as possible and, first of all, to base the regeneration of all food for the human on the autotrophic processes of primary biosynthesis. Agricultural plants perform this function fairly well, regenerating carbohydrates, vegetable oils, and proteins. However, up to half or more of the biomass synthesized by plants is inedible, being a deadlock product. Ways of returning it to the material cycling are yet to be found.

A temptingly easy solution to the problem for space life support systems and for the expanding Earth's population would be to turn to a vegetarian diet, to give up eating animal food. The efficiency of agricultural production would at least double; and the possibility of living on a vegetarian diet has been proved by the experience of many people.

In our opinion, however, this cannot be recognized as an optimal way to solve the food problem either in space or on Earth, since it is founded on a considerable limitation of human physiological and psychological requirements for most people, who are not vegetarians.

Modern science is able to propose another way—to use biotechnological methods in synthesizing an absolute biochemical equivalent to the normal human diet, preserving the organoleptic properties people are used to.

The production of animal protein or its biochemical equivalent remains the most difficult problem. In terms of amino acid composition, 300–400 g of algal biomass contain sufficient amounts of all essential amino acids, including the most scarce—sulfur-containing—amino acids. In this case, however, the diet would include excess quantities of protein, to 150–200 g, nucleic acids, to 15–20 g, excessive pigments, potassium, magnesium, and sulfur. Similar products can be obtained by cultivating hydrogen-reducing or other chemoautotrophic bacteria.

By controlling the direction of biosynthesis parametrically, it is feasible to bring the composition of the produced biomass much closer to human food requirements. To illustrate, the protein content of *Chlorella* can be reduced more than twice, to 20–25%; amounts of nucleic acids, chlorophyll, and mineral substances can be also decreased. However, a gap between human requirements and the products of algal biosynthesis is still wide, and attempts to make it narrower lead to the reduced productivity of photosynthesis.

Another way is the physicochemical transformation of the primary biomass or extraction of parts that can be used for food. Development of this line has started only recently, but it may be expected that a large part of the synthesized biomass can become edible. This is the most realistic approach used today.

Bright prospects are held out by genetic engineering and selection, ways to control the program of biosynthesis so as to attain the maximum correlation between the composition of the primary produced biomass and human food requirements. The closed ecosystem offers favorable conditions for selection–genetic work. The feasibility of having 5–6 generations of wheat a year or autoselection processes going on in microalgal and bacterial continuous cultures are the factors that can permit significant directed shifts in the programs of their biosynthesis to be attained within reasonable time.

Genetic engineering offers prospects for constructing a genome of the organism that would be an absolute metabolic opposite to the human. Hopefully, it would be possible to create such a controlled system with the human that would in essence make the human an autotroph, exchanging only energy with the surroundings. However, this is still a matter for the distant future.

Limiting ourselves to the prospects attainable by currently available methods, we can state that using parametric and selection–genetic methods of controlling the direction of biosynthesis together with the physicochemical processing of the products of biosynthesis may bring us much nearer to the complete closure of the human food cycle in the system.

The second objective that became evident in experiments with the ecosystem is the maintenance of its biological stability. Besides the technical issues discussed above, two biological problems revealed themselves—trace elements and concomitant potential viruses and bacteria.

The migration of trace elements in the experimental system has turned out to be a hardly controllable process, as trace elements enter into the material cycle from system construction materials, are captured and multiply concentrated by system inhabitants in a different way depending on the body part, on the stage of life, and on the state of the organism. Note that this is not an engineering problem even though trace elements are supplied by the system's construction materials. They are not released by passive erosion, but rather due to the diverse biochemical activities of the system microflora exhibiting versatile aggressiveness. Further migration of trace elements involved in element cycling is also determined by biological processes in the system. This threat to system balance was revealed in experiments with closed systems. To find ways of controlling the dynamics of trace elements in closed ecosystems is an important objective of future investigations.

We can show the importance of bacterioflora as a possible source of system disturbance using its contribution to the mobilization of trace elements as an example. The rich gene pool of concomitant microflora constitutes a potential threat that new, unplanned processes would take place in the closed ecosystem.

In Chapter 10 we discussed the dynamics of the microflora in the ecosystem in detail and came to the conclusion that during the experiment it did not reach an absolutely steady state. This was true for human flora in particular. This proves that the system's equilibrium can be really upset by bacterial processes. Viruses may also function as a genetic nuisance, though we had no chance to see any evidence of their activity in the system. To find ways of controlling the state of concomitant microflora still remains an urgent task.

The position of microflora in the ecosystem is different from the natural situation in several ways. We would like to place special emphasis on some conclusions.

Neither the closure of the experimental ecosystem, nor its small size, nor its bacteriological isolation from the outer environment, nor the intensification of its metabolism under the conditions of extremely simplified trophic structure, nor the large flows of microflora from one link to another caused uncontrollable, destructive-for-the-system outbreaks of microorganisms, nor dangerous changes in their species composition or metabolic properties. It seems that in the small closed ecosystem, at least in the versions we have investigated experimentally, the mechanisms ensuring the stability of the microbial biota remain efficient. This renders us optimistic about the sustainability of closed life support systems.

It would be quite interesting to find out to what extent the size of the system could be reduced without losing efficiency of stochastic mechanisms ensuring stability of biocenoses. Experimental closed ecosystems consisting of microorganisms only—microcosms—are convenient objects for such research.

The four-month experiment with the system "human–higher plants" was conducted in the system protected from the invasion of microorganisms from the outside, by pressurizing the cabin with sterile air. The experiment has demonstrated that the system microflora remains essentially the same. Hence, the conditions and capacity of the system are adequate for microflora to sustain itself without any contact with the outer environment.

Another significant regularity of this manmade ecosystem is that despite intensive exchange between links no shifts in their microflora occur, and microbial landscapes remain specific for each link. Thus, the ecological factor remains the dominant one in controlling microflora in the manmade ecosystem. This circumstance is essential for substantiating the choice of the method for controlling microbiological conditions in the closed ecosystem. The choice is either to establish sterile intra-system barriers or to allow nearly free exchange limited only by hygienic measures precluding the seeding of living quarters with fungal spores and bacteria that may be hazardous to humans.

The results of the experiments argue for the ecological method of microflora control, i.e. for controlling the composition and numbers of microflora by maintaining proper environmental conditions. The alternative method—the establishment of sterile barriers between ecosystem links—is difficult to realize technically in a life support system of limited size and possibilities. Sterile barriers may also be more dangerous than an ecological balance: if the barrier is broken by accident, which can hardly be avoided, the microflora, which has been developing in isolation until this moment, can change quickly and profoundly.

Until now we have discussed an experimental closed system with human inhabitants. The demand for such systems has been created by the advance of cosmonautics, but the results of the work can be applied in different spheres, and the ecosystems for such applications need not necessarily include humans.

Investigations of the closed ecosystem may be of theoretical significance because it is a fundamentally new formation, possessing the properties of a biocenosis in terms of species composition, the properties of the organism in terms of the principle of control, and, finally, an absolutely unique property of exchange closure, which has no analog in nature, except for the biosphere as a whole. This object is exceptionally interesting for studying some fundamental properties of living systems. The closed ecosystem, a microworld created by the experimenter, offers an opportunity to reveal the regularities of the existence of closed ecological systems as models of the Earth's biosphere, to define the limits of sustainability, and to observe the transition processes of restoration after the system has been disturbed. The rate of cycling in these ecosystems can be thousands of times higher than in the biosphere. This affords new opportunities for experimental ecology and physiology, for modeling genetic-population and evolutionary processes in the small biosphere, every part of which is transparent, visible, and accessible for analysis and interference. The introduction of the human into the closed ecosystem makes it possible to reveal all human metabolic relations with the environment, including vanishingly small ones, the effect of which can be seen only if the cycles are repeated many times. These opportunities must be primarily used to investigate the physiology and biochemistry of the exometabolism of both normal and sick humans and, probably, to subject them to curative treatment. Those are some research opportunities of experimental ecosystems.

However, in spite of the importance of the objectives, only five closed biological human life support systems have been created by now or are being created: two in Russia—the Ground Experimental Complex in Moscow and the Bios in Krasnoyarsk; one in the NASA Johnson Center—BIO-Plex; one in Europe (Spain—MELISSA, ESA); and one in Japan (CEEF). It would not make sense to construct more of such systems in the near future as they are all very costly, experiments with them are complicated, and information is accumulated slowly. It may be more expeditious to develop miniature closed ecosystems—mini-CES. The International Center for Investigations of Closed Ecological Systems of the Institute of Biophysics has devised such a desk-top CES, a one-tenth scale model of the full-sized CES with a human, but also including the human metabolism. The rate of data accumulation in the CES depends more on the number of material cycles in it than on the length of the period of operation in terms of solar time. Therefore, such a mini-system as a research tool is as many times more efficient than a full-scale system as the rate of the cycle in it is higher than the rate of the cycle in the full-scaled system. The cost of manufacturing and operating such a system is tens, or even hundreds, of times lower than that of the full-scale system. It can be easily cloned and used as a research tool in many laboratories. We may say that in our field we respond to the call of D. Golden, the Head of NASA Administration, to work to make space research "better, quicker, cheaper." In the present-day situation, when it is not only Russian science that lacks material support, this is of great importance.

It would be reasonable to keep full-scale CES as unique centers for periodic, probably not very frequent, key experiments that would demand the presence of humans in the system. The world-wide significance of such experiments and their high cost make it imperative that centers possessing such experimental systems should be officially recognized as open laboratories—international centers. The Center for Investigations of Closed Ecosystems of the Institute of Biophysics, Siberian Branch, Russian Academy of Sciences, in Krasnoyarsk, has the status of an international center and is open to scientists from all over the world.

Quarantine measures on the planets and other celestial bodies to be visited by humans can be better provided for by closed life support systems than by open ones. The closed ecosystems will not release into the outer environment any living material, particularly microorganisms, capable of autocatalytic colonization of space, and, therefore, will not disturb the intact state of the planet. This is the way to protect nature from destructive consequences, which often accompanied the expansion of civilization in the past.

A warning from the past can be provided by the example of Europeans visiting the isolated islands of Oceania and introducing infections against which the aborigines had no immunity and, therefore, died. Of course, we cannot seriously expect a meeting with anthropomorphous inhabitants of other worlds, but even microbial contamination of the celestial bodies could be rather dangerous and, anyway, undesirable.

This objective of cosmonautics can be defined as the creation of a system optimally providing the human requirements with minimum expenditures and without environmental pollution. But isn't it the most vital current problem of humanity on Earth? The similarity of these problems lends great importance to the creation of manmade ecological systems for humans.

It seems unlikely that anyone can predict now when we are going to need closed life support systems outside Earth, but they can find terrestrial applications even now.

For people living in the most severe conditions on our planet it may be more reasonable to design essentially closed ecosystems and maintain the optimal environment inside them than to condition the environment in the open habitat. For northern settlements systems of cascade energy use can be of interest.

According to the project of ecological habitation for northern territories devised at the Institute of Biophysics, an energy cascade with an initial power of 12 kW could provide for a house inhabited by a family of 5–7 members with lighting and heating, as well as supply people during the northern winter and even polar night with fresh vegetables and berries; it will also absorb human wastes without polluting the environment (Bartsev *et al.* 1996d).

Another promising application of closed ecosystems is related to the Sick Building Syndrome (SBS). This is a serious problem not only for such completely sealed products of high technology as spacecrafts and submarines but also for considerably less sealed living and working spaces, where people inhabiting medium and northern latitudes spend the major part of their lives.

The problem has unexpectedly become much more acute after the energy crisis of the 1970s, caused by irregular supplies of oil. In an attempt to reduce the energy expenditures for the heating and conditioning of buildings, additional sealing was done. The better it was performed technically, the higher was the probability of SBS in the inhabitants of these buildings. A flow of papers dealing with the SBS problem ensued, but the mechanism of its origination has not been determined exactly. Against this background the results of our experiments in the Bios-3, which revealed no signs of SBS in our test subjects who stayed for months in the sealed space regenerated by plants, are very important and can be used to overcome SBS in modern highly sealed energy-efficient buildings.

Having covered some distance, it may be well to remember the start.

In the chapter “Ecology of Spaceflight” written by G. Cooke for the famous book “Fundamentals of Ecology” by E. Odum, published several times in the 1970s (Odum, 1971), the author analyzes the prospects of creating biological life support systems, which at that time were still to be devised. G. Cooke pinpointed the difficulties and obstacles to be overcome on the way to creating a biological system that regenerates the environment for a human. He based his opinion on his experience as an ecologist well aware of the properties of open, non-closed ecological systems.

We will quote G. Cooke here and see if his predictions have come true. Let us compare Cooke’s forecasts with later experimental results.

1. The system with microalgae is difficult to control.

“The experiment of Eley and Myers (1964), who closed an algae–mouse system for gas exchange for 82 days, represents one of the longest couplings of a mammal and a support organism. On the basis of ecological theory, one can predict that such oversimplified systems, much like early "bloom" stages in nature, will be difficult to stabilize."

As has been shown in our long-duration experiments, the system is reliable and rather simple to control. Control takes not more than a quarter of the energy provided by photosynthesis.

The reason is that intraorganismal processes form a steadily autocorrelated metabolic network, and if, measuring them, we can make the right choice of one parameter to be monitored, we will be able to control the whole system. For a microalgal population, this parameter is the growth rate, which can easily be measured continuously by an increase in the optical density. Knowing the correlation between the biosynthesis rate and the requirement for biogenic elements, measurements of growth rate will suffice to maintain steady-state conditions for algal cells in continuous culture for an indefinitely long time.

Another important factor of control is the closure of the system. The two links of the system—human and microalgae—consume each other's gaseous exometabolites, exchanging O_2 and CO_2 in respiration and photosynthesis, and thus controlling the photosynthetic rate by limiting CO_2 inflow. So, a feedback system is established between the human and microalgae. Under the conditions of a closed gas cycle, microalgal productivity is regulated through this system.

This method of regulation has been verified in our experiments for many years, in which thousands of microalgal generations succeeded each other in continuous culture, providing humans with air for respiration and showing no signs of inhibition or degeneration.

2. *Algal culture is unreliable.*

"A number of difficulties have been encountered during experiments with algae–mammal couplings; among them have been invasion of competitors or predators (bacteria or zooplankton), accumulation of algal toxins, and the presence of algal mutants with lowered rates of metabolism. These problems are to be expected of an oversimplified ecosystem without adequate internal controls."

We recorded none of the predicted phenomena. In our opinion, the reason is that in our experiments we did not deal with axenic culture. It was a biocenosis with symbiotic relations between the algae and the bacteria.

As for predators and competitors, there is no way that they can be introduced into the closed system. Finally, the main complication predicted—spontaneous emergence of slowly growing mutants—has been combated by continuous turbidostat cultivation. Under such conditions, similar to natural selection in essence, quick-growing cells and their progeny have a higher chance to survive than "slow" mutants, which will be naturally washed out of the cultivator, having no chance to multiply and become dominant in the cultivator. Darwinian selection is inexorable. Experiments described in Chapter 6 demonstrate the efficient action of this mechanism in the event of UV-injured *Chlorella*. Even if for some unpredictable biological or (much more likely) technical reasons the whole algal population does die, only a phenotype will disappear. It is not very difficult to restore the population by the conserved genotype. The necessary information is built into each *Chlorella* cell (or any other organism) measuring only 30 μm. To be on the safe side, the gene pool can be stored in many replications.

3. *Human–algal gas exchange cannot be steadily balanced due to a difference between RQ and AQ.*

"No investigation to date has shown that the two-species ecosystem can maintain an AQ/RQ balance for long periods, even with the use of great numbers of external controls. ... The basic objection to the use of a two-species (man and support organism) life-support system is thus one of reliability."

It has been shown in Section 2 of Chapter 9, that a balance between algal AQ and human RQ is achievable on the basis of the steady controllable growth of algae on a balanced nutrient medium and the correction of the human diet.

In conclusion it should be said that G. Cooke's predictions advanced a priori were quite legitimate and founded on the thorough ecological analysis of the situation, based on the knowledge available by that time. This knowledge led him to the pessimistic conclusion that use of such simple ecosystems imposes severe hazards on the astronaut, and their development as the basis of a life support system should be discontinued.

Nevertheless, we did not give up this work almost 40 years ago, and now, a posteriori, we see that these forecasts have not come true. By investigating the functioning of such systems, we have verified their reliability experimentally and, what is important, we know which stable biological mechanisms act in these systems. They are described in many parts of this book. Our conclusion, supported by this knowledge and its experimental verification, is the opposite to G. Cooke's conclusion, and it is quite optimistic: the creation of closed ecological human life support systems, beginning with the simplest one—with human–microalgal gas exchange, is feasible, and they are at least as reliable as physicochemical systems performing the same function. And they have already been tested. Methods of control developed for them proved to be applicable to higher-plant cultivation. On this basis we devised mulispecies systems, such as Bios-3, which can be efficiently controlled without enormous expenditures. This opens up possibilities for their space applications. Our confidence has been essentially supported by the results of recent experiments on the *Mir* station performed by the joint Russian–American research team. They have shown that plants (wheat, at least, which is most important) can grow almost normally in microgravity, if the technologically controlled environment is suitable (Salisbury *et al.*, 1995).

However, neither future space flights nor the potentials of terrestrial applications of manmade closed ecosystems seem to us so significant as their application as teaching aids demonstrating the vital importance for humans of preserving the biosphere. We started our book off with the discussion of this problem, and we would like to finish it in the same way.

At the same time, the last thing the authors want is that their warnings about the destiny of the biosphere could be understood by readers as unrestrained "green" alarmism. On the contrary, we see the way out of the collision between humanity and the biosphere, which has produced it, in using scientifically substantiated methods for controlling the biosphere. An unchecked advance of scientific knowledge in the twentieth century gives us reasons to hope that the twenty-first century can become the period of harmonization of relations between humans and the biosphere. However, this can be only attained on the condition that humanity would realize soon enough the inadmissibility of further wastage of huge material and intellectual resources for the satisfaction of political ambitions, for national clashes, for wars between each other and with nature. Only a coordinated strategy of interaction with nature, pursued by humanity as a whole, can ensure the transition from the impending danger of the chaotic exploitation of the biosphere by humans to a controlled noosphere. It is hardly possible to find a more vital and universal objective of present-day science than to comprehend the mechanism of the everlasting existence of the biosphere and to propose to society a scientifically substantiated strategy for controlling the biosphere with the aim of preserving this mechanism. This knowledge must become the foundation of our outlook on the world for the generations to come.

The people responsible for strategic political decisions must realize that now, in the twenty-first century, we are in a fundamentally new situation: relations between humanity as a whole and the biosphere rather than human interrelations are becoming critical for the future of both humanity and the biosphere.

It is essential to have a clear understanding of this new problem and it is essential to have a new strategy for further technological development under the conditions of the limitation of the material cycle by biospheric processes. First of all, it is essential to acquire new knowledge, it is essential that the best intellects strive to discover the conditions that limit sustainability and to discover the laws of biosphere functioning, and finally to formulate the principles of control at the stage of the noosphere. As for statesmen, they should have the strong political will to subordinate the interests of one nation to the dominant task common for all humanity—to preserve the biosphere and the humans in it.

We prefaced our book with the epigraph—a quotation from V. Vernadsky. At the end we also want to refer to this great prophet who foresaw the problem encountered by humanity half a century after his death and who found the principal problem of its development. Shortly before his death (he died on January 6, 1945), Vernadsky wrote: “Humanity as a whole is growing into a powerful geological force. And now humanity, with its intellect and labor, faces the problem of restructuring the biosphere for the benefit of freethinking humanity as a single whole. This new state of the biosphere, which we approach unaware of it, is the noosphere.” The authors will be more than adequately rewarded if their work of thirty years contributes, even slightly, to the advance of humanity in this direction.

REFERENCES

A

Aach, H., 1955, *Kosmos* (BRD), **51**, No.8.

Abakumova, I.A. and Fofanov, V.I., 1972, Comparative characterization of biological value of unicellular algae—candidates to be included in closed ecological systems. In *Proceedings of VII All-Union Workshop on Material Cycling in the Closed System,* edited by L. Rubenchik, (Kiev: Naukova dumka), pp. 3–5.

Abakumova, I.A., Akhlebininsky, K.S., Bychkov, K.S., Demochkina, N.G., Kondratyev, Yu.I. and Ushakov, A.S., 1965, Some data for the animal link in the closed ecological system. In *Problemy kosmicheskoi biologii* (Problems of space biology), (Moscow: Nauka), **4**, p. 107.

Abdulaev, D.A., 1971, Toad spit in the diet of agricultural animals. In *Kultivirovaniye vodoroslei i vysshikh vodnykh rastenii v Uzbekistane* (Cultivation of algae and aquatic higher plants in Uzbekistan), (Tashkent), p. 139.

Adamovich, B.A., 1975, Life support systems for crews in short-term and medium-term flights. In *Osnovy kosmicheskoi biologii i meditsiny* (Fundamentals of space biology and medicine), Vol. 3, (Moscow: Nauka), pp. 231–249.

Aeronautics and Space Report of the President. Fiscal Year 1993 Activities, 1994, NASA, Washington, DC 20546, 131 pp.

Agre, A.L., Ivanov, V.M. and Trukhachev, V.N., 1966, On possibility of mineralization of water–fecal mixture by wet oxidation. In *Problemy kosmicheskoi meditsiny* (Problems of space medicine), (Moscow: Meditsina).

Aliev, E.A., 1966, Nutrient solutions for growing tomatoes on artificial substrates. *Agrokhimiya*, **3**, p. 95.

Allen, J., 1991, *Biosphere-2, The human experiment*, (Penguin Books, Synergetic Press, Tucson), 156 pp.

Andre, M. and MacElroy, R.D., 1990, Plants and man in space: A new field plant physiology. *Physiologist*, Vol. 33, suppl. 1, pp. 100–101.

Artsikhovsky, V., 1915, On "air cultures" of plants. In *Izvestiya Alekseevsko-Donskogo Politekhnicheskogo Instituta* (Writings of the Alekseevsko-Donskoi Technical University), IV, otd. II.

Ashida, A. and Nitta, K., 1995, Construction of CEEF [Closed Ecology Experiment Facility] is just started. *SAE Technichal Paper Series 951584.*

Averner, M.M., 1981, An approach to the mathematical modeling of a controlled ecological life support system. *NASA-CR-166331.* (NASA Ames Research Center, Moffett Field, Calif.), 77 pp.

B

Balin, B.A., Zarudny, L.B., Lobanov, A.G., Pechurkin, V.A., Sinyak, G.S. and Shorin, S.N., 1967, On devising a unit for thermal waste processing in the life support system on the basis of material cycling. In *Problemy sozdaniya zamknutykh ekologicheskikh sistem* (Problems of creating closed ecological systems), (Moscow: Nauka), p. 144.

Ballard, R.W. and MacElroy, R.D., 1971, Phosphoenolpyruvate. A New Inhibitor of Phosphoribulokinase in *Pseudomonas facilis, Biochemical and Biophysical Research Communications*. **44(3)**, pp. 614–618.

Ballod, A.A., Yegorova, L.E., Koloskova, Yu.S., Mironov, V.A., Nefedov, Yu.G., Petrov, L.P., Savina, V.P., Uspenskaya, V.A. and Chizhov, S.V., 1967, Investigation of methods of water regeneration from the atmospheric moisture condensate by the sorption method. In *Problemy sozdaniya zamknutykh ekologicheskikh sistem* (Problems of creating closed ecological systems), (Moscow: Nauka), p. 113.

Baranov, S.A., 1967, On negentropic tendency in evolution of natural and manmade cenoses. In *Problemy sozdaniya zamknutykh ekologicheskikh sistem* (Problems of creating closed ecological systems), (Moscow: Nauka), p. 217.

Barashkov, G.K., 1963, *Khimiya vodorosley* (Algal chemistry), (Moscow: Izdatelstvo AN SSSR).

Barta, D.J. and Henninger, D.L., 1994, Regenerative Life Support Systems—Why Do We Need Them? *Advances in Space Research*, **14(11)**, pp. 403–410.

Barta, D.J. and Henninger, D.L., 1996, Johnson Space Center's Regenerative Life Support Systems Test Bed. *Advances in Space Research*, **18(1/2)**, pp. 211–221.

Barta, D.J. and Henderson, K., 1998, Performance of Wheat for Air Revitalization and Food Production During the Lunar–Mars Life Support Test Project Phase III Test. *SAE Technical Paper Series 981704*.

Bartosh, 1965, Intensity of photosynthesis of the cultures of *Scenedesmus quadricauda*, in relation to ambient conditions. In *Izucheniye intensivnoi cultury vodoroslei* (Study of intensive algal culture), (Prague), pp. 137–142.

Bartsev, S.I. and Okhonin, N.S., 1993, Optimization and monitoring needs: possible mechanisms of control of ecological systems. *Nanobiology*, **2**, pp. 165–172.

Bartsev, S.I. and Okhonin, V.A., 1994, On the way to biospherics (2): theory of ecological similarity as obligatory tool for applicable CBLSS research and design. *Abstracts of 30th COSPAR Scientific Assembly* (Hamburg, Germany), 11–21 July, F4.3, p. 331.

Bartsev, S.I. and Okhonin, V.A. 1999, Self-restoration of biocomponents as a mean to enhance biological life support systems reliability. *Advances in Space Research*, Vol.24, **3**, pp. 393–396.

Bartsev S.I., Gitelson, I.I., Lisovsky, G.M., Mezhevikin, V.V. and Okhonin, V.A., 1997, Perspectives of different type biological life support systems (BLSS) usage in space missions. *Acta Astronautica*, Vol.39, **8**, pp. 617-622.

Bartsev, S.I., Gitelson, I.I., Mezhevikin, V.V., Okhonin, V.A., 1996a, Principle of lunar base life support system structure and operation regimes optimization. *SAE Technical Paper Series No. 961551,* pp. 1–6.

Bartsev, S.I., Gitelson, I.I., Mezhevikin, V.V., Okhonin, V.A., Doll S.C. and Rao, N.S., 1996b, First Level of Life Support System (LSS) Closure: Optimization of LSS Structure for Different Functioning Times. *SAE Paper 961556*, 26th ICES.

Bartsev, S.I., Mezhevikin, N.S., Okhonin, N.S., Doll, N.S. and Rao, N.S., 1996c, Life Support System (LSS) Designing: Principle of Optimal Reliability. *SAE Paper 961365*, 26th ICES.

Bartsev, S.I., Mezhevikin, V.V. and Okhonin, V.A., 1996d, BIOS 4 as an embodiment of CELSS development conception. *Advances in Space Research*, Vol.18, **1/2**, pp. 201–204.

Batov, V.A., 1967, Temperature effect on the intensity of *Chlorella* biosynthesis under different lighting levels. In *Nepreryvnoye upravlyayemoye kultivirovaniye mikroorganismov* (Continuous controlled cultivation of microorganisms), (Moscow: Nauka), pp. 144–152.

Bazanova, M.I., 1969, Mineral exchange of the biological system "human–*Chlorella*." *D Sc Thesis*, Krasnoyarsk.

Bazanova, M.I. and Kovrov, V.A., 1968, Dynamics of microelements in the ecosystem "human–microalgae." In *Proceedings of the V Workshop on Material Cycle in the Closed System,* (Kiev: Naukova Dumka), p. 13.

Behrend, A.F. and Henninger, D.L., 1998, Baseline crops for advanced life support program. *NASA Johnson Space Center Memorandum EC3-98-066*, (Houston, Texas).

Belyanin, V.N., Sidko, F.Ya. and Eroshin, N.S., 1964, Light distribution in dense alagal suspensions. In *Upravlyayemoye kultivirovaniye mikrovodoroslei* (Continuous microalgal cultivation), (Moscow, Nauka), p. 24.

Bently, M., 1965, *Promyshlennaya gidroponika* (Industrial hydroponics), (Moscow: Kolos).

Berkovich, E.M., 1964, *Energeticheskii obmen v norme i patologii* (Energy exchange in norm and pathology), (Moscow).

Beyers, R.J., 1963, A characteristic diurnal metabolic pattern in balanced microcosms. *Publications of Institute of Marine Sciences University of Texas,* Vol. 9.

Beyers, R.J., 1963, Balanced aquatic microcosms. Their application for space travel. *American Biology Teacher*, Vol. 34.

Blackwell, A.L., 1990, A perspective on CELSS control issues, *Controlled ecological life support systems: CELSS-1989 workshop (NASA TM-102277),* pp. 327–353.

Blackwell, C. and Gitelson, J.I., 1997, Improving the quality of human performance during space exploration missions via control, *Human Control.*

Bluem, V., 1998, The closed equilibrated biological aquatic system project. *Institute of Environmental Science*. July 21–23, 1998 Conference Rokkasho, Japan.

Bluth, B.J. and Helppie, J.I., 1986, *Soviet space stations as analogs.* 2nd ed. Report under NASA Grant NAGW-659, Wash. D.C. NASA.

Bolsunovsky, A. and Zhavoronkov, V., 1996, Prospects for using microalgae in life-support systems, *SAE Technical Paper Series 951496.*

Bongers, L. and Kok, J.I., 1964, *Developments of Industrial Microbiology*, **5**, p. 182.

Bowman, N.I., 1953, The food and atmosphere control problem in Space vessels. Pt. 2. The use of algae for food and atmosphere control. *Journal of British Interplanetary Society,* Vol. 12, pp. 159–167.

Boysen-Iensen, 1932, *Die Stoffproduktion der Pflanzen.* (Jena).

Brasseaux, S.F., Cornwell, J., Dall-Bauman, L., Henninger, D., Laws, B., Ming, D., Verostko, C., Bourland, C., Real, P.O. and Hurbert, K., 1997, Lunar–Mars Life Support Test Project, Phase II Final Report, *JSC-38800.*

Brasseaux, S.F., Graf, J.C., Lewis, J.C., Meyers, J.C., Rosenbaum, J.C. and Supra, J.C., 1998, Performance of the Physicochemical Air Revitalization System During the Lunar–Mars Life Support Test Project Phase III Test, *SEA Technical Paper Series 981703.*

Brockmann, M.and Henick, J.C., 1958, *Food Technology*, **12**, p. 454.

Bubenheim, D.L., 1991, Plants for water recycling, oxygen regeneration and food production. *Waste Management and Research,* pp. 435–443.

Bubenheim, D.L. and Wignarajan, K., 1995, Incineration as a method for resource recovery from inedible biomass in a CELSS. *Life Support & Biosphere Science*, Vol. 1, **3/4**, pp. 129–140.

Bubenheim, D.L. and Wignarajah, K., 1997, Recycle of inorganic nutrients for hydroponic crop production following incineration of inedible biomass. *Advances in Space Research*, Vol. 20, pp. 2029–2035.

Bubenheim, D.L. and Lewis, C., 1997, Application of NASA's advanced life support technologies in polar regions. *Advances in Space Research*, Vol. 20, **10**, pp. 2037–2044

Budyko, M.I., Ronov, A.B. and Yanshin, A.L., 1985, *Istoriya atmosfery* (History of atmosphere), (Leningrad, Gidrometeoizdat).

Bugbee, B.G. and Salisbury, F.B., 1988, Exploring the limits of crop productivity. *Plant physiology*, Vol. 88, pp. 869–878.

Bugbee, B.G. and Salisbury, F.B., 1989, Controlled environment crop production: hydroponic vs. lunar regolith. In *Lunar base agriculture: soil for plant growth.* (Madison), pp. 107–127.

Bugbee, B., Koerner, G., Albrechtsen, R. *et al.*, 1996, "USU-APOGEE," a new high-yielding dwarf wheat cultivar for life-support systems. *Abstracts, 31st Scientific Assembly of COSPAR*, (University of Birmingham, England).

Burnazyan, A.I., Parin, V.V., Nefedov, Yu.G. and Adamovich, Yu.G., 1969, Year-long medical–technical experiment in the ground life support system. *Cosmicheskaya Biologiya i Meditsina,* **1**, p. 9.

Bychkov, V.P., Kondratyev, Yu.I. and Ushakov, A.S., 1967, Investigation of unicellular algae as a possible food source. In *Problemy sozdaniya zamknutykh ekologicheskikh sistem* (Problems of creating closed ecological systems), (Moscow: Nauka), p. 52.

C

Calloway, D.H. and Margen, S., 1968, Investigation of the nutritional properties of *Hydrogenomonas eutropha, NASA-CR-111599.*

Canham, A.E., 1964, Electricity in horticulture, *The Macdonald Technicians and Grafts series.* (London).

Cao, W. and Tibbits, A.M., 1991, Physiological responses in potato plants under continuous irradiation. *Journal of American Society for Horticultural Science*, **116 (3)**, pp. 525–527.

Cao, W. and Tibbits, T.W., 1993, Growth and carbon assimilation in potato plants as affected by light fluctuations. *Horticultural Science*, **28(7)**, p. 748.

Cao, W., Tibbits, T.W. and Wheeler, R.M., 1994, Carbon dioxide interactions with irradiance and temperature in potatoes. *Advances in Space Research*, Vol. 14, **11**, pp. (11)243–(11)250.

Chetverikov, S.S., 1905, *Volny zhizni* (Waves of life). *Journal of Zoological Department of Emperor's Society of Natural Scientists*, Vol. 3, **6**, pp. 103–105

Chizhov, S.V. and Sinyak, Yu.E., 1973, Provision of spaceship crews with water. In *Problemy kosmicheskoi biologii* (Problems of space biology), **24** (Moscow, Nauka).

Chuchalin, A.P., 1980, *Radiatsionnyi rezhim i ispolzovaniye FAR tsenozon pshenitsy v usloviyakh intensivnoi svetokultury* (Radiation regime and use of PAR by wheat cenosis under intensive light cultivation), *D Sc Thesis*, Krasnoyarsk, Institute of Physics.

Chuchkin, V.G., 1967, Spiral greenhouse, *Priroda*, **10**, p. 58.

Chuchkin, V.G., Kostetsky, A.V., Golovin, V.P., Arbuzova, V.P., Rozhdestvensky, V.I. and Tsvetkova, V.I. 1975, Evaluation of the issue of including the higher plant link in closed life support systems. In *Problemy sozdaniya biologo-tekhnicheskikh sistem zhizneobespecheniya cheloveka* (Problems of creating biological–technical human life support systems), edited by I.I. Gitelson (Novosibirsk: Nauka), pp. 5–13.

Clemedson, C.J., 1959, Toxicological aspect of the sealed cabin atmosphere of space. *Astronautik*, **1**, p. 4.

Combs, G., 1952, *Science*, **116**, p. 453.

Controlled Ecological Life Support System: Transportation Analysis, 1982, NASA-CR-166420, November.

Cooke, G.D., 1967, The pattern of autotrophic succession in laboratory microecosystem. *BioScience*, Vol. 17.

Cooke, G.D., 1971, Ecology of space travel. In *Odum E. Fundamentals of ecology*, 3rd ed., (Philadelphia: Saunders College Publishing).

Cooke, G.D., Beyers R.J. and Odum, E.P., 1968, The case for the multispecies ecological system, with special reference to succession and stability. *Bioregenerative Systems,* NASA Special Publication, Vol. 165.

Corey, K.A. and Wheeler, R.M., 1992, Gas exchange in NASA's biomass production chamber. *BioScience,* Vol. 42, **7**, pp. 503–509.

Curuc, R., 1967, *Isotopes plant nutrition and physiology*, (Vienna).

D

Dadykin, V.P., 1968, Use of higher plants for regeneration of food, water and atmosphere in closed systems. *Selskokhozyaistvennaya biologiya*, **3 (1)**, p. 137.

Dadykin, V.P., 1968a, *Kosmicheskoye rastenievodstvo* (Space plant-growing), (Moscow: Znaniye).

Dadykin, V.P., 1970, Some problems of space plant-growing. In *Proceedings of the 4th meeting on development of K.E. Tsiolkovsky's scientific legacy. 1969 Group of problems of space biology and medicine,* (Moscow), p. 26.

Dadykin, V.P., 1970a, Contribution of V.N. Sukachev's studies on biogeocenosis to development of closed systems. *Izvestiya AN SSSR, biology,* **2**, p. 229.

Dadykin, V.P., 1976, Some underdeveloped problems of space plant growing. In *Proceedings of the IX meeting devoted to development of K.Tsiolkovsky's ideas (Problems of space biology and medicine),* (Moscow), pp. 173–186.

Dadykin, V.P. and Nilovskaya, N.T., 1969, Problems of creating phytocenoses for manmade ecological systems. In *Abstracts of All-Union Meeting on Investigation of Plant Interactions in Phytocenoses,* (Minsk), p. 9.

Dadykin, V.P. and Nikishanova, T.I., 1969a, On the possibility of using sweet potato as the main source of carbohydrates in human diet during the space flight. *Kosmicheskaya biologiya i meditsina,* **1**, pp. 59–63.

Dadykin, V.P., Lebedeva, E.V. and Dmitrieva, I.V., 1967, Vegetable conveyer on hydroponics. *Vestnik selskokhozyaistvennoi nauki,* **7**, p. 65.

Dadykin, V.P., Stepanov, L.N. and Ryzhkova, V.E., 1967, Some data on volatile (oxygen-containing) emissions of some vegetable crops. *Kosmicheskaya biologiya i meditsina,* **6**, p. 48.

Davtyan, G.S., 1967, Investigations in the field of hydroponics. In *Soobshcheniya Instituta agrokhimicheskikh problem i gidroponiki* (Reports of the Institute of agrochemistry and hydroponics), **7** (Yerevan), p. 3.

De Chardin P. Teihard, 1943, Geobiologie et Geobiologia. *Geobiologia,* Vol. 1, pp. 1–5.

De Chardin P. Teihard, 1987, *Human phenomenon* (Russian translation), (Moscow: Nauka).

Dillon, J.C. and Phan, P.A., 1993, *Spirulina as a source of proteins in human nutrition. Spirulina, algae of life,* edited by F.Doumenge *et al.*, Monaco Musee Oceanographique, Bull., **12**, pp. 103–108.

Dreschel, T. and Sager, J., 1989, Control of water and nutrients using a porous tube: a method for growing plants in space. *Horticultural Science,* Vol. 24 (**6**), pp. 944–947.

Drigo, Yu.A., Ivanov, V.M., Kuznetsov, S.O., Novikov, V.M., Sinyak, Yu.E., Trukhachev Yu.E. and Farafonov, N.S., 1967, Mineralization of human and plant wastes by wet oxidation. In *Problemy sozdaniya zamknutykh ekologicheskikh sistem* (Problems of creating closed ecological systems), (Moscow: Nauka), p. 150.

E

Eckart, P. *Spaceflight Life Support and Biospherics,* 1996, (Micro Press and Kluwer Academic Publishers).

Edeen, M.A. and Barta, D., 1996, Early Human Testing Initiative Phase I Final Report, *JSC 33636.*

Edeen, M.A. and Pickering, K.D., 1998, Biological and Physical–Chemical Life Support Systems Integration—Integration Results of the Lunar Mars Life Support Phase III Test, *SAE Technical Paper Series 981708.*

Edeen, M.A. *et al.*, 1998, Lunar–Mars Life Support Test Project Phase III Final Report, *JSC 39114*.

Eley, J.H. and Myers, J., 1964, Study of a photosynthetic gas exchanger: A quantitative repetition of the Priestly experiment. *Texas Journal of Science*, Vol. 16.

Eroshin, N.S., Sidko, F.Ya. and Belyanin, V.N., 1964, On light distribution in *Chlorella* suspension illuminated by different sources of radiant energy. In: *Upravlyayemoye kultivirovaniye mikrovodoroslei* (Controlled microalgal cultivation), (Moscow, Nauka), p. 28.

F

Feoktistova, O.I., 1965, Investigation of seasonal periodicity in *Chlorella* development, depending on cultivation conditions. *Fiziologiya rastenii*, **12 (5)**, p. 888.

Feynman, R.P., 1989, Appendix F: Personal Observations on the Reliability of the Shuttle. In: *What do you care what other people think*, (Bantam books), pp. 220–237.

Fisher, J.W., Pisharody, S., Wignarajah, K., Lighty, J.S., Burton, B., Edeen, M., Davis, K.A. and Finger, B.W., 1998, Waste Incineration for Resource Recovery in Bioregenerative Life Support Systems, *SAE Technical Paper Series 981758*.

Flynn, M. and Borchers, B., 1997, The influence of power limitations on closed environment life support system applications, *SAE Technical Paper Series 972356*.

Folsom, C.E., 1982, *Origin of microcosm life* (in Russian) (Moscow: Mir).

Folsom, C. E. and Hanson, J.E., 1986, The emergence of materially closed system ecology. In *Ecosystem Theory and Application*. (J. Wiley and Sons, LTD), pp. 269–288.

Ford, M. and Thorne, G., 1967, *Annual Botany*, **31 (124)**, p. 629.

Forrester, J.W., 1971, *World Dynamics* (Cambridge, Massachusetts: Wright–Allen Press, Inc.).

Foster, J.F. and Litchfield, J.H., 1967, Engineering requirements for culturing of *Hydrogenomonas* bacteria, *NASA Technical Reports: NASA-CR-90111*.

Friend, D., 1964, *The growth of cereals and grasses*, (London), 181 pp.

Fundamentals of Space Life Sciences, 1 and 2, edited by Churchill, S., 1997, (Malobar Florida: Krieger Publ. Co.), 335 pp.

G

Gafford, R.D. and Craft, C.E., 1959, A photosynthetic gas exchanger capable of providing for respiratory requirement of small animals. *USAF SAM Report*, (Texas: Brooks AFB), pp. 58–124.

Galkina, N.V., 1964, Toad spit as forage plant. *Uzbekskii biologicheskii zhurnal*, **1**.

Galkina, T.B., Kashkovsky, I.N., Kurapova, O.A. *et al.*, 1967, Some characteristics of growth and gas exchange of the alga *Anacistis* in intensive culture. In *Problemy kosmicheskoi biologii* (Problems of space biology), 7, (Moscow: Nauka), p. 480.

Galloway, D.H. and Kumar, A.M., 1969, Protein quality of the bacterian *Hydrogenomonas eutripha. Applied Microbiology*, 17, **1**, p. 176.
Gazenko, O.G., 1968, Space biology and medicine. In *Uspekhi SSSR v issledovanii kosmicheskogo prostranstva* (USSR's advance in space exploration), (Moscow: Nauka), p. 321.
Genin A.M., 1964, Some principles of forming artificial habitat in spaceship cabins. In *Problemy kosmicheskoi biologii* (Problems of space biology), **3**, (Moscow: Nauka), p. 59.
Genkel, P.A., 1954, *Soleustoichivost rastenii i puti ee napravlennogo povysheniya* (Plant resistance to salinity and ways to its purposeful increase), (Moscow: Izdatelstvo AN SSSR).
Gitelson, J.I. and Okladnikov, G.M., 1994, Man as a Component of a Closed Ecological Life Support System. *Life Support & Biosphere Science*, Vol.1, **2**, pp. 73–81.
Gitelson, J.I. and Okladnikov, Yu.N., 1997, Human functions in a biological life support system, *SAE Technical Paper Series 972513.*
Gitelson, I.I., Terskov, I.A. and Batov, V.A., 1964, Automation of cultivation of unicellular organisms to be used in the closed biological system. In *Problemy kosmicheskoi biologii* (Problems of space biology), **3**, (Moscow: Nauka), p. 427.
Gitelson, I.I., Terskov, I.A., Kovrov, B.G., Voitovich Ya.V. and Sadikova, G.I., 1964, On forms of *Chlorella* nitrogen nutrition under continuous cultivation. In *Upravlyayemoye kultivirovaniye mikrovodoroslei* (Controlled microalgal cultivation), (Moscow: Nauka), p. 47.
Gitelson, I.I., Lisovsky, G.M., Terskov, I.A., 1966, On competitive ability of algal culture versus higher plants. In *Upravlyaemyi biosintez* (Controlled biosynthesis), (Moscow: Nauka), pp. 68–75.
Gitelson, I.I., Kuzmina, R.I. and Bazanova, M.I., 1967, *Chlorella* requirement for biogenous elements and effect of their concentration in the background medium on biosynthesis rate. In *Nepreryvnoye upravlyayemoye kultivirovaniye mikrovodoroslei* (Continuous controlled microalgal cultivation), (Moscow: Nauka), p. 126.
Gitelson, I.I., Kirensky, L.V., Terskov, I.A., Lisovsky, G.M., Kovrov, B.G., Sidko, F.Ya., Okladnikov, Yu.N., Antonyuk, M.P., Belyanin, V.N. and Rerberg, M.S., 1970, Experimental biological life support system on the basis of continuous microalgal cultivation , and experiment with human stay in this system for many days. *XIX International Astronautical Congress,* **4**, (Bioastronautics. Pergamon Press), p. 91.
Gitelson, I.I., Fish, A.M., Chumakova, A.M. and Kuznetsov, A.M., 1973, The maximal rate of bacterial reproduction and feaasibility of determining it. *Doklady Akademii Nauk SSSR*, Vol. 211, **6**, pp. 1453–1455.
Gitelson, I.I., Kovrov, B.G., Lisovsky, G.M., Okladnikov, Yu.N., Rerberg, M.S., Sidko, M.S. and Terskov, I.A., 1975, *Eksperimentalnyye ekologicheskiye sistemy, vklyuchayushchie cheloveka. Problemy kosmicheskoi biologii* (Experimental ecological systems with human. Problems of space biology), Vol. 28, (Moscow: Nauka), 312 pp.

Gitelson, I. I., Terskov, I.A., Kovrov, B.G. *et al.*, 1976, Life support system with autonomous control employing plant photosynthesis, *Acta astronautica,* Vol. 3, **9.10**, pp. 633–650.

Gitelson, I.I., Manukovsky, N.S., Pankova, I.M., Somova, L.A. and Tirannen, L.S., 1981, *Mikrobiologicheskiye problemy zamknutykh ekologicheskikh sistem* (Microbiological problems of closed ecological systems), (Novosibirsk: Nauka).

Gitelson, J.I., Terskov, I.A., Kovrov, I.A., Lisovskii, G.M. *et al.*, 1989, Long-term experiments on man's stay in biological life support system. *Advances in Space Research,* Vol.9, pp. 65–71.

Gitelson, I.I., Blum, V., Grigoriev, A.I., Lisovsky, G.M., Manukovsky, N.S., Sinyak, Yu.E. and Ushakova, S.A., 1995, Biological–physical–chemical aspects of a human life support system for a lunar base. *Acta Astronautica*, **37**, pp. 385–394.

Golueke, C.G., 1960, The ecology of a biotic community consisting of algae and bacteria. *Ecology*, Vol. 41.

Golueke, C.G. and Oswald, W.J. 1963, Closing an ecological system consisting of a mammal, algae, and non-photosynthetic microorganisms. *American Biology Tescher*, Vol.25, pp. 522–528.

Gusarov, B.G., Drigo, Yu.A., Novikov, V.M., Samsonov, N.N., Farafonov, N.S., Chizhov, S.V. and Yazdovsky, V.I., 1967, On devising laboratory installations for the utilization link. In *Problemy sozdaniya zamknutykh ekologicheskikh sistem* (Problems of creating closed ecological systems), (Moscow: Nauka), p. 141.

Gusarov, B.G., Sinyak, G.S., Dagaeva, L.V. and Chizhikova, G.I., 1973, On minearlization of the gaseous phase released in processing of organic wastes. In *Eksperimentalnoye i matematicheskoye modelirovaniye iskusstvennykh i prirodnykh ekosistem* (Experimental and mathematical modeling of manmade and natural ecosystems), (Krasnoyarsk, Institute of Physics), p. 18.

Gustan, E. and Vinopal, G.I., 1986, Controlled Ecological Life Support System: Conceptual Design Option Study, *NASA-CR-111599.*

Gyurdzhian, A.A., 1961, Some issues of providing life support in the space flight. *Uspekhi sovremennoi biologii,* **51 (1)**, p. 74.

H

Harvey, R.B., 1922, *Science*, **56**, p. 1448.

Henry, J., 1966, *Biomedical aspects of space flight*, (N.Y.).

Hewitt, E.,J. 1952, *Sand and water culture methods used in the study of plant nutrition,* (Bristol Commonwealth Agricultural Bureaux).

Hill, W.A., 1984, *Selection of root and tuber crops for space missions. The sweet potato for space missions,* edited by Hill, W., Loretan, P., Bonsi, K., (Carver Research Foundation of Tuskegee University), pp. 3–12.

Hoff, G.E., Howe, G.I. and Mitchell, C.A., 1982, Nutritional and cultural aspects of plant species selection for a controlled ecological life support system. *NASA Contractor Report 166324* (NASA, Ames Research Center, Moffett Field, CA).

Humphries, W.R., Seshan, P.K. and Evanich, P.L., 1993, Physicochemical life support systems, Chap. 15 In *Space Biology and Medicine*, edited by

Nicogossian, A. *et al.* (Washington–Moscow: Joint US/Russian Publ.), pp. 331–356.

Humphries, W.R., Seshan, P.K. and Evanich, P.L., 1994, Physicochemical life support systems. In *Life support and habitability,* edited by Sulman, F. and Genin, A. (Amer. Institute of Aeronautics and Astronautics, Washington D.C.), Vol. II, pp. 461–498.

Huttenbach, R.C., Kaiser, A.R., Radford, J.D. and Binot, R.A., 1988, Physical–chemical atmosphere revitalization: the qualitative and quantitative selection of regenerative designs. In *Proceedings of the 3rd European Symposium on Space Thermal Control and Life Support Systems,* (Noordwijk, the Netherlands), (ESA SP-288, December 1988), pp. 57–64.

I

Ierusalimsky, N.D. and Neronova, N.M., 1965, Quantitative relatioship between metabolite concentration and growth rate of microorganisms. *Doklady AN SSSR,* **161 (6)**, p. 1437.

Imshenetsky, A.A., 1967, Biological cycle of hydrogen. *Vestnik AN SSSR,* **6**, pp. 39–44.

J

Jackson, J., Bonura, M. and Putnam, D., 1968, Evaluation of a closed-cycle life-support system during a 60-day manned test. *Douglas Paper*, p. 5105.

K

Kalacheva, L.I., 1971, Leaf:root ratio in different organogenesis phases in various morpho-physiological varieties of radish. In *Morfogenez ovoshchnykh rastenii* (Morphogenesis of vegetable plants), (Novosibirsk: Nauka).

Kalinkevich, A.F., 1964, In *Rol mineralnykh elementov v obmene veshchestv i produktivnosti rastenii* (Role of mineral elements in material exchange and plant productivity), (Moscow: Nauka), p. 73.

Kesler T.G., Trubachev, I.N., Voitovich, Ya.V. and Sidko, F.Ya., 1983, Growth of hydrogen-reducing bacteria on urine as nitrogen source. *Prikladnaya biokhimiya i mikrobiologiya*, Vol. IX, **3**, pp. 480–483.

Khomchenko, G.P., Chizhov, S.V., Sinyak, Yu.E. *et al.*, 1971, Extraction method of water regeneration in closed ecological systems. *Izvestiya TSKA,* **1**, p. 183.

King, R. and Evans, L., 1967, *Australian Journal of Biological Science*, **20 (3)**, p. 623.

Kirby, G.M., 1997, Bioregenerative planetary life support systems test complex: facility description and testing objectives, *SAE Technical Paper series 972342.*

Kirensky, L.V., Terskov, I.A., Gitelson, I.I., Lisovsky, G.M., Kovrov, B.G. and Okladnikov, Yu.N., 1967, Continuous microalgal culture as a link of a closed ecological system. *Kosmicheskaya biologiya i meditsina,* p. 19.

Kirensky, L.V., Terskov, I.A., Gitelson, I.I., Lisovsky, G.M., Kovrov, B.G., Sidko, F.Ya., Okladnikov, Yu.N., Antonyuk, M.P., Belyanin, V.N. and Rerberg, M.S.,

1967a, Gas exchange between human and microalgal culture in the 30-day experiment. *Kosmicheskaya biologiya i meditsina,* **1 (4)**, p. 23.

Kirensky, L.V., Terskov, I.A., Gitelson, I.I., Lisovsky, G.M., Kovrov, B.G. and Okladnikov, Yu.N., 1967b, Experimental biological life support system. In *Continuous cultivation of algae as a link of a closed ecosystem. Proceedings of the Open Meeting of Working Group V of the Tenth Plenary Meeting of COSPAR*, pp. 33–38.

Kirensky L.V., Terskov, I.A., Gitelson, I.I., Lisovsky, G.M., Kovrov, B.G. and Okladnikov, Yu.N., 1967c, Experimental biological life support system. II. Gas exchange between man and microalgae culture in a 30-day experiment. *Proceedings of the Open Meeting of Working Group V of the Tenth Plenary Meeting of COSPAR*, pp. 38–40.

Kirensky, L.V., Terskov, I.A., Gitelson, I.I., Lisovsky, G.M., Kovrov, B.G., Sidko, F.Ya., Belyanin, V.N., Kuzmina, R.I., Okladnikov, Yu.N., Antonyuk, M.P. and Rerberg, M.S., 1970, Closed water exchange in the two-link biological–technical human life support system. *XIX International Astronautical Congress,* **4**. Bioastronautics, (Pergamon Press), p. 51.

Kirensky, L.V., Gitelson, J.I., Terskov, I.A., Kovrov, B.G., Lisovsky, G.M. and Okladnikov, Yu.N., 1971, Theoretical and Experimental Decisions in the Creation of an Artificial Ecosystem for Human Life-Support in Space. *Life Science and Space Research*, **IX**, pp. 75–80.

Klyushkina, N.S. and Fofanov, V.I., 1967, On the value of proteins of unicellular algae. *Kosmicheskaya biologiya i meditsina,* **1 (6)**, p. 52.

Klyushkina, N.S., Fofanov V.I. and Troitskaya, I.T., 1967, Determination of biological value of proteins of unicellular algae and soybean on 4 generations of white rats. *Kosmicheskaya biologiya i meditsina,* **1 (4)**, p. 33.

Knott, W.M., Sager J.C. and Wheeler, R., 1992, Achieving and documenting closure in plant growth facilities. *Advances in Space Research*, Vol. 12, (5), pp. (5)115–(5)123.

Kolchinsky, E.I., 1990, *Evolyutsiya biosfery* (Evolution of the biosphere), (Leningrad: Nauka).

Koptyug, V.A., 1997, *Nauka spaset chelovechestvo* (Science will save humanity), *Summary of the UN Conference on Environment and Developmentm, Mir Nauki*, (Novosibirsk), **4**.

Kordyum, V.A., 1969, *Mikrobiologicheskiye problemy zamknutykh ekologicheskikh sistem* (Microbiological problems of closed ecological systems), *D Sc Thesis*, (Kiev).

Korogodin, V.I., 1991, *Informatsiya i fenomen zhizni* (Information and phenomen of life), (Acad. of Sci. USSR, Pushchino).

Korotaev, M.M., Kustov, V.V., Meleshko, G.I., Poddubnaya, L.T. and Shepelev, Ye.Ya., 1964, Toxic gaseous substances emitted by *Chlorella*. In *Problemy kosmicheskoi biologii* (Problems of space biology), (Moscow: Nauka), **3**, p. 204.

Kovrov, B. G., 1992, Manmade microecosystems with closed material cycle as models of the biosphere. *Biofizika kletochnykh populiatsii i nadorganizmennykh sistem* (Novosibirsk: Nauka), pp. 62–70.

Kovrov, B.G. and Shtol, A.A., 1969, Design of algal cultivators with curved light-receiving surface. In *Upravlyaemyi biosintez i biofizika populyatsii* (Controlled biosynthesis and biophysics of populations), (Krasnoyarsk), p. 48.

Kovrov, B.G. and Lisovsky, G.M., 1972, Quantitative estimation of gas exchange of continuous culture of higher plants as a link of the life support system. *Kosmicheskaya biologiya i meditsina,* **5**, pp. 17–21.

Kovrov, B. G. and Fishtein, G.N., 1980 Distribution of biomass in synthetic closed microbicenoses depending on species structure. *Izvestiia SO AN SSSR, Seriya biologii,* **5**, pp. 35–40.

Kovrov, B.G. and Martynenko, L.L., 1981, Development of the vitamin greenhouse as a biological element of the life support system. In *K.E. Tsiolkovsky i sovremennyye problemy kosmicheskoi biologii* (K.E.Tsiolkovsky and modern problems of space biology), *Proceedings of the XIV meeting,* (Moscow), pp. 99–103.

Kovrov, B.G. and Fishtein, G.N., 1985, Microecosystems and experience of their using for studies of protists in a community of microscopical organisms. *Zhurnal obshchei biologii* (Journal of General Biology), Vol. 46, **3**, pp. 336–344.

Kovrov, B.G., Melnikov, E.S., Belyanin, V.N., Terskov, I.A., Sidko, F.Ya., Eroshin, N.S., Gitelson, I.I., Filimonov, V.S. and Shtol, A.A., 1967, A cultivator for intensive continuous microalgal cultivation. In *Nepreryvnoye upravlyayemoye kultivirovaniye mikrovodoroslei* (Continuous controlled microalgal cultivation), (Moscow: Nauka), p. 14.

Kovrov, B.G., Belyanin, V.N. and Shtohl, A.A., 1969, A calculation of algae cultivatir for life support system. *Proceedings of the 8th International Symposium of Space Technology and Science,* (Tokyo), pp. 1183–1187.

Kovrov, B.G. *et al.*, 1985, Artificial closed ecosystems "man–plants" with a full regeneration of atmosphere, water, and ration vegetable. *Proceedings of the XXXVI International Astronomy Congress.* 7–12 Oct 1985, Stckholm, Sweden. (New York: Pergamon press).

Krauss, R., 1962, *American Journal of Botany*, **49**, p. 425.

Krauss, R., 1964, *Conference on Nutrition in Space and Related Waste Problems*, (Washington D.C., NASA), pp. 289–297.

Kreps, E.M., Verzhbinskaya, P.A., Chenykaeva, E.Yu., Chirkovskaya, E.V. and Gavurina, U.K., 1956, On adaptation of animals to chronic hypoxia. *Physiologicheskii Zhurnal SSSR,* **42 (1)**.

Kurapova, O.A., 1969, Dynamics of build-up of organic substances in the culture fluid in prolonged cultivation of *Chlorella* on repeatedly used medium. In *Upravlyaemyi biosintez i biofizika populyatsii* (Controlled biosynthesis and biophysics of populations), (Krasnoyarsk), p. 20.

Kursanov, A.L., 1966, Preface. In *Fiziologo-biokhimicheskiye osnovy vzaimnogo vliyaniya rastenii v fitotsenoze* (Physiological–biochemical basis of mutual effect on plants in a phytocenosis), (Moscow: Nauka), p. 5.

Kustov, V.V. and Tiunov, L.A., 1969, Toxicology of metabolites and their contribution to formation of artificial atmosphere of sealed space. In *Problemy kosmicheskoi biologii* (Problems of space biology), (Moscow: Nauka), p. 11.

Kuzmina, R.I. and Kovrov, B.G., 1967, Continuous *Chlorella* cultivation on repeatedly used medium. In *Problemy sozdaniya zamknutykh ekologicheskikh sistem* (Problems of creating closed ecological systems), (Moscow: Nauka), p. 91.

L

Lamarck, J.B., 1935, *Philosophy of Zoology* (in Russian), Vol. 1 (Moscow–Leningrad).

Landolt, E., 1957, *Botanical Geography*, p. 67.

Laptev, V.V. and Nilovskaya, N.T., 1968, Sealed installation for investigation of plant gas- and water-exchange. *Selskokhozyaistvennaya biologiya*, **3 (6)**, p. 892.

Latyshev, D.I., 1967, *Vyrashchivaniye ovoshchei v sovkhoze "Teplichnyi"* (Growing of vegetables at the sovkhoz "Teplichnyi"), (Moscow: Kolos).

Lebedeva, E.V., Rusakova, G.G. *et al.*, 1969, Table beet as the autotrophic link in the biological life support system. In *Upravlyaemyi biosintez i biofizika populyatsii* (Controlled biosynthesis and biophysics of populations), (Krasnoyarsk), p. 155.

Leman, V.M., 1961, *Kurs svetokultury rastenii* (A course in plant light culture), (Moscow: Vysshaya shkola).

Le Roy, E., 1927, *L'exigence idealiste et le foit d'evolution*, (Paris).

Le Roy, E., 1928, *Les origines humaines et l'evolution de l'intelligence,* (Paris: Alcan).

Levinskikh, M.A. and Sychev, V.N., 1989, Growth and development of unicellular algae in space flight within the system "algobacterial cenosis–fish." *Kosmicheskaya biologiya i aviakosmicheskaya meditsina*, Vol. 5, pp. 32–35.

Lewis, W.C., 1989, LSA for Mars class missions, Space manufacturing. In *Proceedings of the Ninth Princeton/AIAA SSI Conference* (May 10–13, Washington, D.C.20024), pp. 274–278.

Life Support and Habitability Manual. August 1991, Volume I, ESA PSS-03-406 Issue I.

Lisovsky, G.M., 1972, Objectives and possibilities of selection improvement of higher plants as a link of life support systems. In *Kosmicheskaya biologiya i aviakosmicheskaya meditsina* (Space biology and aerospace medicine), Abstracts, Kaluga, **1**, p. 210.

Lisovsky, G.M. and Sypnevskaya, E.K., 1969, Sodium chloride influence on the composition of *Chlorella vulgaris* in continuous cultivation. *Prikladnaya biokhimiya i mikrobiologiya,* **5 (3)**, p. 369.

Lisovsky, G.M. and Shilenko, M.P., 1970, Cultivar differences in radish reaction to the duration of artificial lighting. *Informatsionnyi bulleten koordinatsionnogo soveta po fiziologii rastenii,* 7, (Irkutsk), pp. 39–40.

Lisovsky, G.M. and Shilenko, M.P., 1971, Culture of cereals as a possible component of the autotrophic link in life support systems. *Kosmicheskaya biologiya i meditsina,* **5 (1)**, p. 22.

Lisovsky, G.M., Gitelson, I.I. and Terskov, I.A., 1965, Biosphere and closed biological systems. In *Abstracts of All-Union Meeting on Controlled Biosynthesis and Biophysics of Populations,* (Krasnoyarsk).

Lisovsky, G.M., Yan, N.A., Sytsnevskaya, E.K. *et al.*, 1966, Productivity of various algal forms in continuous cultivation. In *Upravlyaemyi biosintez* (Controlled biosynthesis), (Moscow: Nauka), pp. 276–282.
Lisovsky, G.M., Gitelson, I.I. and Terskov, I.I., 1967, Biosphere and closed biological systems. In *Problemy sozdaniya zamknutykh ekologicheskikh sistem* (Problems of creating closed ecological systems), (Moscow: Nauka), p. 44.
Lisovsky, G.M., Melnikov, E.S., Sakash, V.G. *et al.*, 1968, Sealed phytotron for investigation of dynamics of biomass increment in planting. *Informatsionnyi bulleten koordinatsionnogo soveta po fiziologii i biokhimii rastenii,* **3**, (Irkutsk), p. 133.
Lisovsky, G., Kovrov, B., Terskov I. and Gitelson, I., 1969, Methods and technique of wheat continuous culture as a link of life support system. *Proceedings of Eighth International Symposium on Space Technology and Science,* (Tokyo), p. 1189.
Lisovsky, G.M., Parshina, O.V., Ushakova, S.A., Bayanova, Yu.N. and Sirotinina, L.A., 1979, Productivity and chemical composition of some vegetable crops grown under "lunar" photoperiod. *Izvestiya SO AN SSSR (biology)*, **1**, pp. 104–108.
Lisovsky, G.M., Prikupets, L.B., Sarychev, L.B., Sidko, F.Y. and Tikhomirov, A.A., 1983, Experimental evaluation of efficiency of light sources in plant light culture. *Svetotekhnika*, **4**, pp. 7–9.
Lovelock, J.E., 1972, Gaia as seen through the atmosphere. *Atmospheric environment*, **6**, pp. 579–580.
Lovelock, J.E., 1990, *The ages of Gaia,* (New York, Toronto: Bantam Books).
Lovelock, J. E., 1991, *Scientists on Gaia,* (England: Mit Press), (First edition published in 1979.)
Luckey, T.O., 1966, Potential microbic shock in manned aerospace systems. *Aerospace Medicine*, Vol. 37, pp. 1223–1228.

M

MacElroy, R.D. and Smernoff, D.T., 1987, Controlled ecological life support systems. *Advances in Space Research*, Vol. 7, **4**, pp. 1–53.
MacElroy, R.D., Johnson, E.J. and Johnson, M.K., 1971, Control of ATP-Dependent CO_2 Fixation in Extracts of *Hydrogenomonas facilis*: NADH Regulation of Phosphoribulokinase. *Archives of Biochemical Biophysics*, **131,** p. 272.
MacElroy, R.D., Tibbits, T.W., Tompson, B.G. and Volk, T., 1989, Natural and experimental ecosystems. *Advances in Space Research*, Vol. 9, **8**, pp. 1–202.
Maksimov, N.A., 1925, Plant culture under electric light and its application for seed control and selection. *Nauchno-agronomicheskii zhurnal*, **7–8**, p. 395.
Manned Mars Mission Study group. "Proposed concept for a manned Mars mission program." 1988, *Proceedings of the Seventh Annual International Space Development Conference,* (Denver, Colorado), (San Diego: Univelt), pp. 411–435.
Manukovsky, N.S., Kovalev, V.S., Zolotukhin, I.G. and Rygalov, V.E., 1996, Biotransformation of plant biomass in closed cycle, *SAE Technical Paper Series 961417*.

Margalef, R., 1963, Successions of populations. *Advances in Frontiers of Plant Science,* (New Delhi), Vol. 2.

Margalef, R., 1963, On certain unifying principles in ecology. *Amer. Nat.*, Vol. 97.

Marshak, I.S., Vasilyev, V.I. and Tokhadze, I.A., 1963, Small-size ballast-free water cooled xenon tube DKsTV-6000. *Svetotekhnika*, **11**, p. 13.

Masachito, T., 1964, *Proceedings of the 5th International Symposium of Space Technology and Science,* (Tokyo), p. 1115.

Masyuk, N.P., 1966, Mass culture of carotene-bearing alga *Dunaliella salina. Ukrainskii botanicheskii zhurnal*, **23 (2)**, p. 12.

Mattson, H., 1966, *International Science and Technology*, **V**, p. 12.

Mayers, J.I., 1958, Study of photosynthetic regenerative systems on green algae, *SAM 58-117, Randolph AEB Texas USAF School of Aviation Medicine Report*, Vol. 58, p. 117.

Meadows, D.L. *et al.*, 1974, *Dynamics of Growth in a Finite World,* (Cambridge, Mass.: Wright Allen Press Inc.).

Meleshko, G.I., 1967, Some characteristics of *Chlorella* population as a link of a closed ecological system. In *Problemy sozdaniya zamknutykh ekologicheskikh sistem* (Problems of creating closed ecological systems), (Moscow: Nauka), p. 73.

Meleshko, G.I. and Krasotchenko, L.M., 1965, Conditions of carbon nutrition in intensive culture. In *Problemy kosmicheskoi biologii* (Problems of space biology), Vol. 4, (Moscow: Nauka), pp. 676–682.

Meleshko, G.I., Lebedeva, E.K., Kurapova, O.A. and Ulyanin, Yu.N., 1967, Prolonged cultivation of *Chlorella* with direct returning of medium. *Kosmicheskaya biologiya i meditsina,* **1 (4)**, p. 28.

Meleshko, G.I., Shepelev, Ye.Ya., Guryeva, T.S. *et al.*, 1991, Embryonal devepoment of birds in microgravity. *Kosmicheskaya biologiya i aviakosmicheskaya meditsina*, Vol. 1, pp. 37–39.

Meleshko, G.I., Shepelev, Ye.Ya., Averner, M.M. and Volk, T., 1993, Biological life support systems, Chapter 16. In *Space Biology and Medicine*, edited by Nicogossian, A. *et al.* (Washington–Moscow: Joint US/Russian Publ.), pp. 357–394.

MELISSA Final report for 1997 activity ESA/EWP-1975, edited by Lasseur, Ch., 1998.

MELISSA Final report for 1998 activity ECT/FG/MMM/97.012, edited by Lasseur, Ch., Fedole, I., 1999, ESA.ESTEC/MCL/26 77. CHL.

Mesarovic, M. and Pestel, E., 1974, *Mankind at the Turning Point* "The Second Report to the Club of Rome." (N.J.: E.P.DUTTO and Co., Inc. Readers Digest Press).

Meyers, K.E., Staat, D.G., Tri, T.O. and Smith, F.D., 1997, Lunar–Mars Life Support Test Project, Phase IIA: A Crew's 60 Day Experience, *SAE Technical Papers 072340.*

Mezhevikin, V.V., Okhonin, V.A., Bartsev, S.I. and Gitelson, J.I., 1994, Indications and counterindications for applying different versions of closed ecosystems for space and terrestrial problems of life support. *Advances in Space Research*, Vol.14, pp. 135–142.

Milko, E.S., 1963. *Izucheniye fiziologii i pigmentoobrazovaniya zelenoi vodorosli Dunaliella* (Investigation of physiology and pigment-formation in the green alga *Dunaliella*), *D Sc Thesis*, 1963.

Milov, M.A. and Balakireva, K.A., 1975, On selecting higher plants for the biological life support system. In *Problemy sozdaniya biologo-tekhnicheskikh sistem zhizneobespecheniya cheloveka* (Problems of creating biological–technical human life support systems), edited by Gitelson, I.I. (Novosibirsk: Nauka), pp. 13–19.

Mizrakh, S.A., Fomin, O.V., Lisovsky, G.M. and Stepanov, G.V., 1969, Effect of relative air humidity on intensity of wheat planting gas exchange. In *Upravlyaemyi biosintez i biofizika populyatsii* (Controlled biosynthesis and biophysics of populations), (Krasnoyarsk), pp. 102–103.

Mizrakh, S.A., Lisovsky, G.M. and Terskov, I.A., 1973, Plant growth and development under lunar photoperiod, *Doklady AN SSSR,* Vol. 210, **2**, pp. 475–477.

Moiseev, A.A., Koloskova, Yu.S., Sinyak, Yu.E. and Chizhov, S.V., 1967, Provision of the crew with water in the space flight. In *Problemy kosmicheskoi biologii* (Problems of space biology), **7**, (Moscow: Nauka), p. 389.

Moiseev, N.N., 1979, Systems analysis of dynamic processes of the biosphere. Systems analysis and mathematical models, *Vestnik AN SSSR,* **1**.

Moiseev, N. N., 1990, *Chelovek i noosfera* (Human and Noosphere), (Moscow: Molodaia gvardiia).

Moiseev N.N. and Svirezhev, Yu.M., 1979, Systems analysis of dynamic processes of the biosphere. Conceptual model of the biosphere, *Vestnik AN SSSR,* **2**.

Moiseev, N.N., Krapivin, Yu.M., Svirezhev, Yu.M. and Tarko, A.M., 1980, systems analysis of dynamic processes of the biosphere. In *Chelovek i biosfera* (Human and the biosphere), **4**, edited by Fedorov, V.D. (Izdatelstvo Moskovskogo Universiteta), pp. 228–258.

Mokhnachev, I.G. and Kuzmin, M.P., 1966, *Letuchiye veshchestva pishchevykh produktov* (Volatiles of food products), (Odessa).

Monod, I., 1950, *Annales de l'Institut Pasteur*, p. 79.

Moore, W.E. and Holdeman, L.V., 1974, Human fecal floras. *Applied Microbiology*, Vol. 27, pp. 961–967.

Moshkov, B.S., 1966, *Vyrashchivaniye rastenii pri iskusstvennom osveshchenii* (Growing of plants under artificial light), (Leningrad: Kolos).

Moshkov, B.S., 1973, *Rol luchistoi energii v vyyavlenii potentsialnoi produktivnosti rastenii* (Role of radiant energy in revealing potential productivity of plants), (Moscow: Nauka).

Muzafarov, A.M., Taubaev, T.T. and Abdiev, M., 1968, Toad spit as valuable vitamin-bearing food for poultry. *Uzbekskii biologicheskii zhurnal*, **3**.

Muzafarov, A.M., Taubaev, T.T. and Abdiev, M., 1971, Toad spit as forage plant and methods of its large-scale cultivation in open pools. In *Kultivirovaniye vodoroslei i vysshikh vodnykh rastenii v Uzbekistane* (Cultivation of algae and aquatic higher plants in Uzbekistan), (Tashkent), p. 117.

Myers, D.I., 1958, Study of photosynthetic regenerative systems on green algae. *USAF School of Aviation Medicine Report*, Vol. 58, p. 117.

Myers, J., 1954, Basic remarks on the use of plants as biological gas exchangers in a closed system. *Journal of Aviation Medicine,* Vol. 24.

Myers, J., 1958, Study of a photosynthetic gas exchanger as a method of providing for the Respiratory requirements of the human in a sealed cabin. *Air University School of Aviation Medicine,* Texas, **58**, p. 117.

Myers, J., 1960, The use of photosynthesis in a closed ecological system. In *The physics and medicine of the atmosphere and space.* (N.Y.: John Wiley and Sons, Inc.).

N

Nakamura, H., 1961, *Chlorella in future* (in Russian), (Tokyo).

Nakamura, H., 1961a, *Reptort of Microalgae Research Institute,* (Japan), **11 (1)**.

Nakamura, H., 1963, *Biological knowledges on species of Chlorella and Scenedesmus,* (Tokyo).

Nakamura, H., 1963a, *Studies on microalgae and photosynthetic bacteria,* (University of Tokyo Press), 197 pp.

Nefedov, Yu.G. and Zaloguev, S.N., 1967, On the problem of habitability of spaceships. *Kosmicheskaya biologiya i meditsina,* **1**, p. 30.

Nelson, M., 1993, Bioregenerative life support for space habitation and extended planetary missions. In *Space life sciences*, edited by Churchill, S., Chapter 22 (Malabar: Orbit Books).

Nelson, M., 1993a, Importance of Biosphere 2 for investigaiton of ecosystem processes. *Vestnik Rossiiskoi Akademii Nauk,* **11**, pp. 1024–1034.

Nelson, M. and Dempster, W.F., 1995, Living in space: results from Biosphere 2's initial closure, an early tested for closed ecological systems on Mars. *Life Support & Biosphere Science*, Vol.2, pp. 81–102.

Nichiporovich, A.A., 1955, *Svetovoye i uglerodnoye pitaniye rastenii* (Light and carbon plant nutrition), (Moscow: Izdatelstvo AN SSSR).

Nichiporovich, A.A., 1956, Photosynthesis and theory of obtaining high yields. In *Proceedings of the XV meeting in honor of Timiryazev,* (Moscow: Izdatelstvo AN SSSR).

Nichiporovich, A.A., 1963, Creation of the human habitat in future space flights. In *Kosmos* (Space), (Moscow: Izdatelstvo AN SSSR), p. 25.

Nichiporovich, A.A., 1966, Objectives of investigations of photosynthetic plant activity as a factor of productivity. In *Fotosinteziruyushchiye sistemy vysokoi produktivnosti* (Photosynthesizing systems of high productivity), p. 7.

Nikishanova, T.I., 1977, Plants for space exploration. *Priroda*, **10**, pp. 105–117.

Nikolaeva, A.M., 1956, Duckweed as food for water birds. *Ptitsevodstvo*, **6**.

Nilovskaya, N.T., 1968, Photosynthesis and respiration of some vegetable plants at different carbon dioxide content in the gaseous environment. *Fiziologiya rastenii*, **15 (6)**, p. 1015.

Nilovskaya, N.T., 1973, *Izucheniye gazoobmena i produktivnosti rastenii v fitotronakh* (Investigation of plant gas exchange and productivity in phytotrons), *D Sc Thesis*, Kiev.

Nilovskaya, N.T. and Bokovaya, M.M., 1967, Air regeneration by higher plants. In *Problemy sozdaniya zamknutykh ekologicheskikh sistem* (Problems of creating closed ecological systems), (Moscow: Nauka), p. 108.

Nilovskaya, N.T. and Razorenova, T.A., 1977, Estimation of potential wheat productivity in phytotrons. *Doklady VASKhNIL*, **5**, pp. 8–9.

Nitta, K., 1998, *Proceedings of International Committee for Material Circulation in Geo-Hydrosphere*, (Institute for Environmental Sciences, Rokkasho, Aonneri, Japan), July 21–23, 1998.

Nitta, K., Ashida, A. and Orsubo, R., 1996, Closed ecology experiment facilities construction planning and present status. *Life Support & Biosphere Science*, Vol. 3, pp. 101–115.

Nosatovsky, A.I., 1957, Wheat biology. In *Pshenitsa v SSSR* (Wheat in the USSR) (Moscow), p. 123.

O

Odum, E.P., 1966, Regenerative systems. In *Human Ecology in Space Flight*. (Interdisciplinary Communications Program, N.Y.).

Odum, E., 1971, *Fundamentals of ecology*, 3rd Edition, Chapter 20, (Philadelphia–London–Toronto).

Odum, E. P., 1989, *Ecology and Our Endangered Life Support Systems*, (Sunderland: Linauer Associates, Inc.).

Odum, H.T. and Hoskin, C.M., 1957, Metabolism of a laboratory stream microcosm. *Publications of Institute of Marine Sciences University of Texas*, Vol. 4.

Okladnikov, Yu.N. and Kasaeva, G.E., 1969, Algal culture as an utilizer of carbon oxide and ammonia in a biological life support system. In *Upravlyaemyi biosintez i biofizika populyatsii* (Controlled biosynthesis and biophysics of populations, (Krasnoyarsk), pp. 137–138.

Okladnikov, Yu.N., Vorkel, Ya.B., Trubachev, I.N., Vlasova, N.V. and Kalacheva, G.S., 1977, Introduction of chufa into human diet as a source of polyunsaturated fatty acids. *Voprosy pitaniya*, **3**, pp. 45–48.

Okladnikov, Yu.N., Vlasova, N.V., Kasaeva, G.E., and Rodionov, V.N., 1979, Human link in experiment (medical–physiological investigations). In *Zamknutaya sistema: chelovek–vysshiye rasteniya* (Closed system "human–higher plants"), edited by Lisovsky, G. (Novosibirsk), pp. 82–99.

Olcott, T. and Conner, W., 1968, Thirty day performance and reliability test of a regenerative life support system. *I.A.F.Paper, B43*, (N.Y).

Olifer, V.A., 1965, Structure of biomass and chemical composition of annual cultured crops on chernozem soils of the Omsk Province. *Izvestiya Sibirskogo Otdeleniya AN SSSR (biology)*, **4 (1)**, pp. 81–86.

Oswald, W.J., 1965, *Sanitation Engineering Division Proceedings of Society of Civil Engineers*, **90 (4)**, p. 23.

Oswald, W.J. and Golueke, C.G., 1964, Fundamental factors in waste utilization in isolated systems. *Division of Industrial Microbiology*, Vol.5, pp. 196–206.

Ovechkin, S.K., 1940, Periodicity in phosphorus nutrition of spring wheats. *Doklady AN SSSR*, Vol. 26, **2**, pp. 186–189.

P

Pankova, I.M., Tirannen, L.S., Somova, L.A., Manukovsky N.S. and Posadskaya, M.N., 1979, System microflora and its dynamics in experiment. In *Zamknutaya sistema: chelovek–vysshiye rasteniya* (Closed system "human–higher plants"), edited by Lisovsky, G. (Novosibirsk), pp. 100–129.

Pankova, I.M., Trubachev, I.N., Kochetova, G.N., Manukovsky, N.S., Bayanova, Yu.I., Abrosov, N.S. and Gribovskaya, I.V., 1985, On using higher edible fungi in manmade ecosystems with humans. In *Mikroorganismy v iskusstvennykh ekosistemakh* (Microorganisms in manmade ecosystems), (Novosibirsk: Nauka), pp. 124–130.

Patterson, M.T., Wignarajah K. and Bubenheim, D.L., 1996, Biomass Incineration as a Source of CO_2 for Plant Gas Exchange: Phytotoxicity of Incineration-derived Gas and Analyses of Recovered Evapotranspired Water. *Life Support & Biosphere Science.*

Pavlov, A.N., 1969, On nitrogen flow from vegetative organs to wheat and corn grain. *Selskokhozyaistvennaya biologiya,* Vol. IV, **2**, p. 230.

Peccei, A., 1977, *The Human Quality,* (Pergamon Press).

Pecoraro, J. and Morris, F., 1972, Progress in regenerative life support systems for a lunar laboratory. *23rd International Astronautical Congress,* (Vienna).

Petrea, V.D., 1968, *Revue Roumaine de biologie. Ser. biologie vegetale,* Vol. 13, **1–2**, pp. 103–106.

Pickering, K.D. and Edeen, M.A., 1998, Lunar–Mars Life Support Test Project Phase III Water Recovery System Operation and Results, *SAE Technical Paper Series 981707.*

Pinevich, V.V., Verzilin, N.N. and Maslov, Yu.I., 1961, Effect of different nitrogen sources on growth and mass accumulation in *Chlorella pyrenoidosa. Vestnik LGU*, **9**.

Podvalkova, P.A., 1959, *Potrebnost yarovoi pshenitsy v azote, fosfore i kalii na razlichnykh stadiyakh razvitiya* (Spring wheat requirements for nitrogen, phosphorus and potassium at different developmental stages), *D Sc Thesis,* (Leningrad, Pushkino).

Pokrovsky, A.A., 1964, On the problem of determining human requirements for food substances, *Vestnik AMN SSSR*, **5**, p. 3.

Polonsky, V.I. and Lisovsky, G.M., 1980, Net production of wheat crop under high PAR irradiance with artificial light. *Photosynthetica*, Vol. 14, **2**, pp. 177–181.

Polonsky, V.I., Lisovsky, G.M. and Trubachev, I.N., 1977, Wheat productivity and biochemical composition at high PAR intensity in the light culture. *Fiziologiya rastenii*, Vol. 24, **4**, pp. 718–723.

Polonsky, V.I., Lisovsky, G.M. and Trubachev, I.N., 1977a, Optimization of PAR intensity during growth period for wheat cenosis. In *Intensivnaya svetokultura rastenii* (Intensive light culture of plants), (Krasnoyarsk, Institute of Physics), pp. 14–34.

Polunin, N. and Greenwald, G., 1993, Biosphere and Vernadsky, *Vestnik RAN*, **2**, pp. 122–126.

Popov, I.G. and Bychkov, V.P., 1994, Cosmonauts' diet. In *Kosmicheskaya biologiya i meditsina* (Space biology and medicine), Vol. II *Obitayemost kosmicheskikh apparatov* (Habitability of spacecrafts), (Moscow–Washington), pp. 313–336.

Posadskaya, M.N., 1976, Peculiarities of biochemical oxidation of wheat straw by active sludge biocenosis. In *Materialy bsesoyuznogo rabochego soveshchaniya po voprosu krugovorota veshchestv v zamknutykh ekosistemakh* (Proceedings of the All-Union Workshop on material cycling in closed ecosystems), (Kiev), pp. 117–119.

Powell, R., Nevels, E. and McDowell, M., 1961, *Journal of Nutrition*, **75**, p. 7.

Pratt, R. and Johnson, E., 1965, *Journal of Pharmacological Science*, **54 (6)**, p. 871.

Prince, R.P. and Knott, W.M., 1989, CELSS Breadboard Project at the Kennedy Space Center. In *Lunar base agriculture, soils for plant growth*, edited by Ming, D.W. and Henninger, D.L., (Agronomy Society of America, Crop Science Society of America, Soil Science Society of America, Madison, WI), pp. 155–163.

Priputina, A.S., Rudenko, A.K. and Litichevsky, N.E., 1964, Characterizatiion of actual feeding and the state of the health of population of some districts of the Kiev Province. In *Materialy XV nauchnoi sessii Instituta Pitaniya AMN SSSR* (Proceedings of the 15 Scientific Session of the Food Institute of AMS USSR), **2** (Moscow).

Proceedings of the 5th International Conference on Indoor Air Quality and Climate. INDOOR AIR'90, Vol.1–5, edited by Walkinshaw, D.S. (Indoor Air Technologies, Ottawa), 1990.

Prokhorov, V.Ya, Shilov, V.M., Akatov, A.K. *et al.*, 1971, Activation of biological properties of staphylococci isolated from humans staying in the sealed cabin for prolonged periods. *Zhurnal Microbiologii,* **9**, pp. 63–68.

R

Rerberg, M.S. and Kuzmina, R.I., 1964, Experience of prolonged stepwise cultivation of protococcal algae in community with bacteria on human wastes. In *Upravlyayemoye kultivirovaniye mikrovodoroslei* (Controlled microalgal cultivation), (Moscow: Nauka), p. 119.

Rerberg, M.S. and Vorobyeva, T.I., 1964, Experience of growing protococcal algae on human wastes under sterile and non-sterile conditions. In *Upravlyayemoye kultivirovaniye mikrovodoroslei* (Controlled microalgal cultivation), (Moscow: Nauka), pp. 124–134.

Rerberg, M.S. and Vorobyeva, T.I., 1967, On sodium chloride influence on biomass growth and chlorophyll synthesis in the protococcal alga *Chlorella*. In *Nepreryvnoye upravlyayemoye kultivirovaniye mikrovodoroslei* (Continuous controlled microalgal cultivation), (Moscow: Nauka), p. 140.

Rerberg, M.S., Vorobyeva, T.I., Kuzmina, R.I. and Barkhatova, I.M., 1965, Human waste treatment by the naturally forming algal–bacterial community. In *Problemy kosmicheskoi biologii* (Problems of space biology), Vol. 4, (Moscow: Nauka), pp. 598–604.

Rerberg, M.S., Popova, M.N., Bazanova, M.I. and Trubachev, I.N., 1968, On the ways of processing human wastes in the biological life support system. In

Proceedings of the V Workshop on Material Cycle in the Closed System on the Basis of Life Activity of Lower Organisms, (Kiev: Naukova dumka), p. 17.

Riely, P.E., Beard, D.B. and Gotts, I., 1966, Effects real and relative of a space type diet on the aerobic and anaerobic microflora of human feces. *Aerospace Medicine*, Vol. 37, **8**, pp. 820–825.

Robertson, A.S., Roberts, K.T., Burge, P.S. and Raw, A., 1990, The effect of change in building ventilation category on sickness absence rates and prevalence of sick building syndrome. In *Proceedings of the 5th International Conference on Indoor Air Quality and Climate. INDOOR AIR'90*, Vol.1–5, edited by Walkinshaw, D.S. (Indoor Air Technologies, Ottawa), Vol.1, pp. 237–242.

Rozov, N.F., 1973, Peculiarities of growing and chemical composition of lettuce, cucumber grass and cress in the phytotron. *Izvestiya Timiryazevskoi Selskokhozyaistvennoi Akademii,* **5**, pp. 141–146.

Rummel, J.D. and Volk, T., 1987, A modular BLSS simulation model. *NASA Conf. Publ. 2480,* (NASA Ames Research Center. Moffett Field, Calif.), pp. 55–56.

Rummel, J.A., Stegemoeller, C.M., Lane, H.W., Nicholson, L.S., Ellis, W.E. and Behrend, A.F., 1998, Advanced life support program plan. *NASA Johnson Space Center Document JSC-39168.*

Rygalov, V.E., 1996, Cultivation of plants in space: their contribution to stabilizing atmospheric composition in closed ecological systems. *Advances in Space Research,* Vol. 18, **4/5**, pp. 165–176.

Rygalov, V.E., Shilenko, M.P. and Lisovsky, G.M., 1995, Minor components composition in closed ecological system atmosphere: mechanisms of formation. *IAF/IAA-95-J4.03*, pp. 1–9.

S

Sager, J.C., 1997, KSC Advanced Life Support Breadboard: Facility Description and Testing Objectives, *SAE Technical Paper Series 972341.*

Salisbury, F.B., Bingham, G.E., Campbell, W.F., Carman, J.G., Bubenheim, D.L., Yendler, B. and Jahns, G., 1995, Growing super-dwarf wheat in Svet on Mir. *Life Support & Biosphere Science*, **2**, pp. 31–39.

Sanders, W.M. and Falco, J.W., 1973, *Ecosystem simulation for water pollution research. Advances in water pollution research.* (N.Y.: Pergamon Press).

Sauer, R.L. and Shea, T.G., 1981, Spacecraft water supply and quality management in Appollo Program. *Water and Sewage Works*, Vol. 118, **7**, pp. 226–233.

Savkin, V.I., Meleshko, G.I. and Adamovich, B.A., 1970, Investigation of processes of regulation of carbon dioxide concentration in the sealed cabin with animals under regeneration of atmosphere by *Chlorella* culture. *Kosmicheskaya biologiya i meditsina*, **4 (5)**, p. 3.

Schaefer, K.E., 1961, A concept of triple tolerance limits based on chronic carbon dioxide toxicity studies. *Aerospace Medicine,* **32**, p. 197.

Schaefer, K.E., 1964, Gaseous requirements in manned space flight. In *Bioastronautics,* (N.Y.–London).

Schieler, L., McClure, L. and Dunn, M., 1953, *Food Research*, **18**, p. 377.

Schulz B., 1962, *Wasserlinsen.* (Leipzig).

Science policy: New mechanisms for scientific collaboration between East and West, 1995, edited by Koptyug, V.A. and Klerk, Joan, *NATO ASI Series, 4. Science and Technology Policy*, Vol. 1, (Kluwer Acad. Publ.), p. 226.
Semenenko, V.E., Vladimirova M.G. and Orleanskaya, O.B., 1967, Physiological characterization of *Chlorella sp.* at high extreme temperatures. *Fiziologiya rastenii,* **14 (4)**, p. 612.
Semenenko, V.E., Vladimirova, M.G. *et al.*, 1969, Intensive culture of *Anocystis nidulans*. In *Biologiya sinezelenykh rastenii* (Biology of blue-green plants), **2**, p. 163.
Seryapin, A.D., Fomin, A.G. and Chizhov, S.V., 1966, Human life support systems in spacecraft cabins using physicochemical methods. In *Kosmicheskaya biologiya i meditsina* (Space biology and medicine), (Moscow: Nauka), p. 298.
Shepelev, Ye.Ya., 1963, Ecological system in space flights. *Aviatsiya i kosmonavtika*, **1**, p. 20–25.
Shepelev, Ye.Ya., 1966, Human life support systems in spaceship cabins on the basis of biological material cycling. In *Kosmicheskaya biologiya i meditsina* (Space biology and medicine), (Moscow: Nauka), p. 330.
Shilenko, M.P., Lisovsky, G.M., Kalacheva, G.S. and Trubachev, I.N., 1979, Chufa as a source of vegetable fats in the closed life support system. *Kosmicheskaya biologiya i avaikosmicheskaya meditsina,* **5**, pp. 70–73.
Shilenko, M.P., Lisovsky, G.M., Trubachev, I.N. and Barashkov, V.A., 1985, Productivity and biochemical composition of wheat under introduction of human liquid waste into the nutrient solution. *Problemy kosmicheskoi biologii i meditsiny i idei Tsiolkovskogo* (Problems of space biology and medicine and Tsiolkovsky's ideas), *Proceedings of XVI–XVII Meeting*, Kaluga, 1981–82, (Publishers of the Institue of Natural History), pp. 135–138.
Shilov, V.M., Bragina, M.P., Borisova, O.K. *et al.*, 1979, Microflora of the organism of the space station crews. In *All-Union Conference on Space Biology and Aerospace Medicine,* (Kaluga), p. 19.
Shtol, A.A., Melnikov, E.S. and Kovrov, B.G., 1976, *Raschet i konstruirovaniye kultivatora dlya odnokletochnykh vodoroslei* (Calculation and design of the cultivator for unicellular algae), (Krasnoyarsk), pp. 95.
Shvarts, S.S., 1976, Evolution of the biosphere and ecological prognostication. *Vesti AN SSSR,* **2**, pp. 61–72.
Sidko, F.Ya, Eroshin, N.S., Belyanin, V.N. and Nemchenko, I.A., 1967, Investigation of optical parameters of unicellular algal populations. In *Nepreryvnoye upravlyayemoye kultivirovaniye mikrovodoroslei* (Continuous controlled microalgal cultivation), (Moscow: Nauka), p. 38.
Sinyak, G.S., Listovsky, P.V., Chizhikova, G.I., Vitashkina, M.A., Karpova, E.I., Gusarov, B.G. and Zabolotsky, L.L., 1971, Catalytic oxidation of some gas exchange products of pyrolysis of human wastes. *Kosmicheskaya biologiya i meditsina*, **5 (5)**, p. 77.
Sinyak, Yu.E. and Chizhov, S.V., 1964, Water regeneration in the cabin of the spaceship. In *Problemy kosmicheskoi biologii* (Problems of space biology), **3**, (Moscow: Nauka), p. 104.
Sinyak, Yu.E., Kuznetsova, L.A., Shikina, M.I., Filchakov, M.I. and Krasnoshchekov, V.V., 1972, Sanitary–hygienic estimation of extraction

technique of water recovery from the condensate of atmospheric moisture. *Kosmicheskaya biologiya i meditsina,* **6 (3)**, p. 22.

Sisakyan, N.M., Gazenko, O.G. and Genin, A.M., 1962, Problems of space biology. In *Problemy kosmicheskoi biologii* (Problems of space biology), **1**, (Moscow: Nauka), p. 17.

Space Biology and Medicine Life support and Habitability, Volume II. 1994, edited by Sulzman, F., Genin, A., *Am. Inst. of Aeronautics and Astronautics,* (Washington D.C., Moscow: Nauka).

Stern, J.A., 1961, *Report 3rd Internat. Sympos. Rockets and Astronautics.* Tokyo.

Stoy, V., 1965, Photosynthesis, respiration and carbohydrate accumulation in spring wheat. *Physiologie Plantarum,* Suppl. IV.

Straight, C. and MacElroy, R., 1990, The CELSS Test Facility: a Foundation for Crop Growth in Space. *Advances in Space Research*, Vol. 12, pp. 575–581.

Straight, C.L., Bubenheim, D.L., Bates, M.T., Flynn, M.T., Belisle W., and Schlick, G., 1993, The CELSS Antarctic Analog Project: A Validation of CELSS Methodologies at the South Pole Station, *SAE Technical Paper Series 932245.*

Straight, C.L., Bubenheim, D.L., Bates, M.T. and Flynn, M.T., 1994, The CELSS Antarctic Analog Project: An Advanced Life Support Testbed at the Amundsen–Scott South Pole Statiion, Antarctica. *Life Support & Biosphere Sciences*, Vol. 1, pp. 52–60.

Strayer, R.F., Alazraki, M.P., Yorio, N. and Finger, B.W., 1998, Bioprocessing Wheat Residues to Recycle Plant Nutrients to the JSC Variable Pressure Growth Chamber during the L/MLSTP Phase III Test, *SAE Technical Paper Series 981706.*

Streshinskaya, G.M., Pakhomova, M.V. and Shevyakova, N.I., 1967, Composition of different strains of the green alga *Chlorella. Prikladnaya biokhimiya i mikrobiologiya*, **3 (4)**, p. 477.

Strogonov, B.P., Kabanov, V.V. and Shevyakova, N.I., 1970, *Struktura i funktsii kletok rastenii pri zasolenii* (Structure and function of plant cells under salinization), (Moscow: Nauka).

Suess, E., 1875, *Die Entstehung der Alpen,* (Wien).

Suess, E., 1909, *Das Entlitz der Erde,* Bd. 3, (Wien, Leipzig).

Sukachev, V.N., 1944, On the principles of genetic classification in biogeocenology. *Zhurnal Obshchei Biologii,* Vol. 5, **4**, pp. 213–227.

Sukhoverkhov, F.N., 1964, Duckweed is cheap and nutritious food. *Rybolovstvo i rybovodstvo*, **2**.

Summary of the UN Conference on the environment and development, 1992, *Mir nauki*, **4**.

Supra, L.N., Reddig, M., Edeen, M.A., Pickering, K.D. and Packham, N.J.C., 1997, Regenerative water recovery system testing and model correlation, *SAE Technocal Paper Series 972550.*

T

Takehashi, Y., 1989, Water oxidation waste management systems for CELSS—the state of art. *Biological Science in Space*, Vol. 3, **1**, pp. 45–54.

Tamponnet, C., Binot, R., Lasseur, C. and Savage, C., 1991, Man in space.—A European challenge in biological life support. *ESA bulletin 67*, pp. 39–41.

Tarko, A.M., 1977, Global role of the system "atmosphere–plants–soil" in compensation of effects on the biosphere. *DAN*, Vol. 237, **1**.

Tashpulatov, R.Yu., Mordvinova, N.B. and Rogunova, K.A., 1971, Microbiological and immunological investigations of human organism in an isolated group. *Zhurnal Microbiologii*, **4**, pp. 68–73.

Taubaev, T.T., Abdiev, M. and Keldibekov, S., 1971, On biological productivity of toad spit in natural water bodies and in the culture. In *Kultivirovaniye vodoroslei i vysshikh vodnykh rastenii v Uzbekistane* (Cultivation of algae and aquatic higher plants in Uzbekistan), (Tashkent), p. 98.

Taubaev, T.T., Neskubo, P.M. *et al.*, 1971, Use of toad spit in fattening of sheep and goats. In *Kultivirovaniye vodoroslei i vysshikh vodnykh rastenii v Uzbekistane* (Cultivation of algae and aquatic higher plants in Uzbekistan), (Tashkent), p. 136.

Tauts, M.I., 1964, Investigation of effect of *Chlorella* metabolites on its growth in intensive culture. *Fiziologiya rastenii*, **2 (2)**, p. 247.

Tauts, M.I., 1966, Effect of changes in the culture medium caused by algal growth on subsequent culture productivity. In *Upravlyaemyi biosintez* (Controlled biosynthesis), (Moscow: Nauka), p. 145.

Taylor, A., 1958, *Food Technology*, **12**, p. 442.

Taylor, G.R., 1976, Medical microbiological analysis of Apollo–Soyus test project crewmembers. In *NASA Technical Memorandum X-58180*, (Houston), p. 22.

Taylor, G.R. *et al.*, 1973, Changes in the fungal autoflora of Apollo astronauts. *Applied Microbiology,* Vol. 26, pp. 804–813.

Terskov, I.A., Lisovsky, G.M., Ushakova, S.A., Parshina, O.V. and Moiseenko, L.P., 1978, Possibility of using higher plants in lunar life support systems. *Kosmichesksya biologiya i aviakosmicheskaya meditsina,* Vol. 12, **3**, pp. 63–66.

Tibbits, T.W. and Alford, D.K., 1982, Controlled ecological support system—use of higher plants. Use of higher plants. *NASA Conference Publication 2231*, (NASA Ames Research Center, Moffett Field, Calif.).

Tibbits, T.W. and Cao, W., 1994, Solid matrix and liquid culture procedures for growth of potatoes. *Advances in Space Research*, Vol. 14, **11**, pp. (11)427–(11)433.

Tikhomirov, A.A., Zolotukhin, I.G. and Sidko, F.Ya., 1976, Influence of light regimes on productivity and quality of radish harvest. *Fiziologiya rastenii,* Vol. 23, **3**, pp. 502–505.

Tikhomirov, A.A., Zolotukhin, I.G., Lisovsky, G.M. and Sidko, F.Ya., 1987, Specificity of reaction of plants of different species to the spectral composition of PAR under artificial lighting. *Fiziologiya rastenii,* Vol. 34, **4**, pp. 774–785.

Tikhomirov, A.A., Lisovsky, G.M. and Sidko, F.Ya., 1991, *Spektralnyi sostav sveta i produktivnost rastenii* (Light spectral composition and plant productivity), (Novosibirsk: Nauka).

Timofeev-Resovsky, N.V. and Tyuryukanov, A.N., 1966, On elementary biochorological divisions of the biosphere. *Bulleten MOIP. Otdeleniye Biologii,* Vol. 71, **1**, pp. 123–132.

Timofeev-Resovsky, N.V., Vorontsov, N.N. and Yablokov, A.V., 1969, *Kratkii ocherk teorii evolyutsii Moskvy* (Outline of the theory of Moscow evolution), (Moscow: Nauka).

Tirranen, L.S., Sizova, T.P. and Pankova, I.M., 1979, *Mikologiya i fitopatologiya* (Mycology and phytopathology), **13 (4)**, pp. 294–298.

Tischer, R., 1960, In *Physics and Medicine of the Atmosphere and Space* (N.Y.–London), p. 397.

Tranquillini, W., 1967, *Angew. Bot.*, **41 (1–2)**, p. 1.

Tri, T.O., 1999, Bioregenerative planetary life support system test comples (BIO-Plex): test mission objectives and facility development, *SAE Technical Paper Series 1999-01-2186.*

Tri, T.O., Edeen, M.A. and Henninger, D.L., 1996, Advanced life support human-rated test facility; testbed development and testing to understand evolution to regenerative life support, *SAE Technical Paper Series 961592.*

Trubachev, I.N., Andreeva, R.I. and Min, Z.V., 1969, Biochemical composition of *Chlorella* and centrifugate in cultivation with addition of human urine and processed solid wastes. In *Upravlyaemyi biosintez i biofizika populyatsii* (Controlled biosynthesis and biophysics of populations), (Krasnoyarsk), p. 21.

Trubachev, I.N., Lisovsky, G.M., Min, Z.V. and Andreeva, R.I., 1970, Biochemical composition of wheat grain grown in the phytotron. *Informatsionnyi Bulleten Koordinatsionnogo Soveta po Fiziologii Rastenii,* **6**, (Irkutsk), pp. 150–151.

Trubachev, I.N., Lisovsky, G.M., Andreeva, R.I., Veber, M.I., Gribovskaya, I.V. and Shilenko, M.P., 1975, Chemical composition of vegetables grown in light culture. *Fiziologiya i biokhimiya rastenii,* Vol. 7, **12**, pp. 185–189.

Trubachev, I.N., Gribovskaya, I.V., Barashkov, V.A. and Kalacheva, G.S., 1979, Biochemical and mineral composition of plants grown in the life support system. In *Zamknutaya sistema: chelovek–vysshiye rasteniya* (Closed system "human–higher plants"), edited by Lisovsky, G. (Novosibirsk), pp. 62–81.

Trukhin, N.V., 1976, On resistance to salinity of *Clorella pyrenoidosa. Botanicheskii zhurnal*, **52 (9)**, p. 1325.

Tsiolkovsky, K.E., 1895, *Grezy o zemle i nebe i effekty vsemirnogo tyagoteniya* (Visions of Earth and sky and effects of gravitation), (Moscow).

Tsiolkovsky, K.E., 1926, *Issledovaniye mirovykh prostranstv reaktivnymi proborami* (Exploration of world space with rockets), (Kaluga).

Tsiolkovsky, K.E., 1964, *Life in interstellar medium* (in Russian), (Moscow: Nauka).

Tsvetkova, I.V., Zamota, V.P. and Maksimova, E.V., 1970, Plant cultivation under the conditions of closed material cycle, using expandedn clay aggregate. *Kosmicheskaya biologiya i meditsina,* **4 (1)**, p. 11.

Turco, R. P., Toon, O.B., Ackerman, T.R. *et al.*, 1983, Nuclear Winter: Global Consequences of Multiple Nuclear Explosions. *Science.* Vol. 222, **4630**, pp. 1283–1300.

U

Ugolev, A.M., 1958, *Estestvennaya tekhnologiya biologicheskikh sistem* (Natural technology of biological systems), (Leningrad, Nauka).

V

Van der Vin, R. and Meyer, G., 1962, *Svet i rost rastenii* (Light and plant growth), (Moscow: Kolos).

Varlamov, V.F., Kozyrevskaya, G.N., Sitnikova, N.N., Koloskova, Yu.S., Chernovich, I.L. and Tsitovich, S.I., 1967, Results of mineralization of human wastes by the biological method under aeration with activated sludge. In *Problemy sozdaniya zamknutykh ekologicheskikh sistem* (Problems of creating closed ecological systems), (Moscow: Nauka), p. 171.

Vernadsky, V.I., 1924, *Biosfera i noosfera* (Biosphere and noosphere), (Moscow: Nauka).

Vernadsky, V. I., 1924, *La geochemie*, (Alcan: Paris).

Vernadsky, V.I., 1926, *Biosphere* (Leningrad), 148 pp.

Vernadsky, V.I., 1927, *Ocherki geokhimii* (Geochemistry essays), (Moscow: Leningrad).

Vernadsky, V.I., 1937, On limits of the biosphere. *Izvestiya Academii Nauk SSSR (Geology)*, **1**, pp. 3–24.

Vernadsky, V. I., 1944, A few words about noosphere. In *Uspekhi sovremennoi biologii* (Advance of modern biology), Vol. 18, **2**, pp. 113–120.

Vernadsky, V. I., 1991, *Nauchnaia mysl kak planetnoe iavlenie* (Scientific thinking as a planetary phenomenon), (Moscow: Nauka), (reprint of the 1938 edition).

Vinberg, G.G., 1960, *Biologicheskaya ochistka i transformatsiya veshchestv zhidkikh vydelenii cheloveka pri uchastii fotosinteziruyushchikh vodoroslei* (Biological treatment and transformation of substances in human liquid wastes, involving photosynthesizing algae), (Minsk).

Vinberg, G.G., Ostapenya, P.V., Sivko, T.N. and Levina, R.I., 1966, *Biologicheskiye prudy v praktike ochistki stochnykh vod* (Biological ponds in sewage water treatment), (Minsk: Belarus).

Vinogradov, M.E. and Shushkina, E.A., 1987, *Funktsionirovaniye planktonnykh soobshchestv epipelagiali okeana* (Functioning of plnakton communities of the ocean epipelagium), (Moscow: Nauka).

Violette, F., Boitean, H. and Bernard, S., 1959, *X International Astronautical Congress* (London).

Vladimirova, M.G., Ignatyevskaya, M.A. and Raikov, N.I., 1966, Characterization of productivity of strains of unicellular algae in intensive laboratory and industrial culture. In *Upravlyayemyi biosintez* (Controlled biosynthesis), (Moscow: Nauka), p. 86.

Vladimirova, M.G., Tauts, M.I., Feiktistova, O.I. and Semenenko, V.E., 1966, Physiological peculiarities of *Chlorella* related to prolonged intensive cultivation of algae. *Trudy Moskovskogo Obshchestva Ispytatelei Prirody. Otdeleniye biologii,* **24**, p. 142.

Vodovotz, Y., 1998, Advanced life support system at Johnson Space Center. *International committee for Material Circultaion in Geo-Hydrosphere and its Applications,* Programs and abstracts, (Japan).

Voitovich, Ya.V., Ponomarev, P.I., Trubachev, I.N., Gitelson, I.I. and Terskov, I.A., 1971, Gas and water balance of the regenerative life support system involving hydrogen-reducing bacteria. *Izvestiya SO AN SSSR (biology)*, **10 (2)**, p. 11.

Volk, T. and Rummel, J.D., 1990, Mass balances for a biological life support system simulation model. *Controlled Ecological Life Support System.* NASA Conf. Publ. 2480. (NASA Ames Research Center, Moffett Field, Calif.), pp. 139–146.

Volk, T., Bugbee, B. and Wheeler, R.M., 1995, An approach to crop modeling with the energy cascade. *Life Support & Biosphere Science,* Vol. 1, pp. 119–127.

Volterra, V., 1931, *Theorie mathematique de la lutte pour la vie.* (Gautiers–Villar).

Voronin, G.I. and Polivoda, A.I., 1967, *Zhizneobespecheniye ekipazhei kosmicheskikh korablei* (Life support of spacecraft crews), (Moscow: Mashinostroyeniye).

Voskresenskii, K.A. and Yurina, E.V., 1965, *Asteromonas gracilis* as the object of large-scale cultivation. *Vestnik MGU. Biologiya, pochvovedeniye*, **2**, p. 29.

W

Walverton, B.C., 1986, Aquatic plants and wastewater treartment. In: *Aquatic plants for water treatment and resource recovery.* (Orlando FL: Magnolia Publ.) pp. 3–5.

Walverton, B.C., 1990, Plants and their microbial assistants. Nature's answer to Earth's environmental problems. In: *Biological Life Support Systems,* edited by Nelson, M. and Soffen, E.A., (Oracle AZ: Synergetic Press), pp. 60–65.

Walwerton, B. and Walwerton, J., 1992, Bioregenerative life-support systems for energy-efficient buildings. *International Conference on Life Support and Biospherics. Proceedings,* pp. 117–126.

Wardlow, J.F., 1967, *Australian Journal of Biological Science*, **2**, p. 20.

Warington, R., 1851, Notice of observation on the adjustment of the relations betwwen animal and vegetable kingdoms. *Quaternary L. Chemical Society*, **3**.

Waslien J., and Galloway, D., 1969, *Nature*, Vol. 229, p. 84.

Watson, D.S., 1952, *Advance in Agronomy,* **4**, p. 101.

Wheeler, R.M., 1998, Bioregenerative life support system at NASA's Kennedy Space Center. *International committee for Material Circultaion in Geo-Hydrosphere and its Applications,* Programs and abstracts, (Japan).

Wheeler, R.M. and Tibbits, T.W., 1986, Growth and tuberization of potato (*Solanum tuberosum L.*) under continuous light. *Plant Physiology,* **80**, pp. 801–804.

Wheeler, R.M. and Tibbits, T.W., 1987, Utilization of potatoes for life support systems in space: III Productivity at successive harvest dates under 12-h and 24-h photoperiods. *American Potato Journal*, pp. 311–320.

Wheeler, R.M., Tibbits, T.W. and Fitzpatrick, A.H., 1991, Carbon dioxide effects on potato growth under different photoperiods and irradiance. *Crop Science*, Vol. 31, **5**, pp. 1209–1213.

Wheeler, R.M., Drese, J.H. and Sager, J.C., 1991a, Atmospheric leakage and condensate production in NASA's biomass production chamber. Effect of diurnal temperature cycles. *NASA Technical Memorandum 103819*, (NASA–KSC, Florida).

Wheeler, R.M., Berry, W.L., Mackowiak, C. *et al.*, 1993, A data base on crop nutrient use, and carbon dioxide exchange in a 20 squalre meter growth chamber: I. Wheat as a case study. *Journal of Plant Nutrition*, **16 (10)**, pp. 1881–1915.

Wheeler, R.M., Corey, K.A., Sager, J.C. and Knott, W.M., 1993a, Gas exchange characterictics of wheat stands grown in a closed, controlled environment. *Crop Science*, Vol. 33, **1**, pp. 161–168.

Wheeler, R.M., Mackowiak, C.L., Sager, J.C., Yorio, N.C., Knott, W.M. and Berry, W.L., 1994, Growth and gas exchange by lettuce stands in a closed, controlled environment. *Journal of the American Society for Horticultural Science*, **119 (3)**, pp. 610–615.

Wheeler, R.M., Mackowiak, C.L., Sager, J.C. and Knott, W.M., 1994a, Growth of soybean and potato at high CO_2 partial pressures. *Advances in Space Research*, Vol. 14, **11**, pp. (11)251–(11)255.

Wheeler, R.M., Mackowiak, C.L., Sager, J.C., Knott, W.M. and Berry, W.L., 1994b, Proximate nutritional composition of CELSS crops grown at different CO_2 partial pressures. *Advances in Space Research*, Vol. 14, **11**, pp. (11)171–(11)176.

Wheeler, R.M., Mackowiak, C.L., Stutte, G.W., Sager, J.C., Yorio, N.C., Ruffe, L.M., Fortson, R.E., Dreschel, T.W., Knott, W.M. and Corey, K.A., 1996, NASA's biomass production chamber: a testbed for bioregenerative life support studies. *Advances in Space Research*, Vol. 18, **4/5**, pp. (4/5)215–(4/5)224.

Wignarajah, K. and Bubenheim, D.L., 1997, Integration of crop production with CELSS waste management. *Advances in Space Research*, Vol. 20, **10**, pp. 1833–1843.

Woodwell, J., Energy cycling in the biosphere. In: *Biosphere*, edited by Gilyarov, M., (Moscow, Mir), pp. 41–59. (see also: The Biosphere, 1970, *Scientific American*, Vol. 223, **3 (IX)**).

Wunder, C.C., 1966, *Life into space,* (Phil.: Davis).

Wydeven, T. and Golub, M.A., 1990, Generation rates and chemical compositions of waste streams in a typical crewed space habitat. *NASA Technical Memorandum 102799.*

Y

Yamakawa, T., 1953, *Bulletin of Japan Society for Scientific Fishery*, **18 (10)**, p. 36.

Yanshin, A.L., 1963, Principle of actuality and the problem of evolution. In *Puti i metody poznaniya zakonomernogo razvitiya zemli* (Ways and methods of comprehending regular development of Earth), (Moscow: Nauka), pp. 1–8.

Yanshin, A.L., 1986, V.I. Vernadsky and his studies of the biosphere and its transformation to the noosphere. In *V.I. Vernadsky i sovremennost* (V.I. Vernadsky and present times), (Moscow), pp. 28–40.

Yanshina, F.T., 1994, V.I. Vernadsky's idea of living material in his works on geochemistry. In *V.I. Vernadsky. Trudy po geokhimii* (V.I. Vernadsky. Works on geochemistry), (Moscow: Nauka), pp. 469–485.

Yazdovsky, V.I., 1966, Principal objectives of space biology and medicine. In *Problemy kosmicheskoi biologii* (Problems of space biology), (Moscow: Nauka), p. 61.

Yurina, E.V., 1966, Experience in cultivation of halobiont algae *Asteromonas gracilis* and *Dunaliella salina*. *Vestnik MGU (Biology, soil science)*, **6**, p. 76.

Z

Zaloguev, S.N., Utkina, T.G., Schin, M.M. and Kareva, N.M., 1971, The microflora of human integument during prolonged confinement. *Life Science Space Research,* Vol. 9, pp. 453–454.

Zamknutaya sistema: chelovek–vysshiye rasteniya (*Closed system "human–higher plants"),* edited by Lisovsky, G.M., 1979, (Novosibirsk: Nauka).

Zharov, S.G., Kustov, V.V., Seryapin, A.D. and Fomin, A.G., 1966, Artificial atmosphere of spaceship cabins. In *Kosmicheskaya biologiya i meditsina* (Space biology and medicine), (Moscow: Nauka), p. 285.

Zhurbitsky, Z.I., 1963, *Fiziologicheskiye i agrokhimicheskiye osnovy primeneniya udobrenii* (Physiological and agrochemical foundations of fertilizer application), (Moscow: Nauka).

Zhurbitsky, Z.I., 1965, Growing of plants on aquatic nutrient solutions. In *Gidroponika v selskom khozyaistve* (Hydroponics in agriculture), (Moscow: Kolos).

Zhurbitsky, Z.I., 1968, *Teoriya i praktika vegetatsionnogo metoda* (Theory and practice of vegetation method), (Moscow: Nauka).

Zolotukhin, I.G., Lisovsky, G.M., Tikhomirov, A.A. and Sidko, F.Ya., 1978, Photobiological investigations of spectral radiation efficiency for wheat. *Svetotekhnika,* **5**, pp. 11–13.

Zweers, T., Preller, G., Brunekreffand, B. and Boleij, M., J.S., 1990, Relationships between health and indoor climate complaints and building, workplace, job and personal characteristics. *Proceedings of the 5th International Conference on Indoor Air Quality and Climate. INDOOR AIR'90*, Vol. 1–5, edited by Walkinshaw, D.S., (Indoor Air Technologies, Ottawa), Vol.1, pp. 495–500.

SUBJECT INDEX